Noise in nonlinear dynamical systems

Volume 3

Experiments and simulations

Noise in nonlinear dynamical systems

Volume 3
Experiments and simulations

Edited by

Frank Moss, *Professor of Physics,*
University of Missouri at St Louis

and

P. V. E. McClintock, *Reader in Physics,*
University of Lancaster

The right of the University of Cambridge to print and sell all manner of books was granted by Henry VIII in 1534. The University has printed and published continuously since 1584.

CAMBRIDGE UNIVERSITY PRESS
Cambridge New York New Rochelle
Melbourne Sydney

Published by the Press Syndicate of the University of Cambridge
The Pitt Building, Trumpington Street, Cambridge CB2 1RP
32 East 57th Street, New York, NY 10022, USA
10 Stamford Road, Oakleigh, Melbourne 3166, Australia

First published 1989

Printed in Great Britain at the University Press, Cambridge

British Library cataloguing in publication data

Noise in nonlinear dynamical systems
Vol. 3, Experiments and simulations.
1. Nonlinear dynamical systems
I. Moss, Frank II. McClintock, P. V. E.
515.3′5

Library of Congress cataloguing in publication data

Experiments and simulations.
(Noise in nonlinear dynamical systems; v. 3)
Includes index.
1. Fluctuations (Physics)–Experiments.
2. Fluctuations (Physics)–Computer simulation.
3. Stochastic processes. I. Moss, Frank, 1934–
II. McClintock, P. V. E. III. Series.
QC6.4.F58N64 vol. 3 003 s [003] 88–9535

ISBN 0 521 35265 7

TM

Contents

Contributors

E. Arimondo
Dipartimento di Fisica
Universita di Pisa
Piazza Toricelli 2
56100 Pisa
Italy

Helmut R. Brand
FB7 Physik
Universität Essen
D4300 Essen 1
W. Germany
and
Department of Physics
Kyushu University
Fukuoka 812
Japan

Leone Fronzoni
Dipartimento di Fisica
Universita' di Pisa
Piazza Toricelli 2
56100 Pisa
Italy

P. Glorieux
Laboratoire de Spectroscopie Hertzienne
Universite de Lille I
59655 Villeneuve d'Ascq cedex
France

D. Hennequin
Laboratoire de Spectroscopie Hertzienne
Universite de Lille I
59655 Villeneuve d'Ascq cedex
France

S. Kai
Department of Electrical Engineering
Kyushu Institute of Technology
Tobata 804
Japan

W. Lange
Institut für Quantenoptik
Universität Hannover
Welfengarten 1
3000 Hannover 1
W. Germany

P. V. E. McClintock
Department of Physics
University of Lancaster
Lancaster LA1 4YB
UK

Riccardo Mannella
Department of Physics
University of Lancaster
Lancaster LA1 4YB
UK

Frank Moss
Department of Physics
University of Missouri at St Louis
St Louis
MO 63121
USA

R. Roy
School of Physics
Georgia Institute of Technology
Atlanta
GA 30332
USA

J. T. Tough
Department of Physics
The Ohio State University
Columbus
OH 43210
USA

A. W. Yu
School of Physics
Georgia Institute of Technology
Atlanta
GA 30332
USA

S. Zhu
School of Physics
Georgia Institute of Technology
Atlanta
GA 30332
USA

Preface

All macroscopic physical systems are subject to fluctuations or noise. One of the most useful and interesting developments in modern statistical mechanics has been the realization that even complex nonequilibrium systems can often be reduced to equivalent ones of only a few degrees of freedom by the elimination of dynamically nonrelevant variables. Theoretical descriptions of such contracted systems necessarily begin with a set of either continuous or discrete dynamical equations which can then be used to describe noise driven systems with the inclusion of random terms. Studies of these stochastic dynamical equations have expanded rapidly in the past two decades, so that today an exuberant theoretical activity, a few experiments, and a remarkably large number of applications, some with challenging technological implications, are evident.

The purpose of these volumes is twofold. First we hope that their publication will help to stimulate new experimental activity by contrasting the smallness of the number of existing experiments with the many research opportunities raised by the chapters on applications. Secondly, it has been our aim to collect together in one place a complete set of authoritative reviews with contributions representative of all the major practitioners in the field. We recognize that as an inevitable consequence of the intended comprehensiveness, there will be few readers who will wish to digest these volumes in their entirety. We trust, instead, that readers will be stimulated to choose from the many possibilities for new research represented herein, and that they will find all the specialized tools, be they experimental or theoretical, that they are likely to require.

Although there is a strong underlying theme running through all three volumes – the influence of noise on dynamical systems – each chapter should be considered as a self-contained account of the authors' most important research in the field, and hence can be read either alone or in concert with the others. In view of this, the editors have chosen not to attempt to impose a uniform style; nor have they insisted on any standard set of mathematical symbols or notation. The discerning reader will detect points of detail on which full concordance appears to be lacking, especially in Volume 1. This should certainly be regarded as the signature of an active and challenging theoretical activity in a rapidly expanding field of study. In selecting the contributors, the editors have made special efforts to include younger authors with new ideas and perspectives along with the more active seasoned veterans, in the confident belief that the field will be invigorated by their contributions.

Finally, it is our pleasure to acknowledge the many valuable suggestions made by our colleagues and contributors, from which we have greatly benefited. Special thanks are due to R. Mannella for constructive criticism and helpful comments at various stages of this enterprise.

Completion of this work owes much to the generous support provided by the North Atlantic Treaty Organization (under grant RG85/0070) and the UK Science and Engineering Research Council (under GR/D/61925).

Frank Moss and P. V. E. McClintock
Lancaster, May 1987

Introduction to Volume 3

As will have become clear to readers of the first two volumes of *Noise in Nonlinear Dynamical Systems*, the theoretical development of the field, already extensive and wide-ranging, is still proceeding at a rapid pace. In fact, although such investigations were initiated as a means towards a better understanding of the observed properties of the physical world, the progress of the theory seems in many cases to have run far ahead of the corresponding experimental studies. Some important questions then arise. Does nature really behave in the manner predicted by the theory? Under what particular physical conditions is any given (approximate) theory to be relied upon? In most cases, these questions have no easy experimental answers, and the aim of this third volume is to survey the limited progress that has been made to date.

Credible experimental tests of stochastic theory have been completed on only a relatively small number of natural systems. These include, particularly, superfluid helium, liquid crystals and lasers, whose diverse properties provide the subject matter of Chapters 1–6.

Heat flow in superfluid helium can result in the formation of a tangle of quantized vortex lines, a particularly simple and well-characterized version of classical turbulence. The process occurs through a continuous instability of the type for which a variety of noise-induced transitions have been predicted. Chapter 1 describes a direct experimental test of these ideas.

The dramatic effect of external noise on liquid crystals is discussed in Chapters 2 and 3, which treat different aspects of the electrohydrodynamic instability (EHD). This is a convective instability which occurs in thin layers of nematic confined between glass plates and subjected to a perpendicular electric field. A variety of convective patterns can form under different circumstances. External noise of appropriate strength and correlation time can serve to *stabilize* the system against the EHD instability, and it is interesting to note that this represents a situation where noise is able to influence directly a spatial structure. Chapter 2 surveys the whole area, but concentrates, particularly, on experimental studies of the influence of noise on the pattern formation process. Chapter 3 provides a description and a critical discussion of experiments and theory for stochastic postponements both of the

EHD and also of higher instabilities. Such postponements were predicted in Fokker–Planck models as early as 1979–80.

Chapters 4–6 all deal with aspects of noise in lasers. Fluctuations in lasers can arise either from internal or external sources. Their influence is difficult to calculate reliably because the noise in question is often colored, so that exact calculations are impossible and it is necessary to employ one of the approximate theoretical approaches discussed in Volumes 1 and 2, and to test their predictions against experiment. Experiments on colored noise in dye lasers are discussed in Chapter 4. The dye laser was, in fact, the first experimental system widely appreciated to demonstrate the effects of colored (as opposed to white) noise dynamics. Chapter 5 deals with optical bistability generally, and with the properties of lasers that have saturable absorbers in their optical cavities. In Chapter 6, a somewhat different approach to the problem is discussed: experiments on transient optical bistability, guided by analogue studies of electronic models, are reviewed. Analogue experiments, more general applications of which are described in Chapters 8 and 9, are shown to provide a very convenient and versatile means of studying the properties of the equations that are believed to describe laser operation.

It will have become apparent to the readers of Volumes 1 and 2 that stochastic theory usually has to be restricted to relatively simple, idealized, model systems. A basic difficulty, therefore, in achieving a meaningful comparison between experiment and theory lies in the identification of whether any observed discrepancies are to be attributed to inadequacies of the theoretical approximations or to significant differences between the idealized model and the real physical system that it is intended to describe. Fortunately there are two supplementary kinds of information that can be sought based, first, on digital simulation and, secondly, on analogue simulation of the relevant stochastic differential equations. In cases where the digital and analogue simulations are both in agreement with the approximate analytic theory, one can have a high degree of confidence that the latter is reliable. There are other situations where no analytic theory, approximate or otherwise, has yet been developed, and where simulation techniques provide the *only* well-characterized means of studying the system in question. Analogue simulations, in particular, offer a strikingly simple and rapid technique for the study of higher dimensional systems, including ones which would require extremely large amounts of computing time for the corresponding digital simulations. Digital simulation techniques are discussed in Chapter 7 and analogue techniques and their application to a wide variety of physical systems are described in detail in Chapters 8 and 9.

1 The effects of colored quadratic noise on a turbulent transition in liquid He II

J. T. TOUGH

1.1 Introduction

Liquid He II represents an important physical system for the experimental study of noise induced dynamical transitions. The basic physics of He II and of He II hydrodynamics are well understood. The flow of heat in this liquid is associated with a kind of turbulence which undergoes a continuous transition as the heat current is increased. At this transition the intrinsic fluctuations and the relaxation time both become large. External noise can be easily added to the driving heat current. Small amplitude noise simply causes the system to fluctuate about its deterministic steady states, but large amplitude noise causes dramatic changes. The stochastic steady states of the system show noise induced bistability.

1.2 He II and superfluid turbulence

When ordinary helium (^{4}He) is cooled below the λ-temperature ($T_\lambda = 2.17$ K at vapor pressure), many of its properties change dramatically. This new liquid is called He II. It has been the object of scientific investigation for over 50 years. A thorough review of the properties of liquid He II has been given by Wilkes and Betts (1987), and shorter discussions can be found in a number of books (McClintock, Meredith, and Wigmore, 1984; Tilley and Tilley, 1974).

The properties of He II can conveniently be discussed within the conceptual framework provided by the two fluid model. This model, based largely upon the theoretical work of Landau (1941), treats the He II as a freely interpenetrating mixture of two separate fluids: the normal fluid and the superfluid. The superfluid has no viscosity or entropy; it is a perfect fluid whose density will be denoted ρ_s. The normal fluid of density ρ_n is rather like a simple conventional fluid with viscosity η and entropy density S. The total fluid density ρ is just the sum

$$\rho = \rho_s + \rho_n. \tag{1.2.1}$$

The superfluid and normal fluid densities vary strongly with temperature subject to this constraint. At T_λ, $\rho_s = 0$, and at the absolute zero $\rho_n = 0$. The

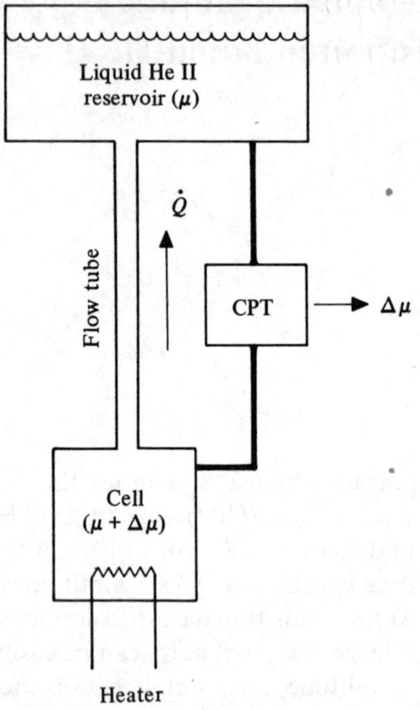

Figure 1.1. Schematic diagram of the apparatus. Heat is carried from the cell to the reservoir through the flow tube. The device labeled CPT detects the chemical potential difference $\Delta\mu$ across the tube.

experiments reported in this chapter were carried out at 1.75 K, where the ratio ρ_s/ρ_n is about 2.6.

He II is a complex hydrodynamic system for several reasons. The superfluid and the normal fluid can move independently with local velocity fields $\mathbf{v}_s$ and $\mathbf{v}_n$ respectively. The equations of motion for the superfluid are:

$$\frac{d\mathbf{v}_s}{dt} = -\nabla\mu \tag{1.2.2}$$

$$\nabla \times \mathbf{v}_s = 0, \tag{1.2.3}$$

where μ is the chemical potential (the Gibbs free energy per unit mass) of the He II. The normal fluid flow satisfies an equation similar to the Navier–Stokes equation for viscous fluids. However, since the flow of the normal fluid is also an entropy current, the local heat current density in the He II can be written

$$\mathbf{w} = \rho S T \mathbf{v}_n. \tag{1.2.4}$$

The significance of this result is that an average normal fluid flow can be imposed on the He II by simply passing a heat current through the system. Consider the apparatus used in the present experiments and shown schematically in Figure 1.1. A tube of area A and length l connects a reservoir of He II to a cell which can be heated at a rate $\dot{Q}$. As a consequence of (1.2.4), there is an *average* normal fluid velocity V_n from the cell to the reservoir given by

$$V_n = \dot{Q}/\rho STA. \tag{1.2.5}$$

Since there is no net mass flow through the tube, it follows that there is a corresponding average superfluid velocity V_s from the reservoir to the cell which satisfies the equation

$$\rho_n V_n = \rho_s V_s. \tag{1.2.6}$$

This particular flow (induced by heat transport without mass transport) is known as thermal counterflow, and is one of the most well-studied hydrodynamic phenomena in He II. In this case the superfluid does not flow independently of the normal fluid, but is constrained by the condition of zero mass transport, (1.2.6). More complex flows, where this constraint is relaxed and the two fluids have independent average flow velocities, are possible, but will not be considered in the present experiments.

For steady state thermal counterflow (1.2.2) gives immediately that

$$\Delta\mu = 0. \tag{1.2.7}$$

The superfluid flows at an average velocity V_s with no chemical potential drop. It is possible to exploit the properties of the He II to construct a device which gives an output proportional to the chemical potential *difference* across its input ports (Yarmchuck and Glaberson, 1979). We have used such a chemical potential transducer (labeled CPT in Figure 1.1) in our experiments (Griswold, Lorenson and Tough, 1987; D. Griswold and J. T. Tough, manuscript submitted to *Phys. Rev. A*; Lorenson, Griswold, Nayak and Tough, 1985). For all values of $\dot{Q}$ less than some onset value $\dot{Q}_0$ we verify that (1.2.7) is obeyed, and the superfluid flows without dissipation. There is of course a temperature difference ΔT associated with the heat flow between the cell and the reservoir. Using thermodynamics to express the chemical potential difference in terms of temperature and pressure differences, it follows from (1.2.7) that the pressure difference ΔP is given by H. London's relation

$$\Delta P = \rho S \Delta T. \tag{1.2.8}$$

It is just this pressure difference which drives the viscous normal fluid at an average velocity V_n.

The simple dissipationless flow of the superfluid described above is limited to small values of the heat current $\dot{Q}$. At larger heat currents the chemical potential difference $\Delta\mu$ is observed to increase approximately as $\dot{Q}^3$. This dissipation is known as superfluid turbulence and has been reviewed recently by Tough (1982). The dissipation results from the interaction of the normal fluid with a tangle of quantized vortex lines in the superfluid (Vinen, 1957). These vortex lines have cores of atomic size around which the superfluid flows with quantized circulation:

$$\kappa = \oint \mathbf{v}_s \cdot \mathrm{d}l = \frac{h}{m}. \tag{1.2.9}$$

Here h is Planck's constant and m is the mass of a helium atom. Glaberson and Donnelly (1985) have provided a thorough review of the theoretical and experimental aspects of quantized vortex lines in He II.

Recently Schwarz (1982) has shown how the self-sustaining vortex tangle of superfluid turbulence can be maintained in a uniform counterflow of the superfluid and normal fluid. Vortex loops grow, like smoke rings in the air, as they move through the normal fluid. This growth is interrupted by vortex reconnections which occur when two loops cross. Computer simulations reveal that an arbitrary initial configuration of vortex lines evolves into a homogeneous random tangle of density L_0 (length of vortex line per unit volume of fluid) approximately proportional to $\dot{Q}^2$. The dissipation associated with the superfluid turbulence is observed as the chemical potential difference $\Delta\mu$, and is given under reasonable approximations (Tough, 1982) by:

$$\Delta\mu = (l/A)(\rho_n/\rho_s)(\kappa B/\rho ST)\, L_0 \dot{Q}, \tag{1.2.10}$$

where B is a dimensionless quantity which characterizes the friction on a quantized vortex line (Barenghi, Donnelly and Vinen, 1983).

Figure 1.2 shows results for the steady state chemical potential difference associated with superfluid turbulence in He II. The chemical potential transducer (CPT) output is given as a function of the heat current $\dot{Q}$. The apparatus is shown schematically in Figure 1.1, and experimental details can be found in Griswold, Lorenson and Tough (1987). We only note here that the length of the flow tube (l) is 1.0 cm, and the internal diameter (d) is 1.34×10^{-2} cm. For heat currents greater than about 140 μW the chemical potential difference increases monotonically, approximately as $\dot{Q}^3$. The results in Figure 1.2 are intended to emphasize the region between 100 μW and 140 μW where there is an obvious change in the dissipation. This change reflects the continuous transition of the superfluid turbulence from a state TI to a state TII (Tough, 1982). The intrinsic fluctuations in the dissipation are strongly enhanced near this TI/TII transition (see Section 1.3). The analysis of these fluctuations (Schumaker and Horsthemke, 1987) introduces a 'paracritical point' at $\dot{Q}_p = 121\ \mu$W where the system relaxation time diverges. It is convenient to scale the heat currents by this value so that

$$\tilde{Q} = \dot{Q}/\dot{Q}_p. \tag{1.2.11}$$

Similarly the reduced chemical potential difference $\Delta\tilde{\mu}$ can be defined in terms of the chemical potential at the paracritical point

$$\Delta\tilde{\mu} = \Delta\mu(\dot{Q})/\Delta\mu(\dot{Q}_p). \tag{1.2.12}$$

Scales for $\Delta\tilde{\mu}$ and for $\tilde{Q}$ are also shown in Figure 1.2.

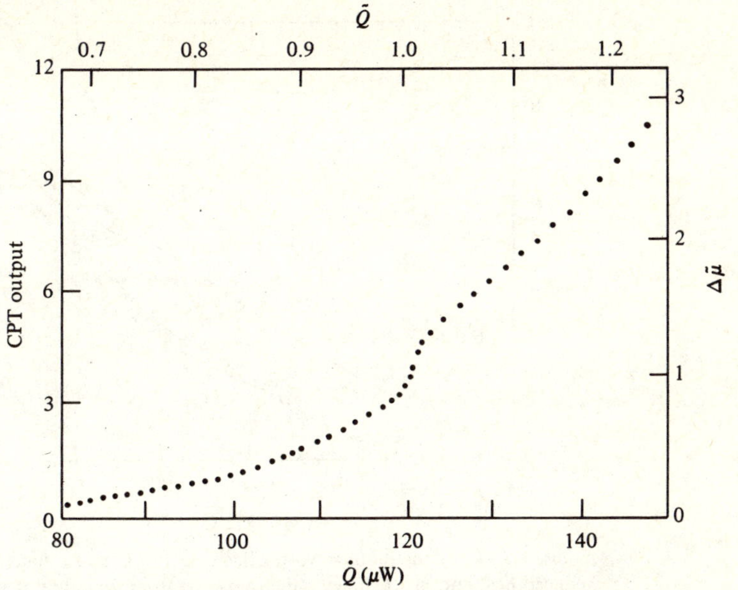

Figure 1.2. The chemical potential difference across the flow tube, as CPT output, as a function of the heat current $\dot{Q}$. The transition from turbulent state TI to state TII occurs near 120 μW. The alternate scales give the reduced chemical potential difference $\Delta\tilde{\mu}$, (1.2.12), and the reduced heat current $\tilde{Q}$, (1.2.11). From Griswold and Tough (1987).

The TI/TII transition is also clearly seen in a plot of the vortex line density L_0 obtained from the $\Delta\mu$ data using (1.2.10). In Figure 1.3 we show the dimensionless vortex line density $L_0^{1/2}d$ as a function of the reduced heat current $\tilde{Q}$. The solid line is calculated from the theory of Schwarz (1982) for homogeneous superfluid turbulence and is seen to agree well with the data at large heat current in the state TII. There is direct experimental evidence that the vortex line density in this state is homogeneous (Awschalom, Milliken and Schwarz, 1984). Evidently, the continuous transition in the vortex line density near the paracritical point represents an evolution from some inhomogeneous distribution of vortex lines in state TI to the homogeneous superfluid turbulent state. It is this TI/TII transition that is the subject of the present chapter.

1.3 Intrinsic fluctuations at the transition

Until quite recently, virtually all that was known about the continuous transition in the vortex line density between the steady states TI and TII was that it depended on the temperature of the He II and the size and geometry of the flow tube (Tough, 1982). These steady state results have now been supplemented by measurements of the intrinsic fluctuations in the dissipation

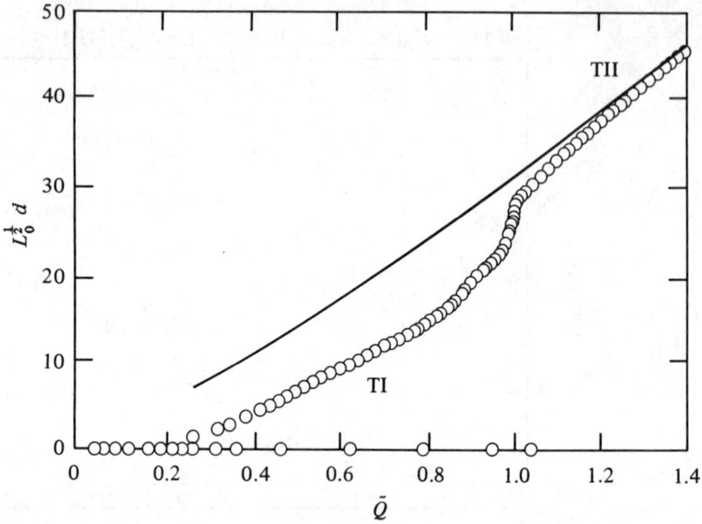

Figure 1.3. The dimensionless vortex line density $L_0^{1/2} d$ as a function of the reduced heat current $\tilde{Q}$ showing the transition from superfluid turbulent state TI to state TII. The solid line is computed from the theory of Schwarz (1982). Adapted from Figure 4 in Griswold, Lorenson and Tough (1987).

near the transition (Griswold, Lorenson and Tough, 1987; Lorenson, Griswold, Nayak and Tough, 1985). The fluctuations have rather featureless power spectra, but show a dramatic gain in amplitude in a small neighborhood of the transition. These data have lead Schumaker and Horsthemke (1987) to propose a dynamical model for the superfluid turbulence in the vicinity of the transition which is in excellent qualitative agreement with all of the experiments.

Before going on to discuss the fluctuations in the chemical potential, it is useful to consider the natural relaxation time of the superfluid turbulence which is known (Ladner, Childers and Tough, 1976; Martin and Tough, 1983) to become long near the transition. The relaxation time τ was determined from the exponential decay following a small step pulse in the heat current. In the experiments (see Figure 1.1), the heat current was stepped from $\dot{Q}$ to $\dot{Q} + \delta\dot{Q}$ ($\delta\dot{Q}/\dot{Q} \leqslant 0.4\%$) and the time response of the chemical potential difference was measured. A computer was used with a programmable power supply driving the cell heater, and the computer was used to collect the time response of the CPT output. A typical measurement would require averaging about 100 pulses. The relaxation times determined by fitting increasing exponentials to these data are shown in Figure 1.4. The results show that the relaxation time goes through a sharp maximum at 121 μW, and this value has been used for the paracritical point $\dot{Q}_p$ in the definition of the reduced heat current $\tilde{Q}$, see (1.2.11). Other methods have been used to determine the relaxation time (Griswold, Lorenson and Tough, 1987) and agree with the pulse data given here.

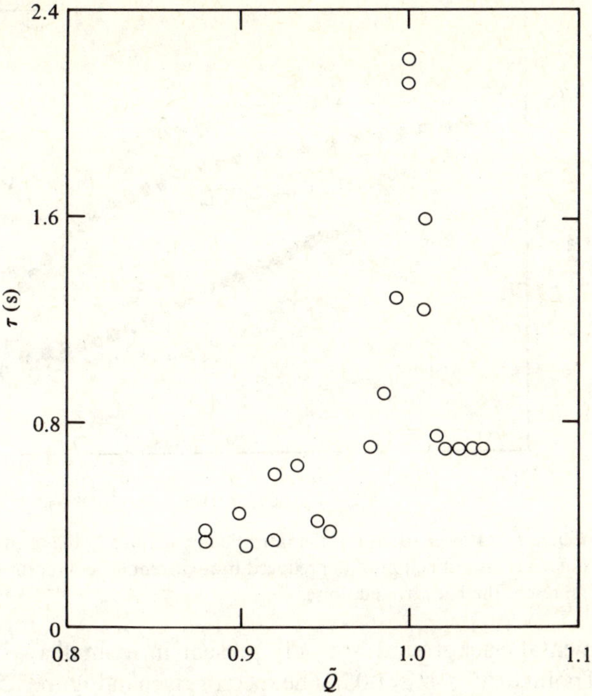

Figure 1.4. The relaxation time of the superfluid turbulence system as a function of the reduced heat current.

Intrinsic fluctuations in the chemical potential are observed in the neighborhood of the paracritical point similar to the phenomenon of enhanced critical fluctuations. Referring to the schematic drawing of the apparatus in Figure 1.1, the CPT output is a signal consisting of a steady part and a fluctuating part corresponding to a chemical potential of the form

$$\Delta\mu = \langle \Delta\mu \rangle + \delta\mu. \tag{1.3.1}$$

The average chemical potential $\langle \Delta\mu \rangle$ was actually used to plot the steady state data shown in Figures 1.2 and 1.3. The fluctuations $\delta\mu$ are studied by passing the CPT output signal through a low pass filter to the computer. The computer collects the time series data, stores it on diskettes, and can be used to determine the distribution function and power series from the stored data. The computer sampling rate and the filter rolloff frequency are set to avoid aliasing as per the Nyquist condition. The sensitivity of the CPT is sufficient to detect fluctuations equal to 0.1% of the steady state value.

The chemical potential time series data are fast Fourier transformed and the real and imaginary parts are squared. The squared parts are then summed and normalized for the sampling bandwidth to produce the power spectrum of the fluctuations at a given reduced heat current $\tilde{Q}$. We find that fluctuations above

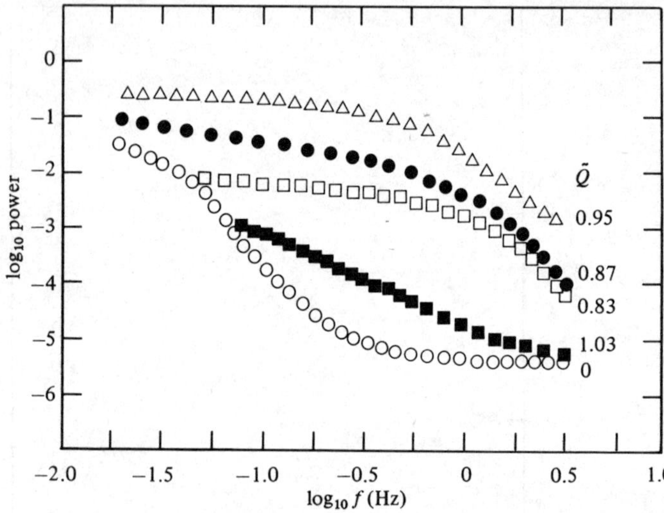

Figure 1.5. Power spectra of the intrinsic fluctuations in the chemical potential difference as a function of the reduced heat current. The data for $\tilde{Q}=0$ represent the background noise.

our experimental background are only present in a small region near the paracritical point ($0.85 \lesssim \tilde{Q} \lesssim 1.05$). The spectra given in Figure 1.5 are typical. The results shown are the average of 10 to 20 individual spectra, each taken over a given bandwidth. Bandwidths of d.c. to 0.1, 2, 5, and in some cases 10 Hz were used. The open circles in Figure 1.5 give the power spectrum of the background fluctuations and show a low frequency contamination due to the temperature regulation of the reservoir. There is little dramatic frequency information in these power spectra. There are no sharp spectral features that might be associated with routes to chaos observed in other dynamical systems. The chemical potential fluctuations appear as simple broad-band noise that can be reasonably fitted to a Lorentzian with a corner frequency determined by the relaxation time τ. Clearly the most dramatic features of these data are the growth of the noise power near the transition and its subsequent decay.

In order to systematically study the dependence of the intrinsic noise power on the reduced heat current $\tilde{Q}$ we have measured the amplitude of the power spectrum as a function of heat current at fixed frequency f_0. For example, referring to Figure 1.5, these measurements will show in detail how the amplitude of the noise power grows as $\tilde{Q}$ is increased from 0.83 to 0.95, and how it decreases again as $\tilde{Q}$ is increased to 1.03. These measurements are based on the fact that the power at any frequency f_0 is proportional to the square of the CPT output associated with that frequency component of the time series. The fluctuating signal from the CPT is passed through a narrow-band filter set at f_0. The filtered signal is then passed through an analog squarer to the computer where more than 10^4 data points are averaged. Since the power

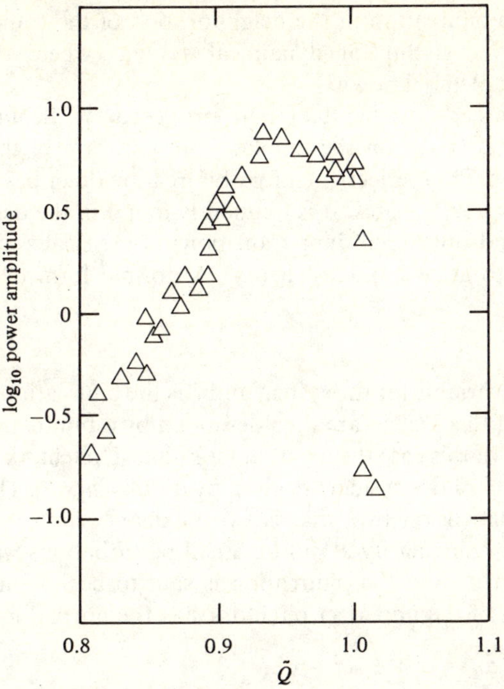

Figure 1.6. The power amplitude at 0.1 Hz as a function of the reduced heat current. Adapted from Figure 9(a) in Griswold, Lorenson and Tough (1987).

spectrum is essentially flat over the narrow bandwidth of the filter, this average is proportional to the power of the fluctuations at the center frequency f_0 of the filter.

The power amplitude data in Figure 1.6 were obtained with a frequency f_0 of 0.1 Hz. The power spectra are virtually flat at this frequency (Figure 1.5), and thus the results in Figure 1.6 should be quite representative of the actual variation of noise power amplitude in the vicinity of the TI/TII transition. The power amplitude has a maximum *slightly below* the paracritical point, and falls dramatically in the state TII. Even at the maximum power amplitude, however, the fluctuations are weak and represent only a few per cent of the steady state dissipation. Accordingly we find that the distribution of chemical potential fluctuations is Gaussian.

The results from the experiments outlined above (Griswold, Lorenson and Tough, 1987; Lorenson *et al.*, 1985) motivated Schumaker and Horsthemke (1987) to propose a model for superfluid turbulence near the TI/TII transition. The model is purely phenomenological, but is successful in giving a unified description of the steady states of the system (Figure 1.3), the relaxation times (Figure 1.4) and the noise power amplitude or the variance of the fluctuations (Figure 1.6). The idea is to identify the underlying bifurcation, and to use the normal form of the bifurcation in the neighborhood of the transition. This approach has proved useful for dynamical systems where the internal processes are incompletely known.

The model suggested by the data and proposed by Schumaker and Horsthemke (1987) is based on the unfolded normal form of the imperfect pitchfork bifurcation. The (perfect) pitchfork bifurcation describes continuous transitions between steady states. It is frequently found in the description of both equilibrium and nonequilibrium transitions in physical systems and is sometimes refered to as a 'Landau Theory'. A normal form of the perfect pitchfork is

$$g(x, \lambda) = -\lambda x - x^3, \tag{1.3.2}$$

where x is the state variable for the system and λ is the bifurcation parameter. The steady states of this system are then described by a bifurcation diagram given by $g(x, \lambda) = 0$. In this case the diagram looks like a 'pitchfork' opening to the left. The dynamics of this system are given by $\mathrm{d}x/\mathrm{d}t = g(x, \lambda)$. The pitchfork bifurcation has codimension two, that is two auxiliary parameters must be adjusted to reach the singularity at $g(0, 0)$. Small perturbations will generally destroy the singularity, and the bifurcation is said to become imperfect. A universal unfolding of the imperfect pitchfork has the normal form

$$p(x, \lambda) = \alpha_0 - \lambda x + \alpha_2 x^2 - x^3. \tag{1.3.3}$$

There are two special cases of the imperfect pitchfork which have codimension one, and are the lowest codimension bifurcations to describe a continuous transition. These are $\alpha_0 = 0$ (the 'transcritical' bifurcation) and $\alpha_0 = \alpha_2^3/27$ (the 'hysteresis' bifurcation).

When α_0 and α_2 have nonzero values the system still undergoes a continuous transition between steady states as λ is varied, but there is no longer a singularity. The stable steady state retains its stability, but that stability is weakened in the transition region. As a result, the magnitude of the fluctuations in the state variable and their relaxation times will be enhanced. The point on the bifurcation diagram where the relaxation time reaches its maximum value is called the paracritical point $(x_\mathrm{p}, \lambda_\mathrm{p})$.

The model of Schumaker and Horsthemke (1987) involves the deterministic equation $\mathrm{d}x/\mathrm{d}t = p(x, \lambda)$ near the transcritical or hysteresis limits, coupled to two sources of noise. The first of these is additive Gaussian white noise intended to represent the internal degrees of freedom of the system. The second is state dependent noise intended to represent the fluctuations in the environment. This term is chosen to be Ornstein–Uhlenbeck noise (Horsthemke and Lefever, 1984) which is coupled linearly to the state variable x. This noise is Gaussian and has a nonvanishing correlation time. If the noise sources are assumed weak, the stochastic differential equations can be linearized about the deterministic steady states, and solved to give the variance of the fluctuations in the state variable.

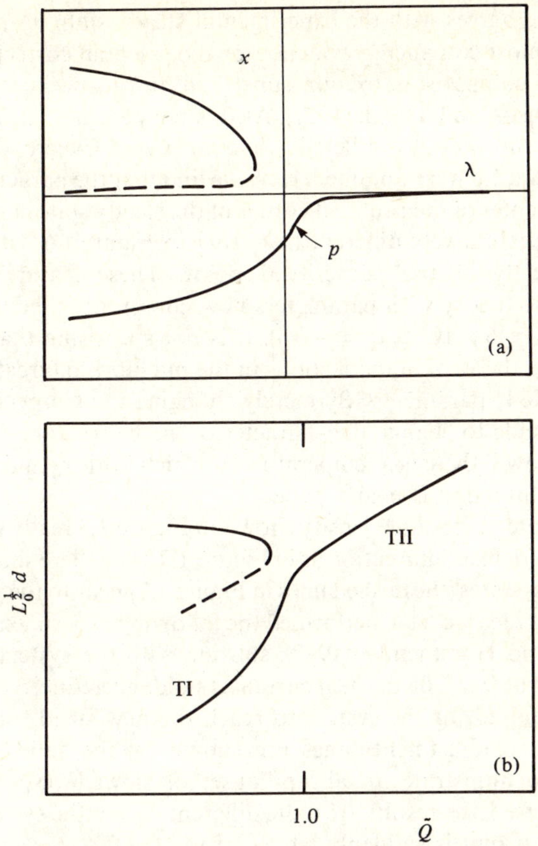

Figure 1.7. (a) Bifurcation diagram for the imperfect pitchfork in the near hysteresis limit. The paracritical point is labeled '*p*'. (b) The physical coordinates $L_0^{1/2}d$ and $\tilde{Q}$ corresponding to the imperfect pitchfork bifurcation shown in (a).

The model described above gives a very satisfactory qualitative representation of the experiments in superfluid turbulence near the TI/TII transition and shown in Figures 1.3, 1.4, and 1.6. Choosing parameters near the hysteresis bifurcation ($\alpha_0 = -8.6 \times 10^{-3}$, $\alpha_2 = -0.5$) gives a noise power or variance for the fluctuations which peaks just below the paracritical point as in the experiments. The steady states for this near hysteresis bifurcation, given by $p(x, \lambda) = 0$, are shown in Figure 1.7(a). These steady states are for the normal form of the imperfect bifurcation, and can be related to the steady states of a more general model by a smooth local transformation about the singular point. Figure 1.7(b) shows the corresponding steady states for the dimensionless vortex line density $L_0^{1/2}d$ as a function of the reduced heat current $\tilde{Q}$, which should be compared to Figure 1.3. The paracritical point is near the inflection point of the steady state curve which agrees with the experimental steady state results.

Probably the most convincing evidence for the essential correctness of the theory involves comparison to our superfluid turbulence data at 1.6 K (Griswold, Lorenson and Tough, 1987). At this temperature all of the He II parameters are substantially different than at 1.75 K where the data in Figures 1.3, 1.4 and 1.6 were obtained. Here we find that the paracritical point is now near the center of the rounded corner of the steady state curve, and the noise power is qualitatively different from that in Figure 1.6 with a second sharp peak exactly at the paracritical point. These features are fully reproduced in the theory with parameters now chosen near the transcritical bifurcation ($\alpha_0 = -6 \times 10^{-4}, \alpha_2 = -0.5$). It is not surprising that the parameters describing the weak imperfections in the pitchfork bifurcation should depend on the He II parameters. By merely changing the temperature of the system, it is possible to change the character of the bifurcation. It would be interesting to know if there is a temperature at which both α_0 and α_2 are very small and the bifurcation is nearly perfect.

If the steady state vortex line density shown in Figure 1.3 really is described by an imperfect hyteresis bifurcation as in Figure 1.7, then there should be a set of unstable steady states (the dashed lines in Figure 1.7) near to the paracritical point. To test this idea we have performed the following experiment suggested by Schumaker and Horsthemke (1987): starting with the system above the paracritical point at $\tilde{Q} = 1.08$ the heat current is suddenly reduced to a value $\tilde{Q}$, and the time required for the system to reach the new steady state at $\tilde{Q}$ is measured. These readjustment times are shown by the solid symbols in Figure 1.8 and demonstrate an abrupt onset of slowed response. If this procedure is reversed, the results are quite different. When the system starts at $\tilde{Q}$ and the heat current is suddenly increased to $\tilde{Q} = 1.08$, the readjustment time depends only weakly on $\tilde{Q}$ as shown by the open symbols. This asymmetry of readjustment times might be a signal of the unstable steady states of the imperfect hysteresis bifurcation which are only sampled when the heat current is reduced from above the paracritical point (Figure 1.7).

1.4 The effects of external noise on the transition

Moss and Welland (1982) were the first to suggest that superfluid turbulence in He II would be an excellent system in which to study noise induced phenomena in nonlinear dynamical systems. They applied the methods of Horsthemke and Lefever (1984) to the dynamical equation proposed by Vinen (1957). This equation is surely an inadequate description of the superfluid turbulence, particularly in the vicinity of the TI/TII transition where the Schumaker and Horsthemke (1987) analysis suggests the bifurcation is an imperfect pitchfork, see (1.3.3). Nevertheless, the early work of Moss and Welland was prophetic. The observation of noise induced bistability in superfluid turbulence (Griswold and Tough, 1987) certainly represents one of

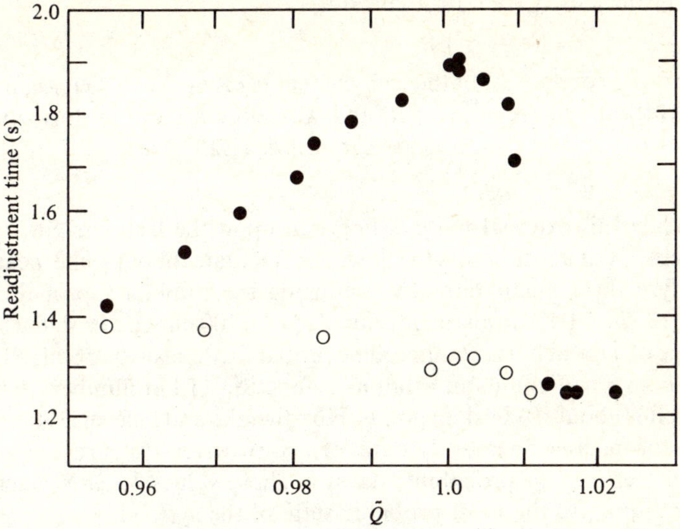

Figure 1.8. The readjustment time observed when the reduced heat current is switched between $\tilde{Q}$ and $\tilde{Q} = 1.08$. The open and closed symbols show the results for increasing and decreasing $\tilde{Q}$, respectively.

the few direct measurements of noise induced phenomena in a real physical system.

The discussion in the preceding sections of this paper indicate the reasons why superfluid turbulence is an excellent candidate for the study of external noise on a nonlinear dynamical system. The noise is easily added to the drive parameter (the heat current $\dot{Q}$), and the fluctuating output is available for computer analysis. The time constants of the system are well understood, and the turbulence appears to be well described by a single variable, which we take as the dimensionless vortex line density $L_0^{1/2}d$. Most important, the continuous TI/TII transition provides a distinctive feature of the system which can be studied as a function of the external noise.

The noise is introduced in the drive parameter by adding a noise component to the heater drive (see Figure 1.1). The output of a voltage noise source is sent through a low pass filter to one input of a summing amplifier. The d.c. component of the noise is zeroed and the output V_0 from a d.c. power supply is added to the noise in the summing amplifier and sent to the cell heater of resistance R. The voltage across the heater is then

$$V_{\mathrm{h}} = V_0 + \delta V, \tag{1.4.1}$$

where δV is Gaussian noise of zero mean and variance σ_v. The heat current driving the superfluid turbulence can then be written as

$$\dot{Q}_{\mathrm{h}} = \dot{Q} + \delta\dot{Q}, \tag{1.4.2}$$

where the steady state heat current is

$$\dot{Q} = (V_0^2 + \sigma_v^2)/R. \tag{1.4.3}$$

The noise component of the heat current $\delta\dot{Q}$ is now non-Gaussian (see below) of RMS amplitude $\sigma^2 = \sigma_v^2/R$, and we define a reduced noise amplitude in terms of the paracritical heat current (see (1.2.11)) as:

$$\tilde{\sigma}^2 = \sigma^2/\dot{Q}_p. \tag{1.4.4}$$

When this external noise is imposed upon the heat current the chemical potential difference across the flow tube fluctuates about some mean value. We analyze these fluctuations by producing the stationary probability density $p(\Delta\tilde{\mu})$. The CPT output range is divided into 100 bins, and a signal falling in the range of a given bin is counted. The probability density is obtained as a plot of the number of counts in a bin as a function of bin number. A typical plot contains about 50 000 data points. Horsthemke and Lefever (1984) have shown that the macroscopic steady states of a noise driven system are associated with the extrema of the probability density. Those values $\Delta\tilde{\mu}_m$ for which $p(\Delta\tilde{\mu})$ is a maximum give the most probable state of the system.

In our experiments (Griswold and Tough, 1987) we fix the noise amplitude $\tilde{\sigma}^2$ and obtain probability densities $p(\Delta\tilde{\mu})$ for different values of the heat current $\tilde{Q}$. From the maxima in each probability density we determine $\Delta\tilde{\mu}_m$ and the equivalent vortex line density L_m (see (1.2.11)). The stochastic steady states for the dimensionless vortex line density $L_m^{1/2}d$ can then be compared with those for the deterministic system given in Figure 1.3. Consider first the results obtained with very weak external noise where we expect the noise induced effects to be insignificant. Figure 1.9 shows the probability densities for several values of the reduced heat current $\tilde{Q}$ using a relative external noise amplitude $\tilde{\sigma}^2$ of 0.33%. The probability density data appear as well-resolved single peaks. Using the maxima in the peaks to define $\Delta\tilde{\mu}_m$ and $L_m^{1/2}d$ as described above gives the points shown on the 'phase diagram' in Figure 1.10. The dashed line reproduces the deterministic steady states from Figure 1.3 for comparison. Clearly the only effect of this weak external noise is to cause the system to fluctuate about its deterministic steady state.

Large amplitude external noise causes major qualitative changes in the bifurcation at the TI/TII transition. Several examples of the probability density obtained with $\tilde{\sigma}^2 = 49\%$ are shown in Figure 1.11. The data at $\tilde{Q} =$ 1.23, 1.29 and 1.39 show broad peaks at values of $\Delta\tilde{\mu}_m$ corresponding to the deterministic TII state. The scale of Figure 1.11 is not sufficient to resolve the peak in the probability density at small values of $\Delta\tilde{\mu}$. An example of a probability distribution where this peak is resolved is given in Figure 1.12. Higher resolution scans show that this lower peak is present in all the distributions shown in Figure 1.11. Clearly the distributions at the higher heat currents have become bimodal, and the external noise has induced bistability.

Figure 1.13 shows a 'phase diagram' for the stochastic steady states with $\tilde{\sigma}^2 = 49\%$. The maxima in the probability densities are used to define $L_m^{1/2}d$

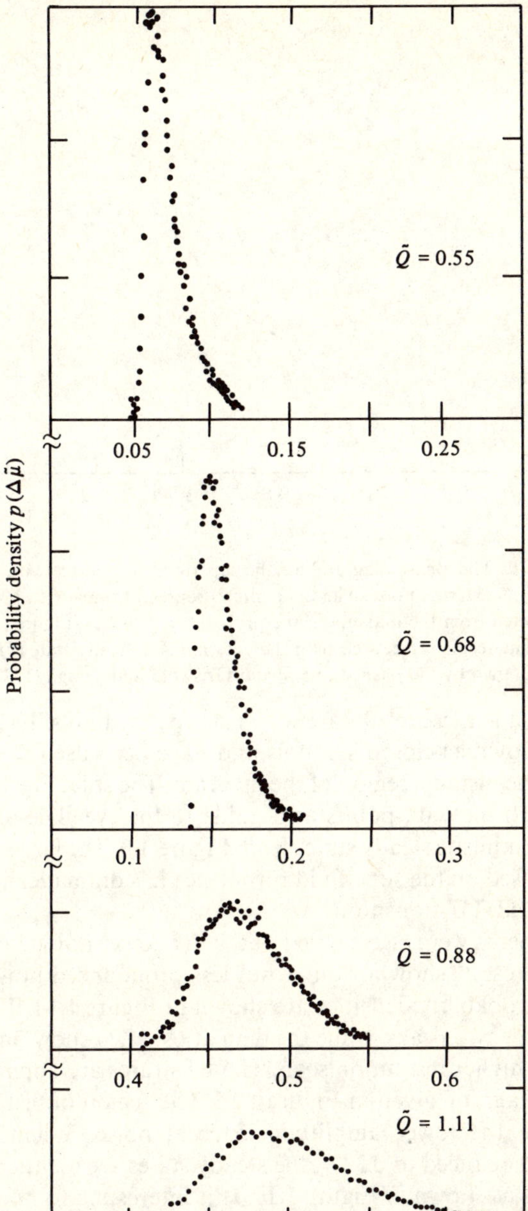

Figure 1.9. Probability densities obtained with low level ($\tilde{\sigma}^2 = 0.33\%$) external noise at several different heat currents. Note that the zero is suppressed on the $\Delta\tilde{\mu}$ scale to reveal the narrow peaks. From Griswold and Tough (1987).

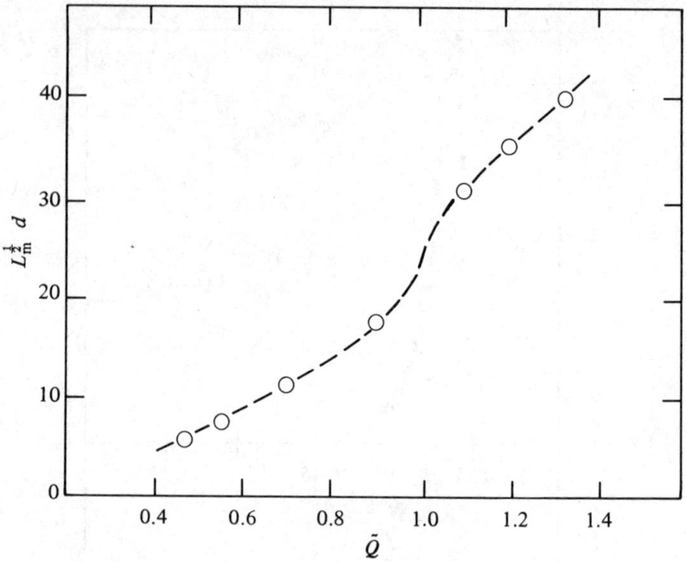

Figure 1.10. The 'phase diagram' for the superfluid turbulence with low level ($\tilde{\sigma}^2 = 0.33\%$) external noise. The open circles represent the vortex line densities L_m computed from the maxima in the probability densities (Figure 1.9) using (1.2.10). The results agree well with the deterministic steady state (Figure 1.3) which is shown by the dashed line. From Griswold and Tough (1987).

as described above, and these results are shown as open symbols. The *minima* in the densities are shown as closed symbols and have been used to define the unstable states in the bistable region of the diagram. The solid line has merely been drawn through the data points as a guide to the eye. The dashed line represents the deterministic steady states from Figure 1.3. The large amplitude external noise imposed on the superfluid turbulence has dramatically changed the features of the TI/TII transition.

Identical experiments were also carried out with a lower noise amplitude of $\tilde{\sigma}^2 = 15\%$, and the results show a similar but less pronounced noise induced bistability. Some probability densities are shown in Figure 1.14. The data at $\tilde{Q} = 1.02$ clearly show two peaks, while the data at $\tilde{Q} = 1.24$ show only a single peak even under a higher resolution scan. These features are apparent in the stochastic 'phase diagram' given in Figure 1.15. The region of bistability has been reduced with the lower amplitude external noise. When the noise amplitude is further reduced to 0.33%, the steady states are identical with the deterministic states as shown in Figure 1.10. It is interesting to note that this behavior of the TI/TII transition in superfluid turbulence is just the reverse of the model bistable system studied by Welland and Moss (1982). In that model, the degree of hysteresis in a deterministically bistable system is reduced as the noise level is increased.

It would clearly be useful to extend the model of Schumaker and

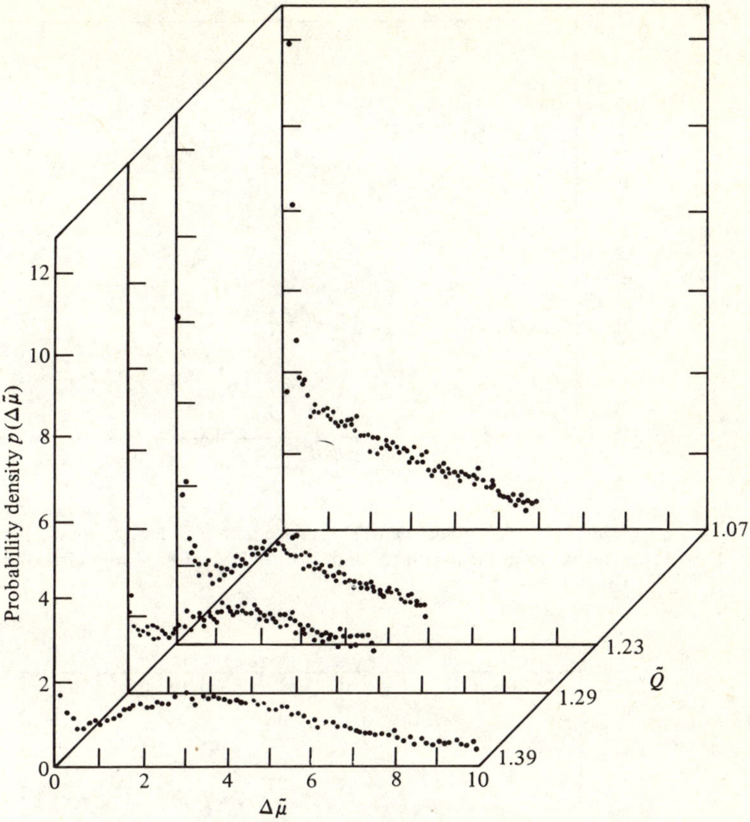

Figure 1.11. Probability densities obtained with high level ($\tilde{\sigma}^2 = 49\%$) external noise at several different heat currents near the paracritical point ($\tilde{Q} = 1$). At the larger values of $\tilde{Q}$ the densities become bimodal, although the lower peak is not resolved in this figure.

Horsthemke (1987) to include the case of external noise on the drive parameter. There are two aspects of our experiment, however, which will make comparison with theory somewhat complicated. Our noise is colored and quadratic. The noise is colored since the noise voltage driving the heater is passed through a low pass filter set at 5Hz. The relaxation time of the system is sufficiently long (see Figure 1.4) that noise of higher frequencies simply would not appear in the He II as uniform fluctuations of the heat current. The system itself 'filters out' higher frequencies. The noise is also quadratic and non-Gaussian. This follows because the noise in the bifurcation parameter (the heat current, (1.4.2)) is obtained by squaring a noise voltage. These experimental difficulties can hopefully be eliminated in a more thorough future study.

In concluding this article on the effects of noise on the TI/TII transition in superfluid turbulence, it is appropriate to mention a very early series of

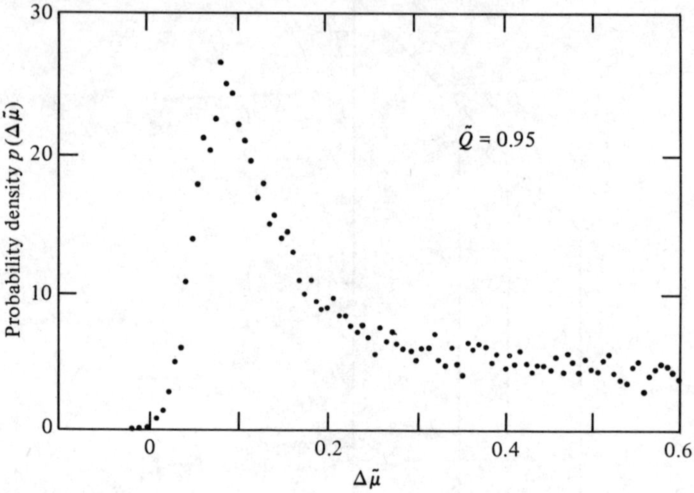

Figure 1.12. Probability density obtained with high level ($\tilde{\sigma}^2 = 49\%$) external noise in which the narrow peak at low $\Delta\tilde{\mu}$ is resolved. From Griswold and Tough (1987).

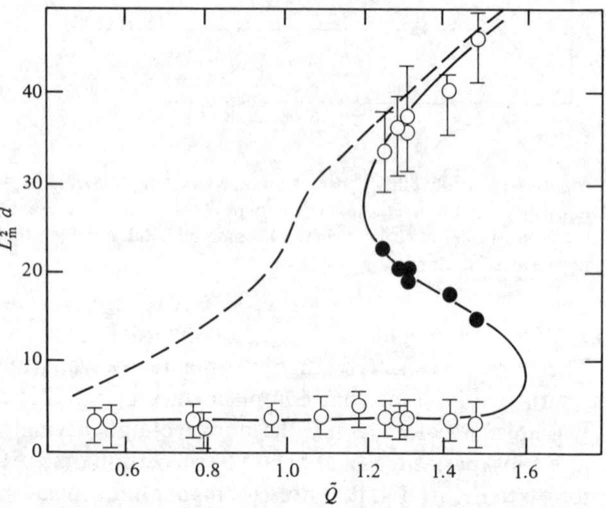

Figure 1.13. The 'phase diagram' for superfluid turbulence with high level ($\tilde{\sigma}^2 = 49\%$) external noise. The open symbols represent values of the vortex line density L_m computed from the maxima in the probability densities (Figures 1.11 and 1.12) using (1.2.10). The solid symbols are obtained in a similar way from the minima, and represent unstable states of the system. The solid line is simply a guide to the eye emphasizing the bistability induced by the noise. The dashed line is the deterministic steady state (Figure 1.3). From Griswold and Tough (1987).

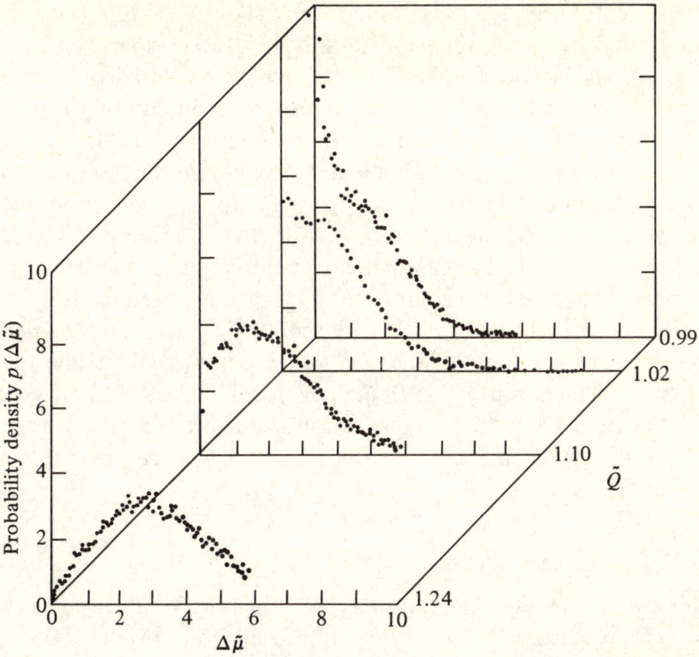

Figure 1.14. Probability densities obtained with intermediate level ($\tilde{\sigma}^2 = 15\%$) external noise at several different heat currents near the paracritical point ($\tilde{Q} = 1$). At the largest value there is no longer any evidence of bistability.

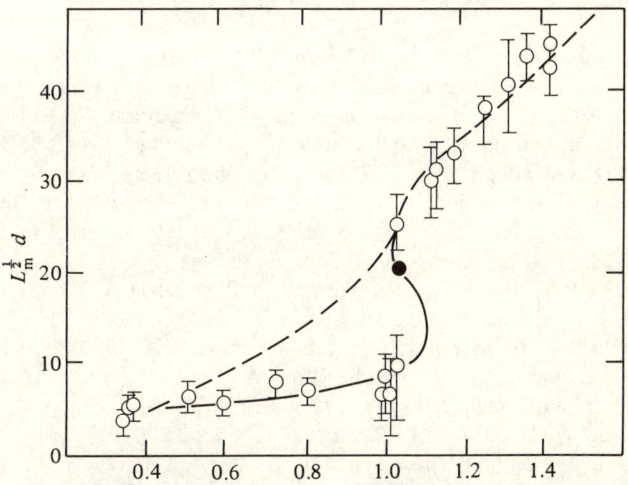

Figure 1.15. The 'phase diagram' for the superfluid turbulence with intermediate level ($\tilde{\sigma}^2 = 15\%$) external noise. The symbols are the same as in Figure 1.13. The noise induced bistability is much reduced compared to the high level noise results in Figure 1.13. From Griswold and Tough (1987).

experiments (Oberly, 1971; Oberly and Tough, 1972) which may now be interpreted as noise induced bistability. In these experiments an oscillatory (*single frequency*) component was added to the driving heat current. The frequency was sufficiently low that a uniform modulation was produced in the He II sample. The dissipation due to the superfluid turbulence was detected as a periodic temperature difference, and the median value was taken to represent the steady state. The experiments showed that a large amplitude oscillatory component in the heat current made the continuous TI/TII transition discontinuous, and increased the 'critical' value of the heat current at the transition. The results are in general qualitative agreement with the present data shown in Figures 1.10, 1.13 and 1.15. These early experiments also showed that the effect of the oscillatory component on the transition was maximized for a particular frequency f_m. This was interpreted as a *spacial* effect, where the viscous penetration depth $(\eta^{1/2}/\pi\rho f_m)$ was equal to the tube radius. The meaning of this result in the context of noise induced transitions is not clear.

Acknowledgement

This work has been supported by the National Science Foundation, Low-Temperature Physics Program, Grant No. DMR8218052.

References

Awschalom, D. D., Milliken, F. P. and Schwarz, K. W. 1984. *Phys. Rev. Lett.* **53**, 1372.
Barenghi, C. F., Donnelly, R. J. and Vinen, W. F. 1983. *J. Low Temp. Phys.* **52**, 181.
Glaberson, W. I. and Donnelly, R. J. 1985. In *Progress in Low Temperature Physics, vol. IX* (D. F. Brewer, ed.), pp. 1–142. Amsterdam: North-Holland.
Griswold, D., Lorenson, C. P. and Tough, J. T. 1987. *Phys. Rev. B* **35**, 3149.
Griswold, D. and Tough, J. T. 1987. *Phys. Rev. A* **36**, 1360.
Horsthemke, W. and Lefever, R. 1984. *Noise Induced Transitions.* Berlin: Springer.
Ladner, D. L., Childers, R. K. and Tough, J. T. 1976. *Phys. Rev. B* **13**, 2918.
Landau, L. D. 1941. *J. Phys. (USSR)* **5**, 71.
Lorenson, C. P., Griswold, D., Nayak, V. U. and Tough, J. T. 1985. *Phys. Rev. Lett.* **55**, 1494.
McClintock, P. V. E., Meredith, D. J. and Wigmore, J. K. 1984. *Matter at Low Temperatures.* New York: John Wiley.
Martin, K. P. and Tough, J. T. 1983. *Phys. Rev. B* **27**, 2788.
Moss, F. and Welland, G. V. 1982. *Phys. Rev. A* **25**, 3389.
Oberly, C. E. 1971. Ph.D. Thesis (The Ohio State University).
Oberly, C. E. and Tough, J. T. 1972. *J. Low Temp. Phys.* **7**, 223.
Schumaker, M. F. and Horsthemke, W. 1987. *Phys. Rev. A* **36**, 354.
Schwarz, K. W. 1982. *Phys. Rev. Lett.* **49**, 283.
Tilley, D. R. and Tilley, J. 1974. *Superfluidity and Superconductivity.* New York: Van Nostrand Reinhold.

Tough, J. T. 1982. In *Progress in Low Temperature Physics, vol. XIII* (D. F. Brewer, ed.), pp. 133–220. Amsterdam: North-Holland.
Vinen, W. F. 1957. *Proc. Roy. Soc. A* **242**, 493.
Welland, G. V. and Moss, F. 1982. *Phys. Lett. A* **89**, 273.
Wilkes, J. and Betts, D. S. 1987. *The Properties of Liquid and Solid Helium.* Oxford: Clarendon Press.
Yarmchuck, E. J. and Glaberson, W. I. 1979. *J. Low Temp. Phys.* **36**, 381.

2 Electrohydrodynamic instability of nematic liquid crystals: growth process and influence of noise

S. KAI

2.1 Introduction

The electrohydrodynamic (EHD) instability in nematic liquid crystals (NLCs) has received considerable attention in recent years for two main reasons. The first is that it enables one to study the formation process of a dissipative structure including the onset of chaos (Hirakawa and Kai, 1977; Kai, Araoka, Yamazaki and Hirakawa, 1979a; Kai and Hirakawa, 1977, 1978; Kai, Wakabayashi and Imasaki, 1986). The second is that it allows experimental access to a multiplicative stochastic process (Schenzle and Brand, 1979) because it is easy to obtain a fluctuating control parameter, the so-called multiplicative noise (Kai, Kai, Takata and Hirakawa, 1979b; Kawakubo, Yanagita and Kabashima, 1981). These two aspects of the EHD instability involve a wide range of interesting new phenomena that need to be accounted for. The onset of dissipative structures (pattern formations) can be observed corresponding to instabilities relaxing from unstable to stable states of thermodynamically excited systems. The dynamical phenomena near such instability points have been studied in various fields, such as hydrodynamic instabilities (Croquette and Pocheau, 1984; Greenside and Coughran, 1984; Greenside and Cross, 1985; Pocheau and Croquette, 1984; Pomeau and Zaleski, 1983; Tesauro and Cross, 1986); phase separations (Furukawa, 1985; Gunton, San Miguel and Sohni, 1983; Komura, Osamura, Fujii and Takeda, 1984); oscillatory instabilities in electrical circuits (Kabashima, Itsumi, Kawakubo and Nagashima, 1975); and in laser radiation (Arecchi and Degiorgio, 1971). The multiplicative noise process has been also investigated in some of these fields (Fox, James and Roy, 1984; Kabashima, Kogure, Kawakubo and Okada, 1979; Linz and Lucke, 1986; Mannella, Moss and McClintock, 1987; Masoliver, West and Lindenberg, 1987; Moss and Welland, 1982; Sancho, San Miguel, Yamazaki and Kawakubo, 1982; Smythe, Moss and McClintock, 1983), but there have been very few experimental studies. In this chapter we review our recent experimental studies of EHD instability in NLCs on the pattern formation and its multiplicative stochastic processes, and introduce a new concept concerning an analogy with a crystal growth.

2.2 Electrohydrodynamics

2.2.1 Nematic liquid crystals and electrohydrodynamics

Nematic liquid crystals (NLCs) are molecules of a rod-like shape whose local orientation is defined as the director (de Gennes, 1982). An optical anisotropy of liquid crystals originates in this orientational order. The director aligns parallel to glass plates in the x direction when one encloses a thin layer (100 μm) of liquid crystals between two glass plates (Figure. 2.1a). This is called homogeneous alignment in terms of liquid crystals. On applying an electric field, an instability takes place when the intensity of the field exceeds a certain threshold (Figure 2.1b). This happens due to the dielectric and conductive anisotropies of liquid crystals, and therefore is called the electrohydrodynamic (EHD) instability. The basic mechanism is schematically illustrated in Figure 2.1.

EHD instability is similar in some ways to Rayleigh–Bénard (R–B) instability (a thermally excited convection), when the applied electric field and its frequency correspond to the temperature difference between the bottom and the top of a fluid layer (the Rayleigh number, R) and the characteristic time of the fluid (the Prandtl number, P), respectively (Kai and Hirakawa, 1978). The properties of the bifurcation and the flow structure in R–B convection can be determined by the parameters R and P and by the aspect ratio Γ of the container, the ratio of the lateral size l to the thickness d of a layer (Gollub and Benson, 1980). In the EHD case, the circumstances are similar; that is, the instability is characterized by the normalized voltage $k = V/V_c$ (or more suitably $\varepsilon = (V^2/V_c^2) - 1$), the normalized frequency $P_e = f_c/f$ and Γ, the aspect ratio, of the cell. Here V_c is a threshold voltage at the first instability, the so-called Williams domain (WD) instability, and f_c is a critical frequency separating two different regimes: the conductive ($f < f_c$) and the dielectric ($f > f_c$). Then f_c is determined by relaxation processes of inhomogeneities (fluctuations) in the system (de Gennes, 1982; Dubois-Violette, de Gennes and Parodi, 1971).

Such fluctuations relax through two different processes: first, that corresponding to the relaxation of space charge q by the anisotropic conductivity $\sigma_a = \sigma_\parallel - \sigma_\perp$, and, secondly, through the relaxation of the director by the anisotropic dielectric constant $\varepsilon_a = \varepsilon_\parallel - \varepsilon_\perp$ and the elastic constant K of liquid crystals. Here '$\parallel$' and '$\perp$' denote the components parallel and normal to the director, respectively. The critical frequency f_c, therefore, is determined by a balance between these two processes and is given by $f_c \sim \tau_c^{-1} = \varepsilon_\parallel/4\pi\sigma$ or, more appropriately, $= \varepsilon_H/4\pi\sigma_H$, where ε_H and σ_H are, respectively, Helfrich's dielectric and conductive parameters containing ε_a, σ_a and the anisotropic viscosities (Helfrich, 1969; Smith *et al.*, 1975). Thus the convective and the oscillatory instabilities occur for $f < f_c$ and $f > f_c$, respectively. The dynamical behavior of the phenomenon can be described by a set of linear parametric

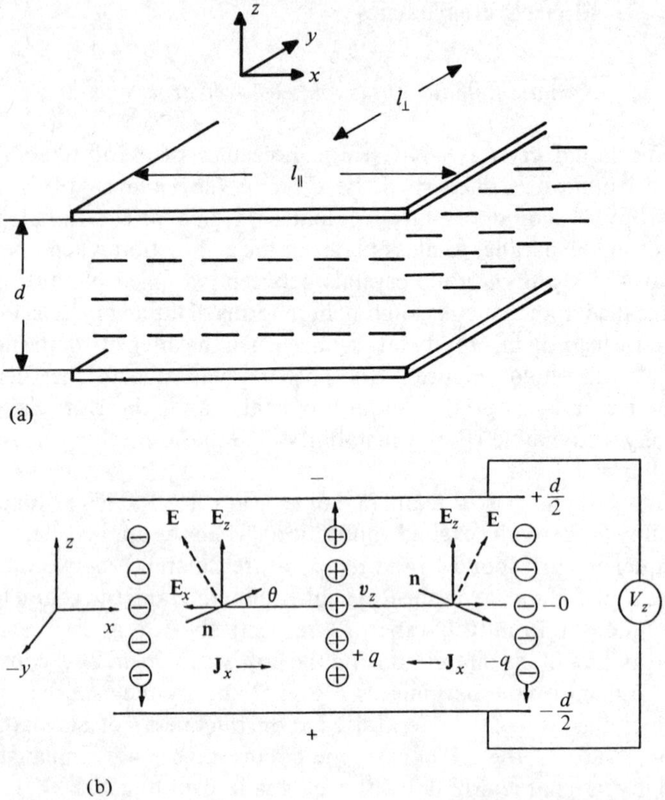

Figure 2.1. Director orientation in a thin (~ 100 μm) nematic film with a negative dielectric anisotropy ($\varepsilon_a = \varepsilon_\parallel - \varepsilon_\perp < 0$). (a) Undisturbed homogeneous alignment at $E = 0$. d is the thickness of the film (~ 100 μm); $l_\parallel$ and $l_\perp$ are the lateral lengths parallel and perpendicular to the director orientation, respectively. From these two lateral scales, we determine two aspect ratios $\Gamma_\parallel = l_\parallel/d$ and $\Gamma_\perp = l_\perp/d$. (b) Periodic deformation of the director at the onset of the electrohydrodynamic (EHD) instability for the applied field $|\mathbf{E}|$ larger than the threshold value $|\mathbf{E}_c|$. $\mathbf{E}_x$ = induced electric field in the x direction; $\mathbf{E}_z$ = externally applied electric field; $\mathbf{E}$ = composite electric field ($= \mathbf{E}_x + \mathbf{E}_z$); $\mathbf{J}_x$ = current flow in the x direction; $\mathbf{n}$ = director. Supposing that fluctuation of the deformation mode shown in the figure occurs when $E > E_c$, which has the largest eigenvalue, the excess space charge q_x starts to be stored in the x direction by the current flow $\mathbf{J}_x$ due to the conductive anisotropy $\sigma_a = \sigma_\parallel - \sigma_\perp > 0$. This excess charge induces the additional electric field $\mathbf{E}_x$ in the x direction. The composite field $\mathbf{E} = \mathbf{E}_z + \mathbf{E}_x$ works to maintain or to increase the deformation because the director always becomes aligned normal to $\mathbf{E}$. Simultaneously, molecules in the region with the space charge q_x move toward the electrodes due to the electrostatic force. This induces a flow v_z which inclines the director further. Thus, the periodic director deformation grows without decaying and is realized as the macroscopic deformation (c). These processes are continued. This feed-back mechanism is called the Carr-Helfrich effect.

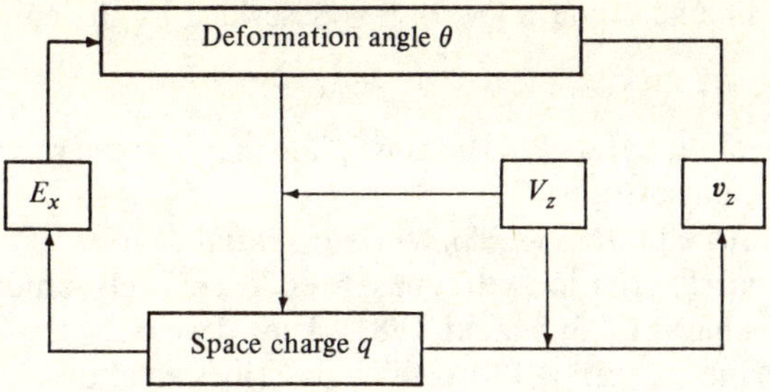

(c)

equations with two variables q and ψ;

$$\dot{q} + q(t)/\tau_c + \sigma_H E_m \psi(t) \cos \omega t = 0 \quad (2.2.1)$$

$$\dot{\psi}(t) + \lambda(E_0^2 + E_m^2 \cos^2 \omega t)\psi(t) + (E_m/\zeta) q(t) \cos \omega t = 0, \quad (2.2.2)$$

given by the Orsay group (Dubois-Violette, de Gennes and Parodi, 1971). Here, $\psi = \mathrm{d}\theta/\mathrm{d}x$ (θ is the director angle with respect to the x axis), and $E_m \cos \omega t$ is the externally applied electric field with an angular frequency $\omega(= 2\pi f)$. ζ and λ are the effective viscous and the effective elastic relaxation coefficients, respectively (see Figure 2.1). Readers requiring more detail on EHD may refer to the following excellent text book and original papers: Dubois-Viollete, de Gennes and Parodi (1971), de Gennes (1982), and Smith *et al.* (1975). Because they are linearized equations, however, (2.2.1) and (2.2.2) can only predict the onset of the first instability and give the threshold value V_c; they cannot describe features of the patterns, the steady-state amplitude of unstable modes and detail of structures above V_c such as a second bifurcations and the transient dynamics of new patterns.

2.2.2 *Macrodynamics on a pattern formation*

If an electric field higher than $E_c(= V_c/d)$ is applied stepwise, the new ordering pattern grows depending upon the frequency of the field. Such a pattern is characterized by the amplitude of the convective velocity ($f < f_c$), or by an oscillatory amplitude of the director ($f > f_c$). The evolution of the pattern may be experimentally obtained from optical measurement because the temporal change of the director angle causes a change in the optical anisotropy. The amplitude equation of motion describing the above nonlinear macrodynamics is in general given by the following time dependent Ginzberg–Landau (TDGL) equation,

$$\dot{Y} = \alpha Y - \beta Y^3 + \nabla^2 Y, \quad (2.2.3)$$

where Y is the amplitude of the director angle θ (Hijikuro, 1975, for EHD; Newell and Whitehead, 1969, for R–B convection). For two-dimensional flow such as in the Williams domain (WD), θ is described by the equation

$$\theta(x,t) = Y \sin k_x x, \tag{2.2.4}$$

where x is the direction parallel to the glass plate and k_x is the wavenumber in the x direction (see also Figure 2.1).

The homogeneous equation (2.2.3), without spatial degrees of freedom but containing thermal fluctuation $\eta(t)$, has been extensively studied (Kubo, Matsuno and Kitahara, 1973; Suzuki, 1981). They developed the fluctuation-enhanced theory and its scaling form which describes an anomalous increase in the variance at the onset time t_{m}. In a later section we will discuss an important physical meaning of t_{m}. When systems have spatial degrees of freedom, however, such dynamical behavior becomes more complicated and difficult to obtain analytically for the entire growth dynamics. The transient kinetics of these inhomogeneous phenomena, therefore, is studied by dividing them into several characteristic time stages. At the beginning, the macroscopic order grows exponentially from thermal fluctuation. Then, in the second stage, nonlinear growth dominates due to the terms Y^3 and $\nabla^2 Y$, the so-called wavenumber selection and phase dynamics for the late stage. The final step of the growth involves the motion of defects; that is, defect dynamics plays an important role in making a finally uniform structure for the entire area.

If a noise term is included in the above equation, it becomes the following type of differential stochastic equation:

$$\dot{Y} = \alpha Y - \beta Y^3 + \nabla^2 Y + \xi(t) Y + \eta(t). \tag{2.2.5}$$

This type of homogeneous equation has been widely studied for the last decade from the viewpoints of both additive and multiplicative noises (Horsthemke and Lefever, 1984). The inhomogeneous equation, however, has not been tackled in much detail (Feigelman and Staroselsky, 1986).

Since Γ is closely related to the degrees of freedom of spatially excited modes, a large number of modes may be enhanced, the behavior of the flow becoming much more complicated for large Γ. But for small Γ only a few modes may grow stably and therefore the flow's behavior may be relatively simple. Accordingly the 'route to chaos' in EHD (as well as in R–B) instability strongly depends on Γ (Kai, 1986). Furthermore, contrary to R–B instability of isotropic fluids, the symmetry is initially broken in EHD instability because of the director orientation due to the strong surface anchoring force. This especially happens in a very thin cell, typically less than $d \lesssim 4\,\mu$m or in a relatively thick cell ($d \lesssim 50\,\mu$m), with a special surface treatment such as polymer-coated glass. Therefore we have to take into account the thickness d as an additional control parameter indicating the interfacial interaction with substrates, even though d is already involved in Γ. Because of this initial symmetry breaking even in a stable state, we have to distinguish between the aspect ratios $\Gamma_{\parallel}$ (parallel) and $\Gamma_{\perp}$ (perpendicular to the director) (Kai, 1986). The onset stage of defects strongly depends upon Γ and an applied field frequency by which the roll size can be controlled. This is an additional feature of EHD instability. Such initial conditions are important factors in the control of the transient kinetics of pattern formations. It is therefore possible to demonstrate a quite new aspect of hydrodynamics phenomena.

2.2.3 Multiplicative stochastic process in electrohydrodynamics

One other interesting feature of EHD instability is that it gives a nonlinear multiplicative stochastic process (MSP) when noise, $\xi(t)$, is superimposed on to a deterministic field (Brand, Kai and Wakabayashi, 1985; Kai *et al.*, 1979b; Kawakubo, Yanagita and Kabashima, 1981). According to preliminary results achieved by Kai *et al.* (1979a, b), the multiplicative noise process in EHD instability has been shown to stabilize the system; that is, the thresholds for the onset of instabilities for both the Williams domain and turbulent flow were shifted upward. This was explained qualitatively by the fact that the external noise could potentially excite many small scale convective modes while the deterministic field excited the most stable mode k_m. It was considered that such modes with higher wavenumbers might cancel the store of space charge necessary to induce the convective instability. Theoretical explanations for such a shift of the threshold were given by Brand and Schenzle (1980), Kawakubo, Yanagita and Kabashima (1981), and very recently by Behn and Müller (1985), and Müller and Behn (1987), who took into account the frequency dependence of the thresholds based on (2.2.1) and (2.2.2) and gave the phase diagram.

In MSPs of liquid crystals, however, an important fact is the existence of a nonlinear effect, i.e. the quadratic noise term coupled with Y, as well as spatial structures. This leads to different noise effects from the linear coupling term $\xi(t)\,Y$ in dynamical properties of EHD instability. The contribution of this type of quadratic term in the Freedericksz transition of a liquid crystal may be described by modifying the original potential profile (Horsthemke, Doering, Lefever and Chi, 1985; Sagues and San Miguel, 1985; Sagues, San Miguel and Sancho, 1984). Experimental evidence appearing to support this will be also observed in EHD. Here the Freedericksz transition is a magnetic-field-induced instability which occurs in the following way. When the field becomes higher than the threshold, H_c, the director alignment is uniformly switched from its initial direction (normal to an applied magnetic field) to one parallel to the field (Figure 2.2). Theoretical results suggest the following facts for this instability (Sagues and San Miguel, 1985). The threshold field H_c increases with an increase in the external magnetic-field noise. The influence on the relaxation time is stronger than that for the threshold shift. Beyond the instability point the mean first passage time becomes shorter with increase in noise intensity, but becomes longer below H_c. These changes can be understood in terms of a modification by noise of the curvature of the potential.

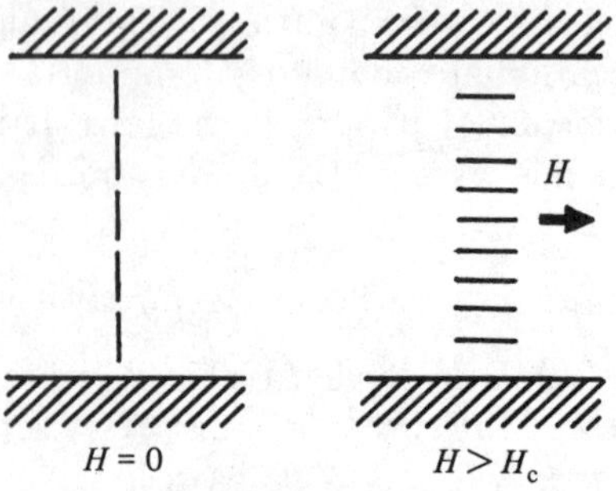

Figure 2.2. Magnetic-field-induced Freedericksz transition.

On the other hand, the steady state distribution at $H > H_c$ takes several forms, depending upon the intensity and the correlation time of applied noises (Horsthemke *et al.*, 1985). Such a multistability has been also obtained theoretically in an optical system (Broggi and Lugiato, 1984; Fedchenia, 1984). One frequently calls this the noise-induced transition. On increasing the correlation time of the noise, the shift of H_c^2 becomes smaller, until no shift occurs for any intensity of the noise at a correlation time comparable to that of the deterministic field (Horsthemke *et al.*, 1985). Finally, noise with a long correlation time can destabilize the system, i.e. H_c^2 decreases with the increase in the noise intensity. We will describe similar behavior of EHD instability later in this chapter. Qualitatively, if the correlation time of the noise approaches the characteristic time of the deterministic field, it may not act as noise but works cooperatively with the deterministic field (Graham and Schenzle, 1982; Kai *et al.*, 1985). The similar experimental fact observed in EHD, therefore, might be basically explained by the above theories in a similar way.

There is, however, a great discrepancy in EHD instability from the Freedericksz transition; that is, EHD instability has a spatial pattern, which makes the problem complicated. For such a spatially inhomogeneous system, a spatially uniform noise does not in fact behave as such when applied to the system in the usual way. The effective spatial property of noise thus depends on that of the object studied. There seems therefore to be no other study of MSPs in systems with spatial degrees of freedom. Interesting points in the present study are as follows: (1) How does the external noise affect dynamical properties of the pattern formation, such as the growth time and the defect motion? (2) How does it change a spatial structure? (3) Is there any universal influence by the noise on the systems?

2.2.4 *Analogy between pattern formation processes in EHD instability and in a first order phase transition*

One well-investigated example with spatial degrees of freedom is the phase-separation process, which is a transient phenomenon in a process going from

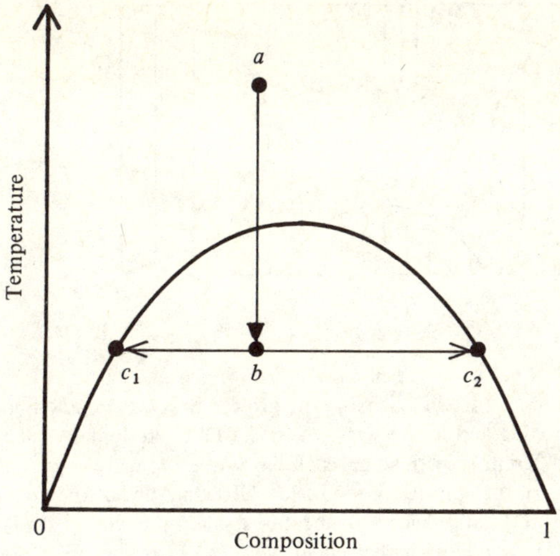

Figure 2.3. Phase separation in a quenching system. A system such as a melt is thermodynamically unstable when it is quenched from a to b. The nucleation or the spinodal decomposition, depending on the quenching point in the phase diagram, takes place by the driving force due to the thermodynamical instability. After sufficiently nucleating and growing, they reach an almost steady state, and then the ripening plays an important role in the growth. That is, the growth kinetics changes.

nonequilibrium to equilibrium. Qualitatively it is somewhat similar to the pattern formation process in EHD instability, except for the conservation law. In spinodal decomposition in alloys, for instance, the order parameter is conserved (Furukawa, 1985),

$$\dot{Y} = \nabla^2(\alpha Y - \beta Y^3 + \nabla^2 Y), \tag{2.2.6}$$

since the integration of (2.2.6) is given by

$$\int \dot{Y}\,\mathrm{d}x\,\mathrm{d}y\,\mathrm{d}z = \int \nabla^2(\alpha Y - \beta Y^3 + \nabla^2 Y)\mathrm{d}x\,\mathrm{d}y\,\mathrm{d}z = 0. \tag{2.2.7}$$

Its growth kinetics therefore are clearly different from the nonconserving case (2.2.3) such as EHD instability. Recently, however, a large increase in electrical fluctuations was observed during a process of crystallization from supercooled liquid glycerol (Tazaki and Yamaguchi, 1984) which resembled transient phenomena in EHD instability (Kai, Wakabayashi and Imasaki, 1986) and in laser radiation (Arecchi and Degiorgio, 1971). An idea for the separation process therefore might be applied extensively to other systems with spatial degrees of freedom (Kai, Wakabayashi and Imasaki, 1986; San Miguel and Sagues, 1987).

Let us consider the situation using an analogy with crystal growth (see Figure 2.3). The state at $t < t_{\mathrm{m}}$ is mainly dominated by the nucleation process of the macroscopic order and the growth process without the interaction.

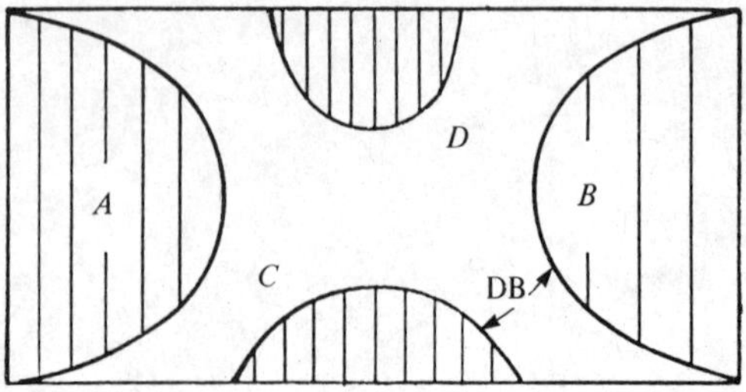

(a)

Figure 2.4. Analogy between convective domain growth in EHD (a) and crystal growth in first order kinetics (b). The growth process is made under relatively deep quenching and for very large Γ in EHD. Each domain, such as *A*, *B*, *C* or *D*, is the different growth stage; each have different wavelengths, phases and amplitudes of the convective modes (a). Growing further they make domain boundaries (DB) and defects (dislocations) when they are in contact (see also Figure 2.21); usually many defects are created. Then the ripening process, when the more stable modes 'eat' the others, takes place and they grow further. For example, if domain *A* has the most stable mode of convection it can grow further and suppress the others (see also Section 2.4). This is somewhat similar to the crystal growth process schematically sketched in (b). The larger crystals are more stable according to the Gibbs–Thomson equation, $C_{eq}(R) = C_{eq}(\infty)\cdot(1 + \alpha'/R)$, where $C_{eq}(R)$ is the equilibrium concentration of a crystal of size R, and α' is the capillary constant. The larger crystals grow by the Ostwald ripening process which is due to this size-dependent solubility and diffusion of mass, i.e. smaller crystals dissolve into monomers which diffuse into larger crystals and are crystallized again. Naturally there are other growth kinetics for further growth, such as direct coalescence (collision) of nuclei.

(We shall discuss t_m later on, in Sections 2.4 and 2.5, in relation to the onset time defined by Suzuki, 1981.) The nuclei are sufficiently produced until t_m is reached (Toschev, Milchev and Stoyanov, 1972; Wong and Meijer, 1982). Each one, however, is in a spatially different growth stage. Then the interaction among nuclei at different locations appears for $t \sim t_m$; for example, through the diffusion process in conserving systems. One calls this stage a ripening stage (cf. Ostwald ripening in a crystal growth in Figure 2.4; see Kahlweit, 1975; Lifshits and Slyozov, 1961). Thus for $t \gtrsim t_m$ each nucleus grows bigger mainly by the ripening process (competitive growth of nuclei). In nonconserving systems, the coalescence (growth by collisions) of nuclei dominates the growth dynamics for $t \gtrsim t_m$ as shown in Figure 2.5. In these stages, (2.2.3) and (2.2.6) are no longer valid. A new kinetic equation would come out corresponding to such ripening processes. For example, when the interfacial thickness of grains is very thin compared to the bulk size of the new ordered phase, the equation of motion on the interface describes the growth (Ohta, Jasnow and Kawasaki, 1982). Thus, one way to seek to attain

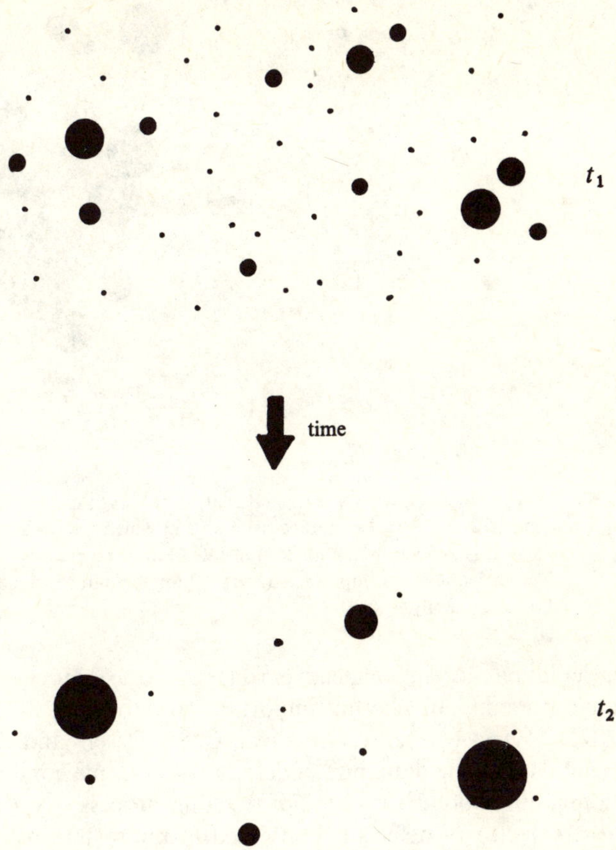

(b)

deep understanding of the transient dynamics very close to the steady state is to search for the dominant macroscopic kinetic equation which is newly derived from bulk kinetics.

More generally, there is no coupling among Fourier components of modes in the linear growth regime. For the late stage, however, coupling exists because of the Y^3 term, and the spatial interaction of modes is produced through the $\nabla^2 Y$ term. The difference in kinetics therefore must be due to spatial inhomogeneity produced by the many domains and the defects. Some modes, however, are removed by mode selection during the final stage, which resembles ripening. This process leads to the possibility of a new macroscopic equation that is closely related to the dynamics of defects and the mode selection (Shiwa and Kawasaki, 1986; Tesauro and Cross, 1986). In EHD as well as in R–B convection, therefore, wavefront motion, pattern selection

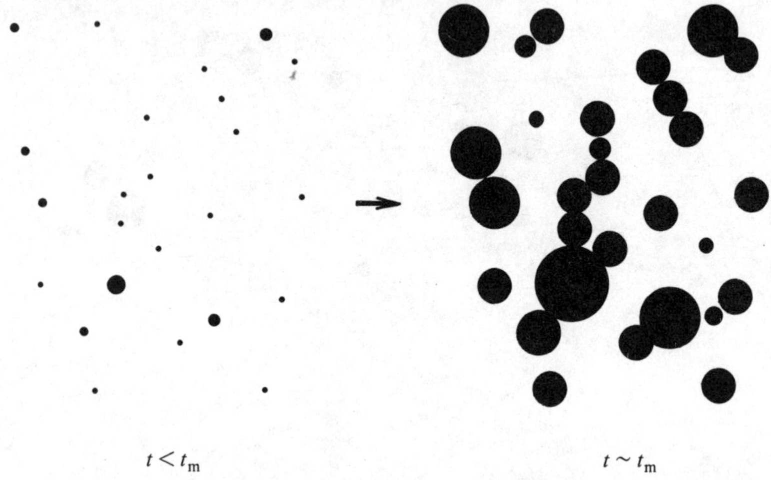

Figure 2.5. Nucleation ($t < t_m$) and coalescence stage ($t \gtrsim t_m$) in nonconserving systems. Here t_m is the characteristic time at which the coalescence of nuclei starts; it is defined by the nucleation rate Θ and the growth velocity v (see Section 2.4.3); for example, $t_m \sim (v^3\Theta)^{-1/4}$ for three-dimensional systems from a dimension analysis.

and defect dynamics in convective rolls, and the competitive growth of convective domains all play an important role to form the final structure (WD) (Kai, 1986; Kai and Hirakawa, 1985; Pesch and Kramer, 1986; Whitehead, 1983). The dynamics of defects, for example combining, climbing and gliding, correspond to a type of ripening process (see Figure 2.4), and the Burgers vector of defects is conserved in the system with a sufficiently large Γ and in systems enhanced at relatively deep quenching (a large step of the applied field). In this case, the Burgers vector may be regarded as an order parameter for the final stage (Kawasaki, 1984a, b, c, 1985; Shiwa and Kawasaki, 1986). There is also a physical analogy with dislocation motions in ordinary crystals (Tesauro and Cross, 1986).

Although the precise analogy between the kinetics of a crystal growth process and the order formation in EHD is still left out here, the concept of phase separation may be useful in qualitatively understanding the kinetics of the system far from equilibrium. A united approach for these different systems is a scaling theory, which may describe those transient kinetics in a simple way by only taking relevant parameters, as done in phase separation systems (Furukawa, 1985; Komura *et al.*, 1984). We will also shortly propose an experimental result from this point of view.

In EHD instability, there are two advantages when investigating these subjects. They are that the characteristic time is much shorter than that in R–B convection, typically by two to three orders of magnitude, and a bifurcation parameter can be easily controlled (Kai and Hirakawa, 1978).

2.3 Experimental preparation

2.3.1 *Sample*

A sample of the material MBBA (*N*-(p-methoxybenzylidene)-p-butylaniline), which shows a nematic phase at room temperature, is sandwiched between two horizontal glass plates on which are deposited SnO_2-conducting electrodes. The thickness d of the sandwiched cell ranges from 4 to 110 μm and the lateral dimensions $l_{\parallel}$ are 1–10 mm, giving rise to an aspect ratio in the range 500 ($l_{\parallel} = 10$ mm, $d = 20\,\mu$m) to 10 ($l_{\parallel} = 1$ mm, $d = 100\,\mu$m). By rubbing the surfaces, homogeneous alignment is attained; a.c. voltages of various frequencies and variable magnitude are applied across the sample. The threshold voltage, V_c, for WD ($f < f_c$) lies the range 6–10 V, and for the oscillatory instability ($f > f_c$) it is between 40 and 300 V, both depending upon the frequency of the field. The temperature during the experiments is controlled to within ± 0.05 degree using a copper container with a double wall.

2.3.2 *The nature of externally applied noise*

Noises with various natures, applied to the sample, are shown in Figures 2.6–2.10, which are realized by using commercially available equipment (NF co. WG722). This equipment generates a quasi-random noise with a period $T_p \sim 510$ days for a 20 μs clock. For this clock a 5 kHz band-width of Gaussian noise is realized.

The interval distribution between two binary pulses can be described by a Poisson distribution (Figure 2.6),

$$N(n, v) = \frac{(v)^n}{n!} \exp(-v), \qquad (2.3.1)$$

where n is the unit time-delay given by $\Delta t/b$, Δt is the time interval between the successive two pulses, and b is the minimum width of the binary pulse which depends on a clock frequency of the shift register; for example, $b = 20\,\mu$s for a 20 μs clock and 200 μs for 200 μs clock ($T_p \sim 14$ years). Since a pulse cannot be generated on superimposition, we cannot observe the pulse in a time-lag Δt shorter than twice the pulse width, $2b$.

In Figure 2.6, the Gaussian distribution,

$$N(\Delta t) = (2\pi)^{-1/2} \sigma_t^{-1} \exp\left(-\frac{(\Delta t - 2b)^2}{2\sigma_t^2} \right), \qquad (2.3.2)$$

has been fitted, with parameters $2b = 40\,\mu$s and $\sigma_t = 50\,\mu$s, as shown by the dotted line.

The power spectrum $P(\omega)$ of a Poisson pulse is given as

$$P(\omega, b) = \frac{K^2 b^2}{2\pi} \left(\frac{\sin(\omega b/2)}{(\omega b/2)} \right)^2. \qquad (2.3.3a)$$

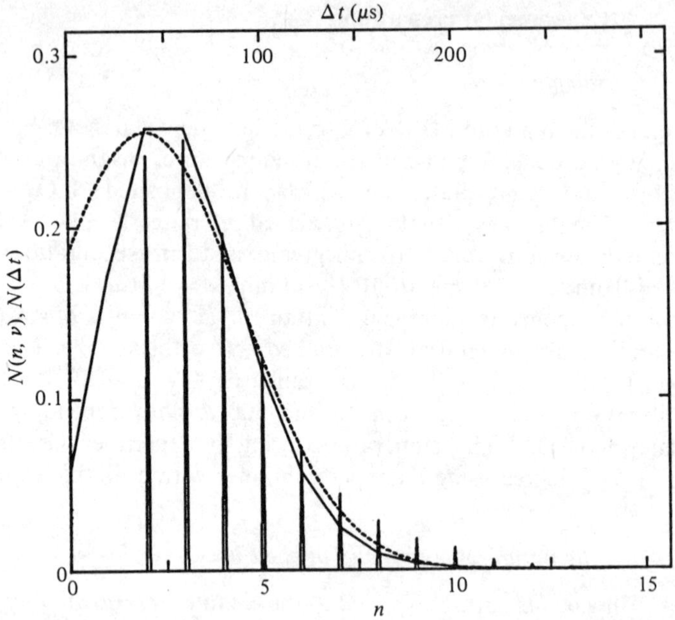

Figure 2.6. Generating probability function $N(n, \nu)$ of a binary pulse for a 20 μs clock. Actual time-lag Δt is given as 20 μs. The solid line indicates the Poisson distribution of (2.3.1) with $\nu = 3$ and $N_0 = 1.15$, and the dotted line is the Gaussian distribution of (2.3.2) with $N_0 = 32$, $\sigma_t = 50$ μs and $2b = 40$ μs.

Here b is the pulse width and K is the pulse height. For the narrowing limit of the width of unit pulse, the power spectrum $P(\omega)$ is given by the equation,

$$P(\omega) = \lim_{b \to 0} \frac{K^2 b^2}{2\pi} \left(\frac{\sin(\omega b/2)}{(\omega b/2)} \right)^2 = \frac{1}{2\pi}, \tag{2.3.3b}$$

namely the white noise spectrum. The experimentally observed $P(f)$ shown in Figure 2.7 (where $f = \omega/2\pi$) agrees well with the white noise spectrum up to about $f = 20$ kHz. Other kinds of noise may be made from this binary noise through use of special electrical filtering techniques. Figures 2.7–2.10 show original signals, power spectra, and probability density functions of amplitudes of all types of noise used here. The profiles of their probability density functions give them the names 'binary', 'Gaussian', 'uniform' and 'binomial'. $1/f$ noise can also be generated by our equipment. The intensity of the noise is determined by $Q = V_N^2 = \langle V(0)^2 \rangle$ using a fast Fourier transform (FFT) analyzer (Iwatsu co. SM-2100A). The correlation time of the noise is determined from the auto-correlation function or the band-width of the power spectrum using this analyzer.

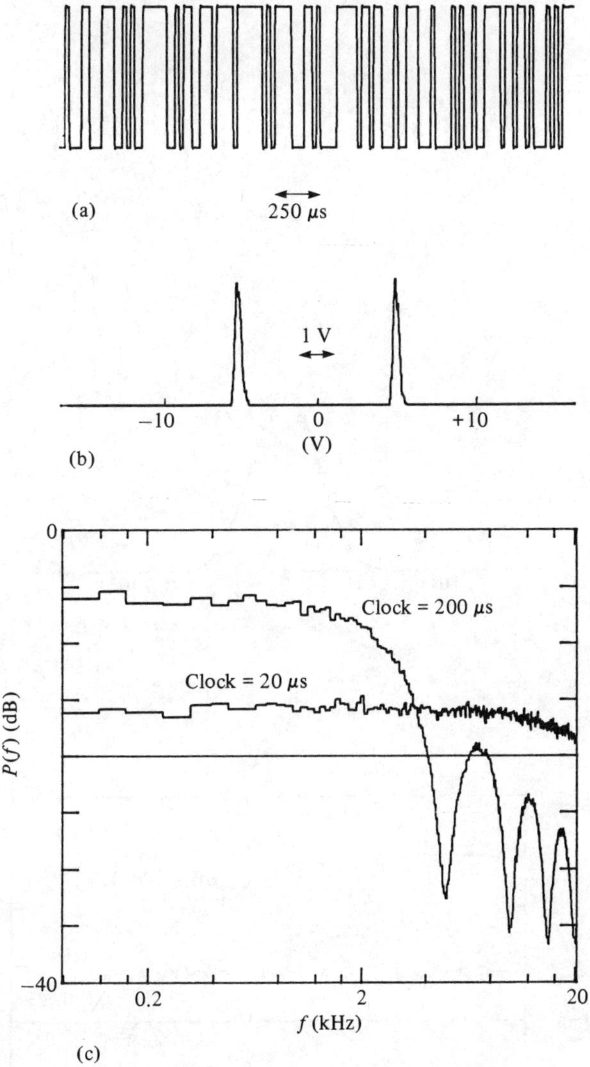

Figure 2.7. Binary noise: (a) original; (b) probability distribution of the amplitude; (c) power spectrum ($V_N = 6.70$ V).

2.3.3 *Convective regime* ($f < f_c$)

Measurement of evolution of director angle using birefringence

We obtained the linear time constant ($\exp(\pm t/\tau_L)$) using the dynamical birefringence technique (Pieranski, Brochard and Guyon, 1973). The director is initially aligned homogeneously with $\theta = 0$. Then, neglecting flow effects,

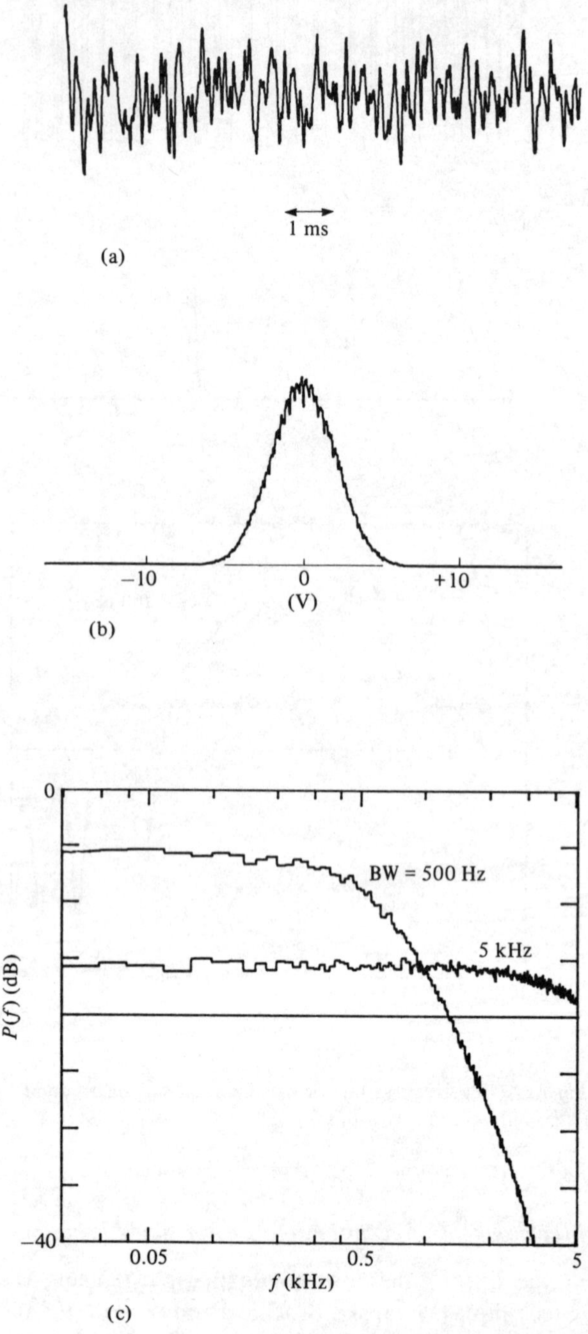

Figure 2.8. Gaussian noise: (a) original; (b) probability distribution of the amplitude; (c) power spectrum (BW: band-width) ($V_N = 2.65$ V).

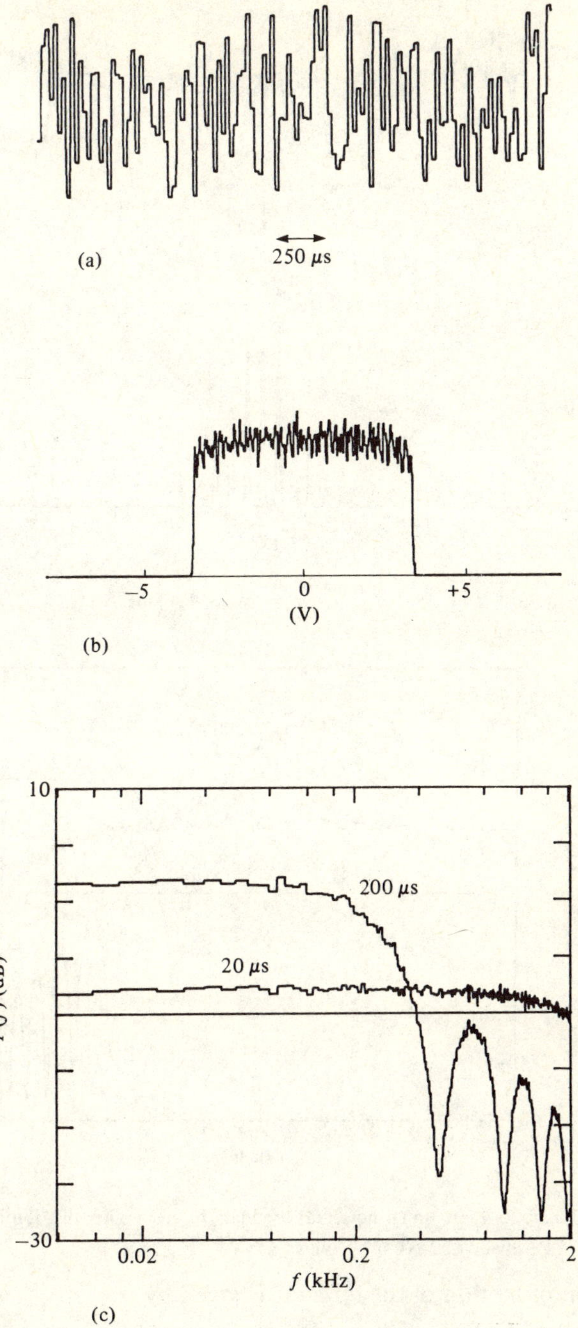

Figure 2.9. Uniform noise: (a) original; (b) probability distribution of the amplitude; (c) power spectrum ($V_N = 2.65$ V).

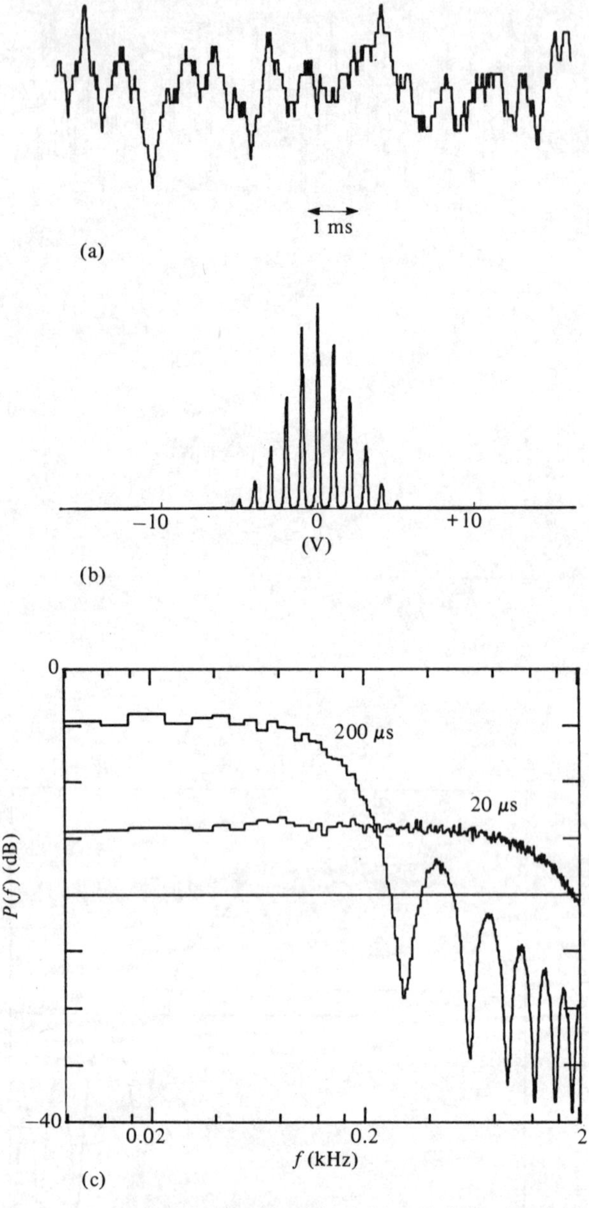

Figure 2.10. Binomial noise: (a) original; (b) probability distribution of the amplitude; (c) power spectrum ($V_N = 2.90$ V).

the equation of motion of the director is given by

$$\frac{d\Phi}{d\theta} = \frac{d\theta}{dt} = \tau_L^{-1}(\varepsilon)\theta - \beta\theta^3, \tag{2.3.4}$$

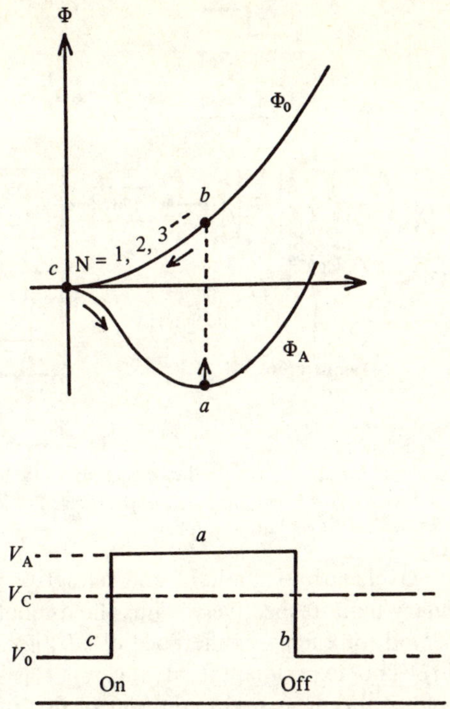

Figure 2.11. Dynamical behavior of the director on a potential. Φ_0 and Φ_A are potential profiles for $V_0 < V_C$ and $V_A > V_C$, respectively.

and

$$V_c^2 = \frac{4\pi^3\varepsilon_\perp}{\varepsilon_a\varepsilon_\parallel}(K_{11} + 4K_{33}\lambda_x^{-2}d^2) = V_c^2(0) + V_c^2(k_x). \tag{2.3.5}$$

Here Φ is the potential, ε_a is the anisotropy of the dielectric constant, K_{ii} is the elastic constant ($i = 1$: splay; $i = 3$: bend), and k_x is the wavenumber of WD in the x-direction ($k_x = 2\pi/\lambda_x$). At $V = 0$, Φ is minimum at $\theta = 0$. The precise expressions of coefficients τ_L^{-1} and β are very complicated even under d.c. field, which is derived from the reductive perturbation method involving flow effects (Hijikuro, 1975). When the external voltage is varied from $V_0 < V_c$ to $V_A > V_c$, the state falls into $\theta \neq 0$, i.e. into a of the curve Φ_A in Figure 2.11, where V_0 is a certain bias voltage lower than V_c to obtain the parallel alignment to the glass plate, and V_A is higher than V_c.

As a liquid crystal has birefringence, the phase difference ϕ between the ordinary and the extraordinary light passing through the sample gradually changes with the temporal evolution of the angle of the director, according to the equation

$$\phi(t) = \left(\frac{\pi d}{\kappa}\right)\theta^2(t)(n_e - n_o), \tag{2.3.6}$$

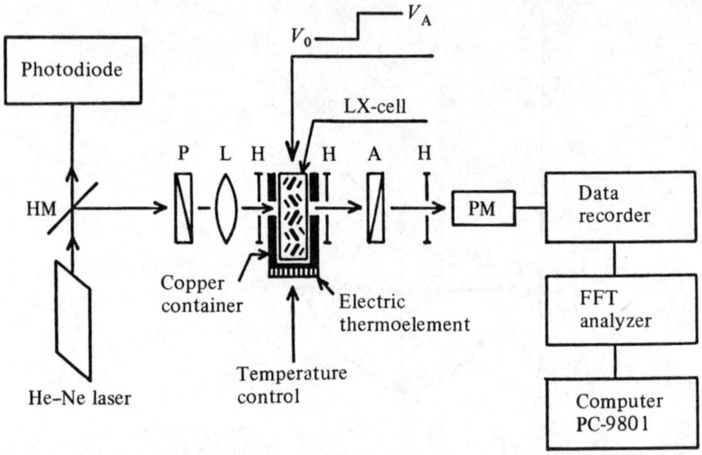

Figure 2.12. Experimental set-up to measure the optical birefringence of nematic liquid crystals (LX). HM = half mirror; P = polarizer; L = lens; H = pin hole; A = analyzer; PM = photomultiplier.

where κ, n_e and n_o are the wavelength of the light and refractive indices for extraordinary and an ordinary light, respectively. Thus the oscillation of the light intensity can be observed for each 2π-difference of $\phi(t)$, following the temporal evolution of $\theta(t)$. The experimental arrangement is shown in Figure 2.12 and an example of the measured evolution of $\theta(t)$ is shown in Figure 2.13, clearly exhibiting an oscillation. Here N shows the Nth oscillation. The inverse of the interval is proportional to the slope of Φ. The time constant is obtained from the interval of these oscillations as follows:

$$\tau_L = \frac{t_{N+1} - t_N}{\ln((N+1)/N)}, \tag{2.3.7}$$

under the assumption $d\theta/dt = \tau_L^{-1}\theta$, neglecting the cubic term in (2.3.4). This holds only for very small N which is always chosen here to be $N = 1$. Qualitatively we can infer from the signal shown in Figure 2.13 that the slope of Φ is very flat near $\theta = 0$ and the steady state, but steep for the transient region (the short interval A in Figure 2.13). In the same manner, we could also measure the relaxation time switching the field from V_A to V_0. Thus the divergent tendencies toward V_c from both sides are clearly observed (Figure 2.14). This dependence on ε can be described by (2.3.4), in which τ_L is proportional to ε^{-1}. Another time constant for the growth is the nonlinear growth time τ_r, which is chosen as the 90%-rising time. It contains the influence of the higher order term in (2.3.4).

Two-dimensional image processing for pattern dynamics

In order to investigate the convective pattern dynamics and the defect motion, image processing techniques are used. The system consists of a digital image

In order to enhance the director oscillation, the electric field is repeatedly (1024) applied in a burst-like way at the chosen value for each fixed $\varepsilon(=(V^2 - V_c^2)/V_c^2)$, and the results averaged (see below). The phase of the applied field for each burst is always the same, being controlled exactly by a computer. Furthermore, a burst width is varied with ε and d, and the interval of the burst is chosen to be the time to recover completely from the modified state to the initially uniform state. A single mode He–Ne laser ($\lambda = 632.8$ nm, 3 mW) and photodiode (NEC Co. LSD-39A) are used as an incident light source and as a detector, respectively (see Kai, Wakabayashi and Imasaki, 1986, for details).

The development of the oscillation is obtained from the envelopes of the oscillatory signal (Figure 2.16). There are 1024 repetitions of the bursts, and the temporal development of each envelope from t_{0i} to the steady state is obtained. Then the most probable path $\bar{Y}_0$ (mean formation route of the order) is defined as the averaged route for all envelopes,

$$\bar{Y}_0(t^j) = \frac{1}{N}\sum_i^N Y_i^j, \quad \text{at } t^j \tag{2.3.8}$$

where N is 1024. The variance is then given by

$$\sigma_0(t^j) = \frac{1}{N}\sum_i^N (\bar{Y}_0 - Y_i)^2 \quad \text{for } t^j. \tag{2.3.9}$$

The experimental procedures for the scaling are basically the same as above (Kai, Wakabayashi and Imasaki, 1986).

2.4 Conductive regime ($f < f_c$)

2.4.1 Convective pattern in EHD

When the field reaches V_c, one finds first a roll pattern called Williams domains (WD) (Figure 2.17a). As the voltage is increased from V_c, these rolls start to fluctuate in time with a characteristic ε-dependence (fluctuating Williams domains, FWD) and a transition to the grid pattern (GP) at V_{GP} (~ 1.5–$2.0\ V_c$) takes place (Figure 2.17b). As the applied voltage is stepped up further, the spatially incoherent regime, usually called the dynamic scattering mode (DSM), is formed (Figure 2.18). Thus many interesting dissipative structures have been found in EHD instability (Hirakawa and Kai, 1977; Kai and Hirakawa, 1977; Kai, Yamaguchi and Hirakawa, 1975). Very recently two new flow structures have been observed: a zig-zag and a skewed varicose in a cell with an extremely large $\Gamma_\parallel$ (> 500) and with a special surface treatment coated by a polymer (Joets and Ribbota, 1986a, b). This procedure induces a very strong surface anchoring effect and forces liquid crystal molecules to align. Examples are shown in Figure 2.19, which shows a zig-zag structure in

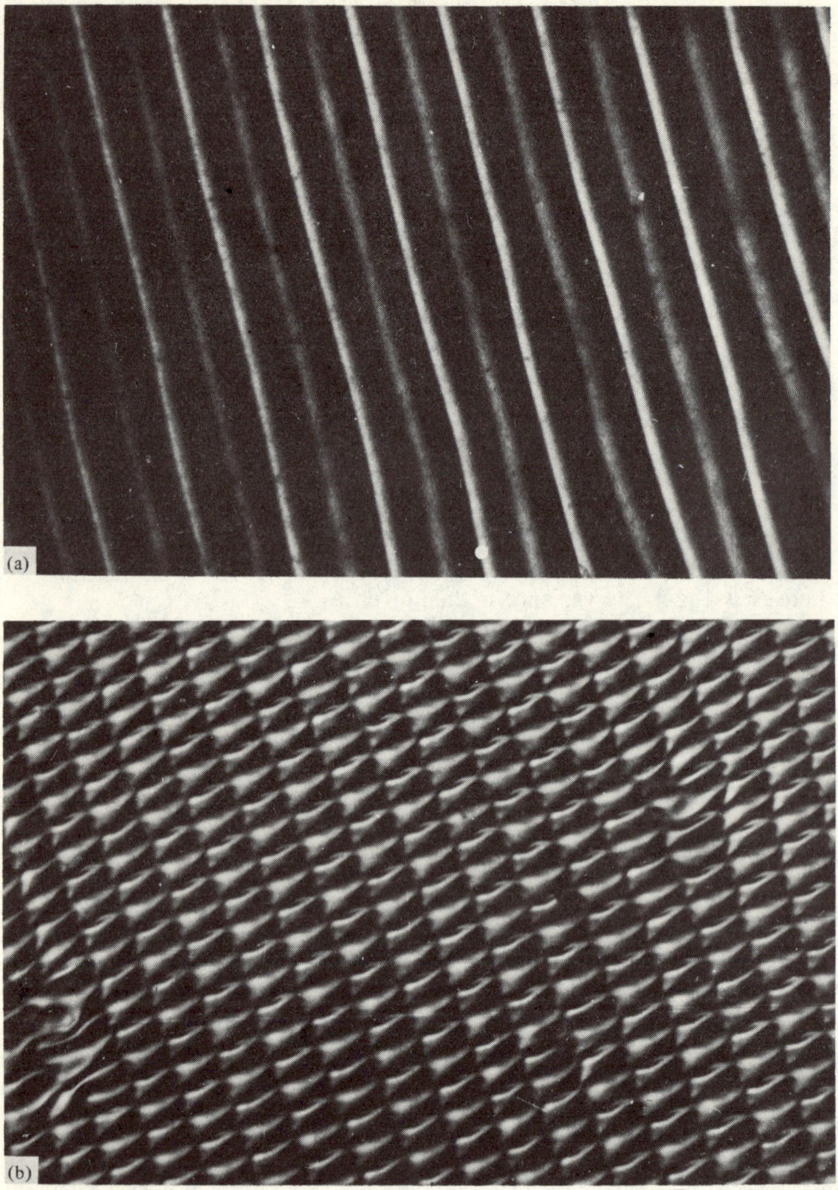

Figure 2.17. Typical convective patterns observed in the conductive region ($f < f_c$). (a) Williams domains (WD) of two-dimensional convective flow. (b) Grid pattern (GP) of three-dimensional cellular convective flow.

EHD instability for a very thin cell ($d = 4\ \mu$m). These types of patterns in EHD instability are theoretically obtained by Zimmerman and Kramer (1985).

The wavelength λ_x in the x direction of these hydrodynamic structures

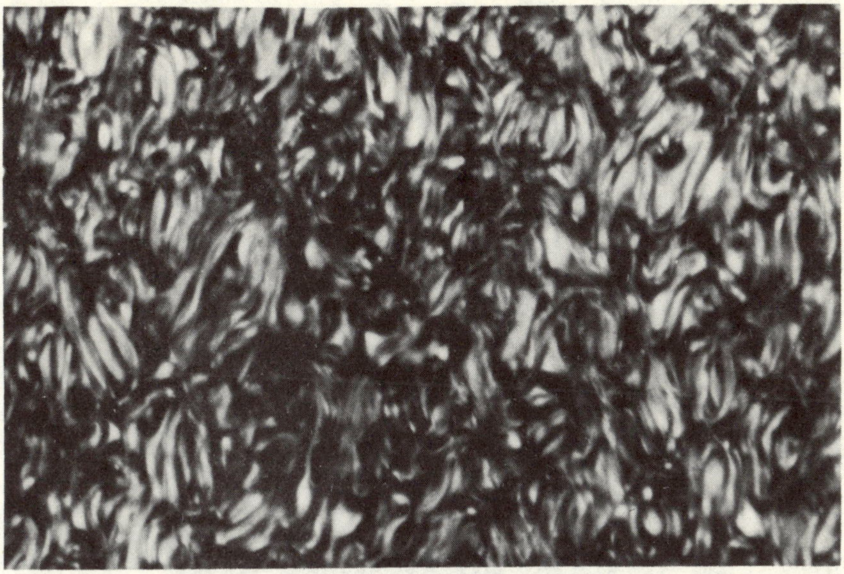

Figure 2.18. Dynamic scattering mode (DSM) corresponding to turbulent flow.

becomes small with increasing applied field frequency, and it goes asymptotically to zero at f_c (see, for example, Figure 2.20 for λ_x of WD). Actually it never quite reaches zero but becomes very small (of the order of micrometers) as shown in Figure 2.20(b). We call this pattern the chevron.

2.4.2 *Transient properties of WD formation*

Pattern selection

Let us consider a one-dimensional situation of the convective pattern formation when applying the field V_A higher than V_c (see Figure 2.21). In EHD instability, all rolls are arranged in parallel because of initial symmetry breaking, unlike in R–B convection. It is therefore relatively simple to describe the phase and the defect dynamics in EHD instability. Here we restrict ourselves to the discussion of the system with a few defects and the situation close to $\varepsilon = 0$. The $\Omega(k)$ dispersion relation for this scenario is shown in Figure 2.22, where Ω is the eigenvalue of the modes. Since all modes with $\Omega > 0$ can grow, the band-like excitation of modes may be realized. The mode $k_c(= 2\pi/\lambda_c)$ with the maximum eigenvalue Ω_c is the fastest growth mode. However, k_c is not always equal to the most stable mode k_m in the steady state, which depends on all boundary conditions. Suppose that convections start from both lateral boundaries with the scale λ_c, determined by the linear stability theory associated with a thickness of the layer under an infinite lateral condition. They grow into the middle of the container, and, after a while, they

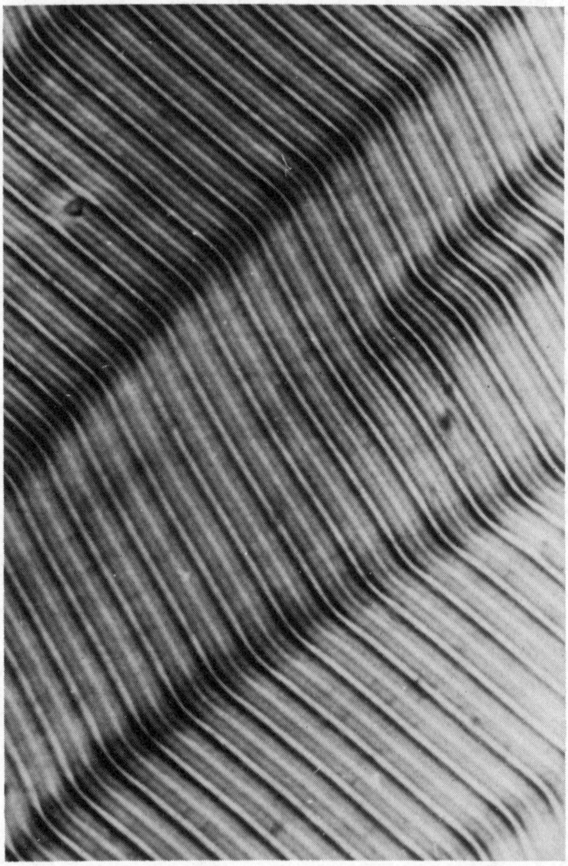

Figure 2.19. Zig-zag pattern observed in a cell with a strong anchoring effect.

are in contact with one another. Then they become 'aware' of their 'misfitting' and start to modify their scale into the most stable scale $\lambda_m (= 2\pi/k_m)$. In this situation, the deviation of k_m from k_c is very small for very large $\Gamma (l \gg d)$, i.e. $\Delta k = k_m - k_c \sim 0$ (see Figure 2.21a). That is, the convective rolls growing from both lateral boundaries may fit together relatively well at the middle of the cell. Accordingly, the formation of defects is not always necessary in this case (Tsuchiya and Horie, 1985), and k_c can continuously change into k_m because of a continuous and small barrier. On the other hand, the barrier is usually large and discontinuous for small Γ. In this case the rolls modify their scales by making defects because of a large 'misfitting force' and energy barrier (see Figure 2.21b). An example of such formation processes is shown in Figure 2.23.

The dynamics from k_c to k_m would happen as defect dynamics and/or phase dynamics depending upon the values of Δk and $\Delta\Omega$. Here, phase

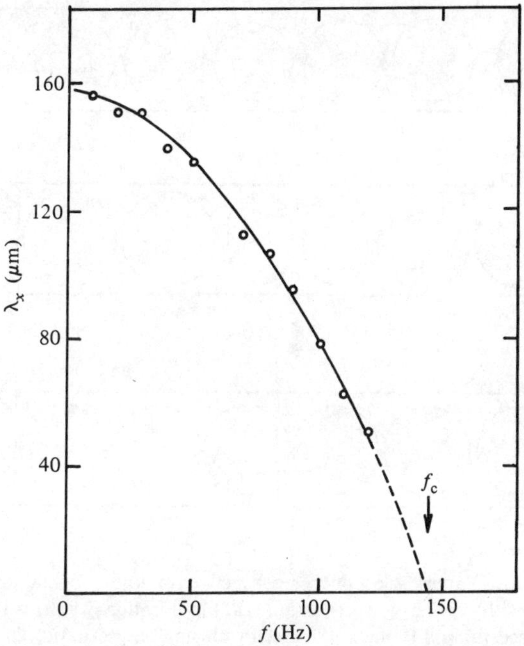

(a)

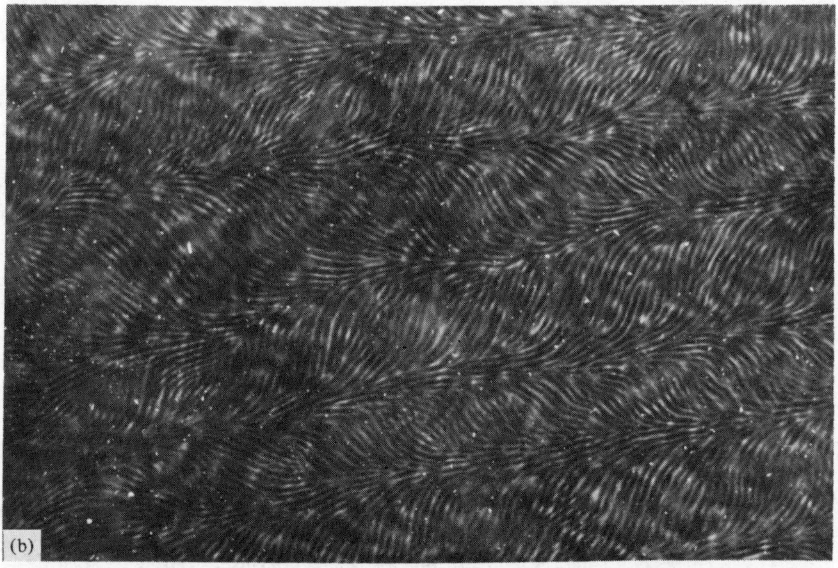

(b)

Figure 2.20. Frequency dependence of spatial scales of a convective pattern. (a) Wavelength λ_x of WD versus applied frequency. (b) Chevron pattern observed in a dielectric regime ($f > f_c$).

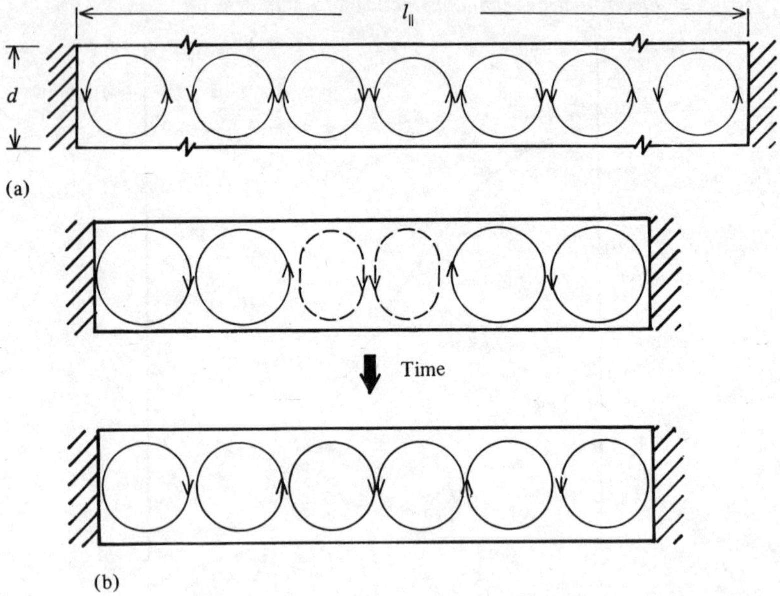

Figure 2.21. Pattern selection in a convective roll formation. (a) Roll formation in a cell with very large aspect ratio $l \gg d$. Modification of rolls is very small in this case. (b) Roll formation in a cell with small aspect ratio. The dotted roll indicates an unstable roll which is awakened when rolls growing from both sides contact. Then a modification over all rolls starts.

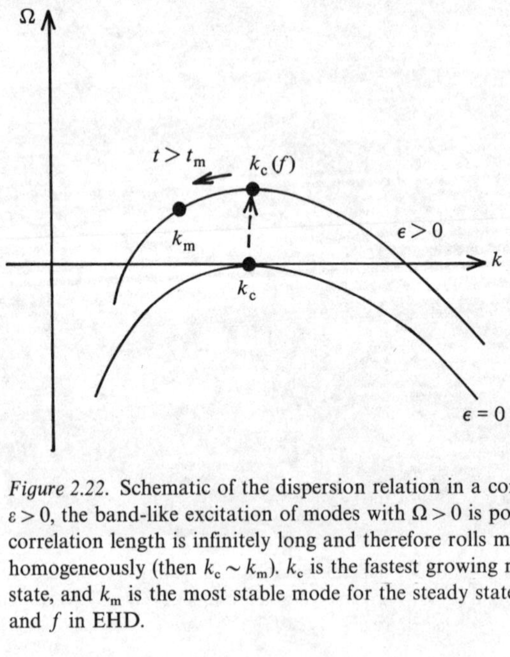

Figure 2.22. Schematic of the dispersion relation in a convective instability. For $\varepsilon > 0$, the band-like excitation of modes with $\Omega > 0$ is possible. For $\varepsilon = 0$, the correlation length is infinitely long and therefore rolls may form homogeneously (then $k_c \sim k_m$). k_c is the fastest growing mode for the transient state, and k_m is the most stable mode for the steady state. Both depend on Γ and f in EHD.

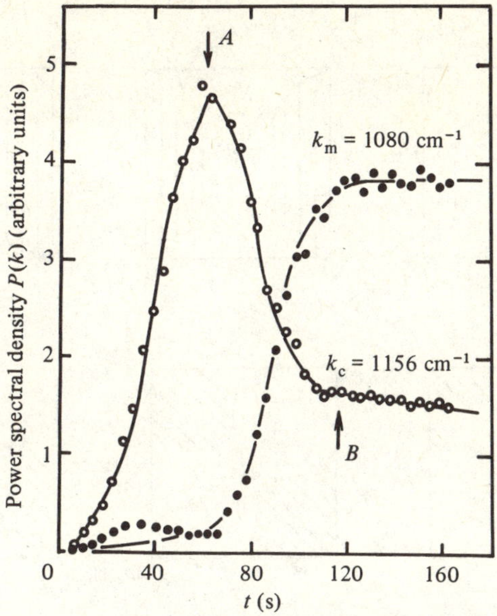

Figure 2.23. Typical mode change in power spectra for roll formation process with small aspect ratio ($\Gamma_{\parallel} = 14, \Gamma_{\perp} = 11, d = 100\,\mu$m, $f = 10$ Hz, $\lambda_c = 54\,\mu$m and $\lambda_m = 58\,\mu$m). The spectrum is taken along the direction perpendicular to the roll axis. At $t \gtrsim t_m$ (~ 50 s), the mode modification from k_c to k_m starts following the contact of two domains. The amplitude of k_c decreases but that of k_m increases with time. This process is influenced by an application of white noise. At A, a defect starts to form. At first k_m tries to grow simultaneously with k_c but it is suppressed in the early stage. After a long period of being only negligibly excited, the k_m mode starts growing rapidly at $t \simeq 70$s. The so-called defect motion starts at B, from which the linear decay of $P(k_c)$ is observed (see also Figure 2.28 for its temporal relation).

dynamics means the dynamics of spatially slow variation modes in convective systems without singular points such as defects. Naturally there must be other origins of the onset of defects for large Γ, e.g. inhomogeneities of boundaries (thickness d), impurities and inhomogeneous external fields. Neglecting these origins and simply discussing the situation on the basis of the idea described above, we may consider that defects form when Δk is relatively large. Then the defect dynamics are closely connected with the mode changing from k_c into k_m (Figure 2.23). On the other hand, when Δk is small the phase dynamics may play a relatively important role for the late stage of the evolution. An external noise can also modify the Ω–k curve and change the dynamical properties of defects, as does Γ. (For example, the onset stage of defects seems to be delayed in the presence of an external noise.)

Wavefront motion

We can observe two different types of growth of convective rolls, as is shown in Figure 2.24 (Kai and Hirakawa, 1985). In Figure 2.24(a), rolls are arranged

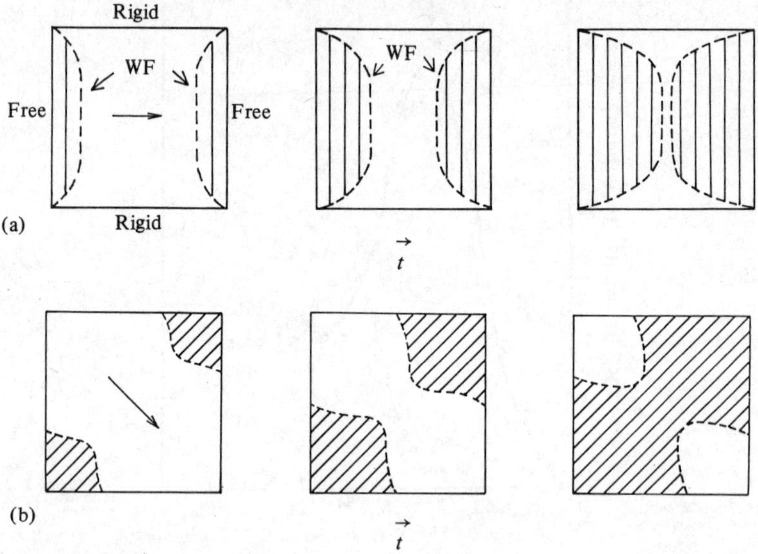

Figure 2.24. Wavefront profiles of WD-roll domain. (a) Stable profile of wavefront. (b) Unstable profile. The arrow shows the initial injection direction of a liquid crystal. The dotted lines show wavefronts (WF).

along the direction normal to the rubbing and to the injected direction of a liquid crystal. The growth of rolls starts from boundaries, and two wavefronts (WF) of the convective domains can be observed parallel to the roll axis. This is the most common case. A WF is rarely observed to be normal to the roll axis (see Figure 2.24b); slip dislocations are often formed in this type. The velocity of this WF, once it forms, seems to be faster than that in (a). Formation process (b), however, does not always appear and is very rarely observed in a cell with relatively large $\Gamma(\sim 50)$ and $d(\sim 50\,\mu\text{m})$. Since the formation process of (b) changes into one of (a) after repeating the experiment several times, it is clearly unstable and is probably due to an anisotropy of NLCs, i.e. due to the initial symmetry breaking by the injection of NLCs. This is an interesting feature of EHD instability. The onset time t_m is then roughly equal to the time at which a WF just contacts another WF.

2.4.3 *Defect motion at the later stages*

An isolated defect motion

The defect always disappears by climbing and gliding motion to the lateral walls. However, the phenomenon is very sensitive to Γ and f since they can change k_c and k_m (see Section 2.4.2). To understand this better, let us choose one example of an isolated defect motion and try to describe general aspects as far as possible. The stable number of roll pairs N_s is shown in Figure 2.25 as

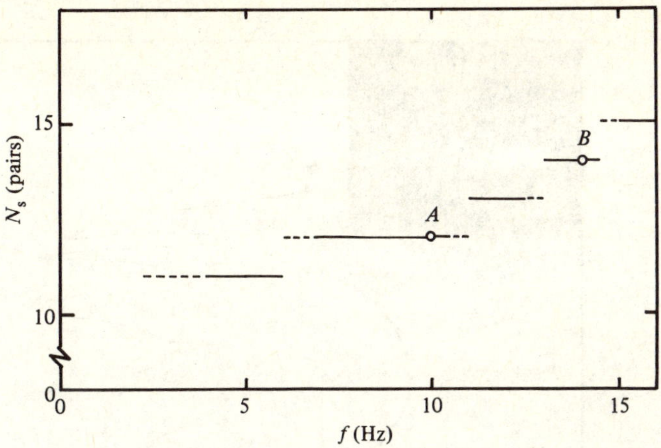

Figure 2.25. Steady state number of total roll pairs N_s in the cell shown as a function of frequency f for $\varepsilon < 0.4$ ($\Gamma_{\parallel} = 14, \Gamma_{\perp} = 11$ and $d = 100\,\mu$m). The state with N_s corresponds to the mode k_m. Motions for A and B are described in the text in detail.

a function of the applied frequency for the cell with $\Gamma_{\parallel} = 14$, $\Gamma_{\perp} = 11$ and $f_c = 20$ Hz. For $f = 10$ Hz, the state of $N_s = 12$ is more stable than $N = 13$ (corresponding to the fastest growing mode k_c). When $f = 12$ Hz, k_c ($N = 13$) is identical to k_m ($N_s = 13$). In this case no defect is observed for ε as long as the system is below the FWD point. For $f = 14$ Hz, the state $N_s = 14$ is the final stable state in spite of the fact that the fastest growing mode is at $N = 15$. Thus the number of rolls at the steady state always decreases from the initial number, i.e. $\Delta k = k_c - k_m$ is positive (Figure 2.25), when ε is smaller than a certain value. Oscillatory gliding happens more frequently with larger amplitude, and the climbing tends to be more suppressed as f and N_s increase. The final disappearance of a defect usually happens by the climbing motion, and so becomes more and more difficult with increasing f. As f approaches f_c, it is more difficult to observe an isolated defect but it is easier to form many defects, and the gliding motion becomes continuously oscillatory without decay. This is clearly related to the chevron pattern. The amplitude of the gliding oscillation depends on f and ε. In contrast to this, the oscillatory dependency for the climbing motion is usually very weak. One possible origin of these oscillatory tendencies is the coupling between the gliding and the climbing motions.

One typical example of climbing motion is shown in Figure 2.26. The velocity increases drastically as the defect approaches the wall. The characteristic distance l^* from the wall to the location at which the speed-up starts is roughly the same as L (the size of one pair of rolls) and/or d. The minimum climbing velocity is always observed where the gliding amplitude is at a maximum. This means that the large gliding velocity leads to a small climbing velocity.

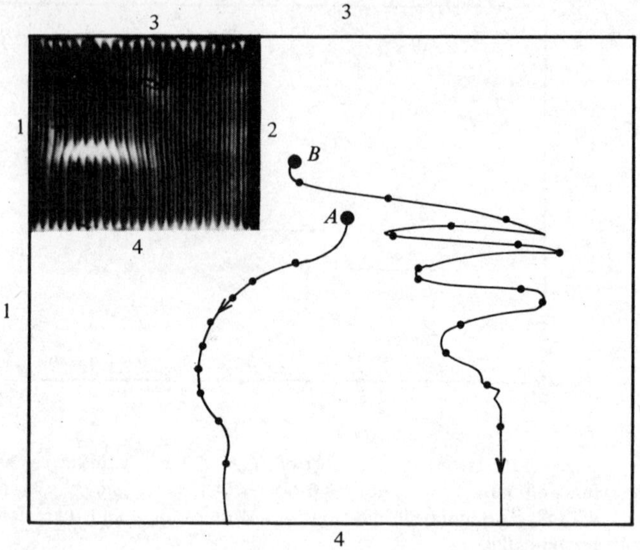

Figure 2.26. Typical examples of a defect motion (A: $f = 10\,\text{Hz}$ ($\varepsilon = 0.199$) and B: 14 Hz ($\varepsilon = 0.132$)). The small closed circle indicates the location of a defect every two minutes for A and every minute for B. The strong oscillatory tendency can be clearly seen for $f = 14\,\text{Hz}$. The inset photograph shows an isolated defect for $f = 10\,\text{Hz}$. A large deformation field is observed. Dust seen in the photograph is not inside the cell. 1, 2: free boundaries; 3, 4: rigid boundaries.

The climbing distance, l_c, from the lateral wall and its velocity, v_c, normalized by L are plotted in Figure 2.27 as an example. As shown in Figure 2.27(a), the defect begins to form around $t = 70\,\text{s}$ ($\sim t_m$) and is complete at $t \sim 120\,\text{s}$ where the transmitted intensity of light is almost steady (Figure 2.28). The temporal change of the total light intensity through the liquid crystal film qualitatively indicates the evolution of the convection. The onset time t_m may be roughly estimated as 64 s from $I(t_m) \sim I(t_{1/2}) \sim \frac{1}{2}I(\infty)$ from its result. Only a short time after $t \sim 120\,\text{s}$, the defect motion starts. The motion however is not steady until 150 s (indicated by the dotted line in Figure 2.27). The so-called defect dynamics therefore is usually studied for the almost steady state after complete formation of the rolls. The maximum velocity is about $1\,L\,\text{min}^{-1}$, and the minimum velocity is about $0.15\,L\,\text{min}^{-1}$, both of which depend on ε.

Figure 2.29 shows the typical velocity oscillation in the gliding motion of an isolated defect which disappears by climbing at 30 min.

One way to describe the defect dynamics is to use an analogy with the dynamics of an electron gas (Kawasaki, 1985; Shiwa and Kawasaki, 1986). The defect motion then is dominated by the image force due to a two-dimensional Coulomb-like interaction. Both original and imaged defects here are attractive

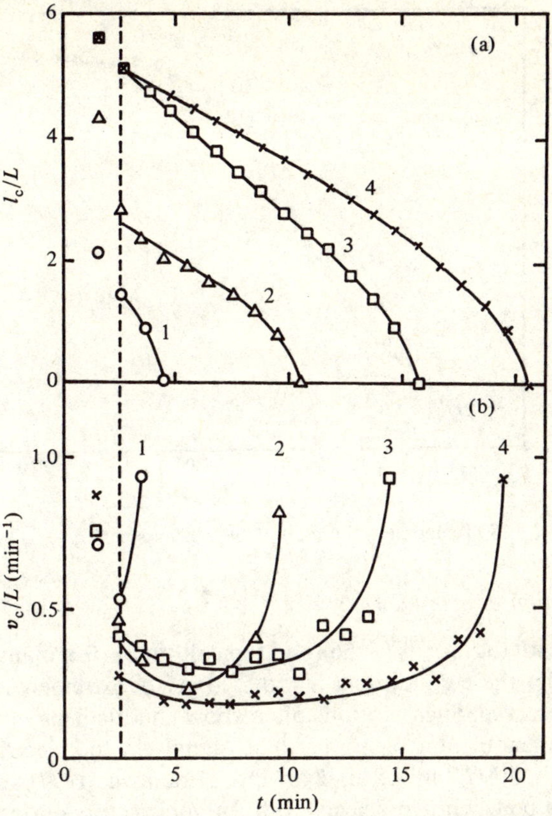

Figure 2.27. Temporal change of climbing distance l_c and its velocity v_c for $f = 10$ Hz. Marks 1 to 4 are different runs. The dotted line shows the time when the roll and the defect are perfectly formed.

with opposite sign (see Figure 2.30). In such a case, the characteristic distance among many of the defects is theoretically predicted to be proportional to $t^{1/2}$ for two-dimensional systems. Even the dynamics of an isolated defect shows such a relation, which suggests conventional diffusive transport kinetics (Yamazaki, Kai and Hirakawa, 1987). Then the transport coefficient, such as $D_c = 3.88 \times 10^{-6}\ \mathrm{cm^2\ s^{-1}}$ for the climbing, coincides with an order of the ionic diffusion constant of NLCs (Galerne, Durand and Veyssie, 1972).

The defect induces the deformation of the neighboring rolls. Such a deformation field changes with time. The width of the deformation, W, is initially small, and then becomes large as the defect moves. The maximum width, W_m, always occurs at the minimum climbing velocity and at the maximum gliding distance (or at the extreme value of the gliding, i.e. at the turn-over point in general). Then finally W becomes narrow as it approaches the lateral wall. Thus W correlates strongly with the gliding motion of the defect. All tendencies described in this section are basically universal.

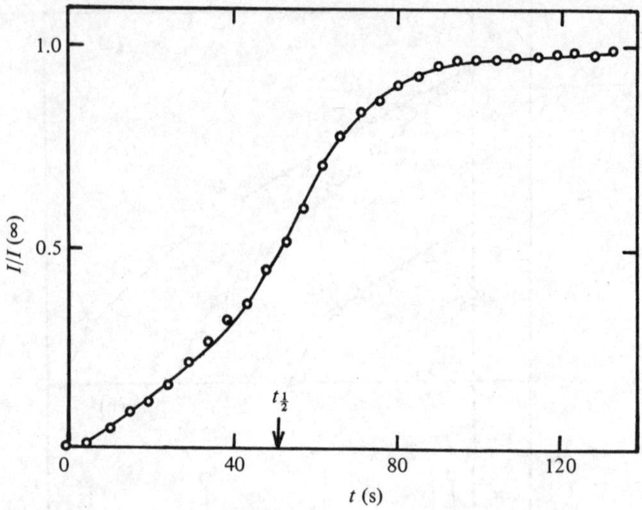

Figure 2.28. Temporal evolution of light transmission $I(t)$.

Motion of many defects

Using the cell with the large $\Gamma(\sim 56)$, and typically at $\varepsilon \sim 0.3$, many defects are produced. Under these conditions, a wide variety of extremely complicated motions are observed. Such motions often show chaotic behavior and cause the transient phenomenon of nonperiodic signals in the electrical current density through NLC film (Kai, Kai and Hirakawa, 1977). One typical situation would be when two defects, with Burgers vectors of opposite signs, combine attractively and disappear. Another example is that of repulsive motion when both defects have equal signs; in this case, the two defects move apart from each other. We can also understand this behavior by employing the idea of Coulomb interactions with image forces (Figure 2.30). Let us turn our attention to the attractive motion of two defects in many, typically 10 to 30, defects. At the beginning, both defects are located in different rolls (Figure 2.31a, b). One defect starts to move first by the gliding motion (Figure 2.31c). As soon as they reach the same roll, they move along the direction of the roll axis, i.e. the climbing starts (Figure 2.31c, d), and they disappear by combining (Figure 2.31e, f). The temporal behavior of two defects is summarized in Figure 2.32, in which we also notice an oscillatory tendency of the gliding motion. In general, it can be said that defects are removed by the climbing motion; this follows gliding, which requires a higher energy.

We summarize here the formation process of rolls and defects for large Γ as follows (Kai and Hirakawa, 1985). The roll structures do not form over the entire area for $t < t_m$. Near $t = t_m$ roll structures form in almost the entire area of the cell, and defects start to be formed; namely, WF just contacts others at $t \sim t_m$. Accordingly t_m represents the characteristic time at which the coalesc-

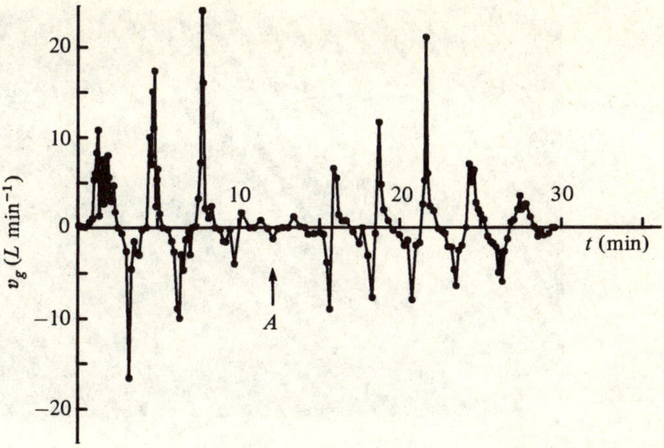

Figure 2.29. Temporal oscillation in the gliding velocity ($f = 14$ Hz, $\varepsilon = 0.152$ and $f_c = 20$ Hz). When the defect comes to the middle of the cell, the gliding velocity is reduced (A). The oscillation period is about 3 min.

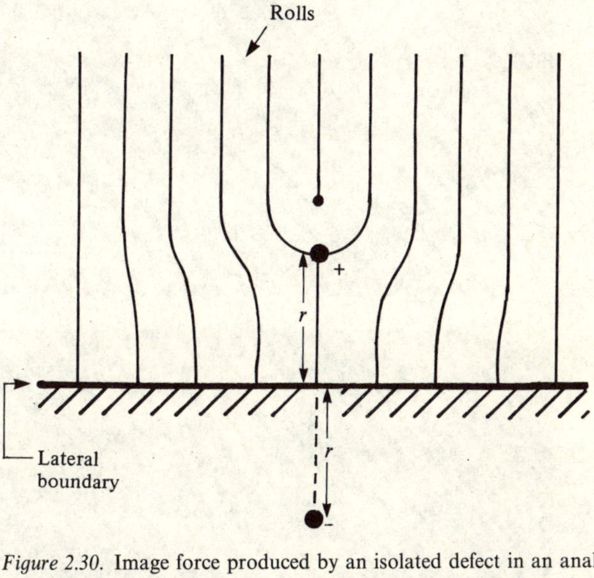

Figure 2.30. Image force produced by an isolated defect in an analogous description with the Coulomb interaction. Symbols + and − show the sign of the Burgers vector. For $r \sim L$, the Coulomb-like force plays a dominant role.

ence of domains by the collision of WF and defect dynamics starts to play an important role in the pattern selection. In other words, using the growth velocity v and the nucleation rate Θ of the domains, we have the following characteristic time:

$$t_m \sim (v^2 \Theta)^{-1/3} \tag{2.4.1}$$

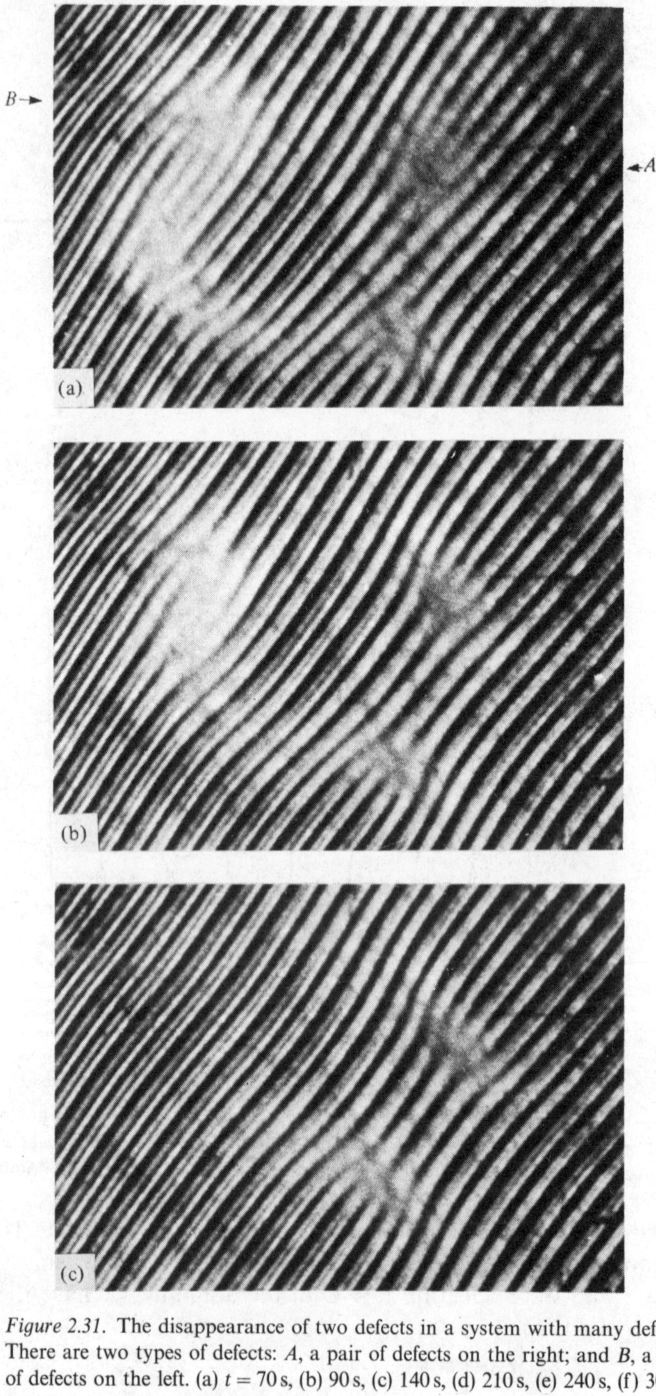

Figure 2.31. The disappearance of two defects in a system with many defects. There are two types of defects: *A*, a pair of defects on the right; and *B*, a pair of defects on the left. (a) $t = 70\,s$, (b) 90 s, (c) 140 s, (d) 210 s, (e) 240 s, (f) 300 s.

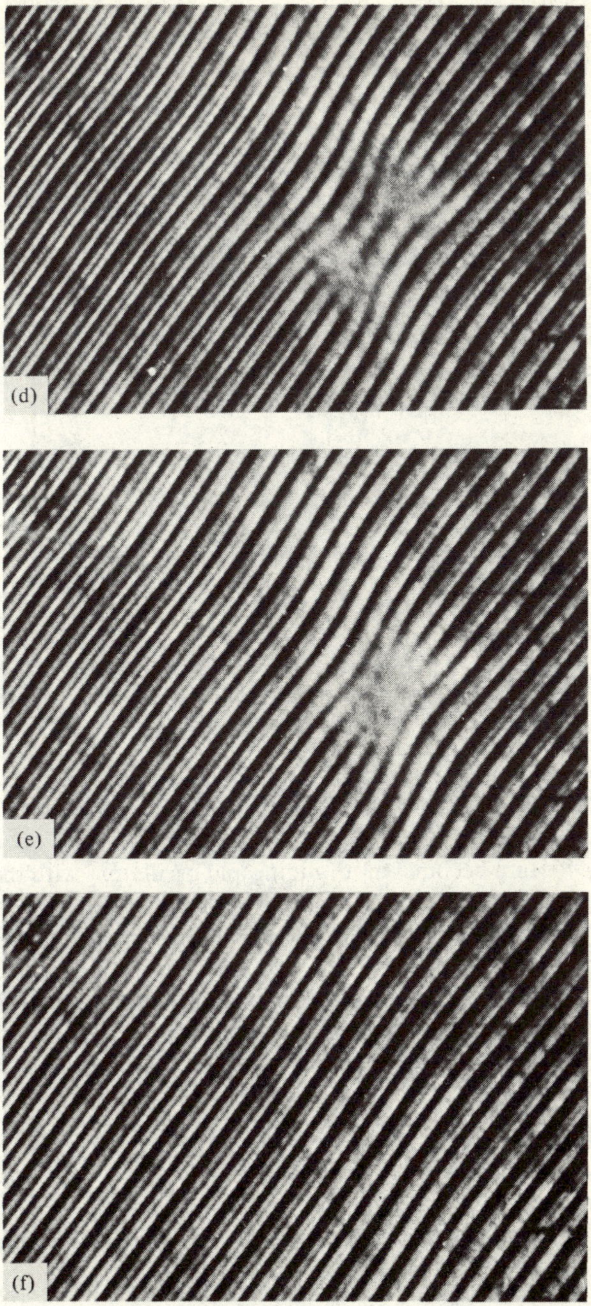
(d)
(e)
(f)

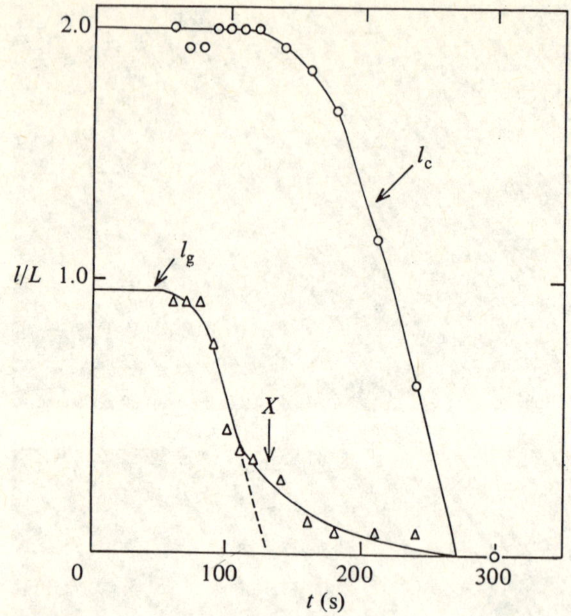

Figure 2.32. Temporal relations between gliding and climbing motions of a defect of type A in Figure 2.31. The counter defect does not move in this case. At X, the defect comes to the same roll by gliding motion. The oscillatory behavior can be observed for the gliding motion.

for two-dimensional systems from dimension analysis (see also Figure 2.5). This is another definition of t_m from an analogy with crystal growth (see Section 2.2.4).

The defect motion becomes more gliding and oscillatory with the increase in ε (typically $\varepsilon > 0.2$–0.3, depending on the applied frequency), without disappearance by combination, and new defects are nucleated. The mechanism of this motion therefore obviously relates to the fluctuating Williams domain (FWD). In FWD, the number of defects changes nonperiodically and it is observed that there is more likely to be an even number of states than an odd number; that is, the defects tend to pair up. Occasionally a 'no-defect state' is observed, but always new defects are being created.

2.4.4 *Multiplicative noise process*

Threshold shifts for the onset of instabilities by application of external noise

We use various types of noise distinguished by the probability density of the amplitude as was described in Section 2.2. For given values of τ_N and intensity, no dependence of the threshold shift on the type of the amplitude distribution

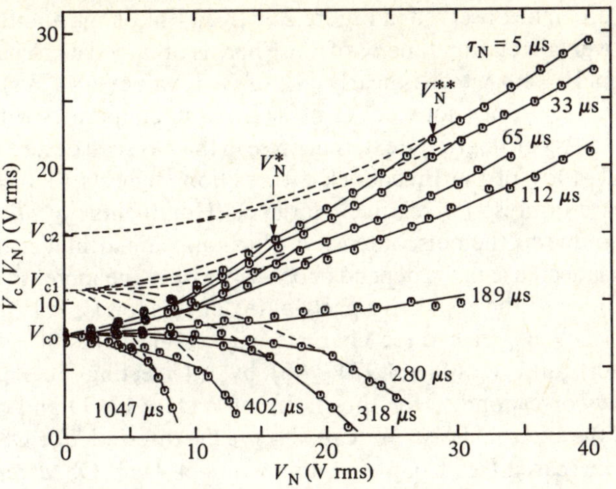

Figure 2.33. Dependence of phase diagram on correlation time τ_N. Note that extrapolations converge into each threshold voltage, even for negative slopes. V_N^* and V_N^{**} are marked for $\tau_N = 5\ \mu s$.

function of noise can be observed. Also there is no dependence on the frequency of a deterministic field, $f < 50\,\text{Hz}(\ll f_c)$.

In Figure 2.33, the typical results of the threshold shift for the first instability are plotted as a function of V_N. Here higher noise can always lead to a stabilization of the periodic structure for $\tau_N \ll \tau_c$. These shifts can be intuitively explained by the fact that the external noise reduces the amount of the space charge (which must be accumulated to cause convection by the Carr–Helfrich effect) by inducing random oscillation of the director. If, in place of one noise, one superimposes a sinusoidal wave with $f > f_c$ on one of $f = 30\,\text{Hz}$, then one observes rather similar shifts, albeit very small ones.

The correlation time τ_N of noise thus strongly influences the structure of the phase diagram. As τ_N becomes shorter, the slope of the noise dependence in the phase diagram becomes steeper. On the other hand, as τ_N becomes closer to the characteristic time of the system, $\tau_c = \omega_c^{-1}(442\ \mu s)$, where the condition $\tau_N \ll \tau_c$ does not hold any more, the slope becomes flatter. Here ω_c is the critical angular frequency separating the dielectric from the conductive regime. Finally, we observe that a threshold shift happens toward the negative direction. Discontinuities of slope are seen here as well as for $\tau_N \ll \tau_c$. When we try to fit each part between the bends to the following equation:

$$V_c(V_N) = aV_N^2 + bV_N + c, \tag{2.4.2}$$

where a, b and c are the coefficients obtained from experimental results by the least-mean-square method, two to three different slopes are clearly distinguished.

The systematic change of the coefficient a in (2.4.2), which indicates the slope of the curve, can be observed in Figure 2.35 (Kai, Fukunaga and Brand, 1987). The crossover correlation time τ_N^* of noise here is obtained as 255 μs in this cell ($d = 100\ \mu$m, $\Gamma = 40$), which is nearly 60% of the τ_c value ($\sim 442\ \mu$s). That is, the noise with $\tau_N^* < \tau_N$ does not work as noise here but cooperates with the deterministic field. Accordingly, the external noise in this case can destabilize systems. Theoretical study of the Freedericksz transition suggests that H_c decreases with increase in noise for relatively large τ_N (Horsthemke *et al.*, 1985).

The bend due to the noise intensity, which appears at higher intensities of noise with a decrease in τ_N, depends upon τ_N as does the increase in the slope. Extrapolations of the respective parts of the curves for $V_N < V_N^*$, $V_N^* < V_N < V_N^{**}$ and $V_N^{**} < V_N$ give three characteristic voltages $V_{c0}(0) = 7.6$ V, $V_{c1}(0)$ ($\sim 1.5\, V_{c0}(0)$), and $V_{c2}(0)(\sim 1.97\, V_{c0}(0))$ by intersecting at $V_N = 0$, which respectively correspond to the thresholds of WD, of FWD and of GP in the absence of the external noise. We can observe the threshold for DSM when V_N is further increased (see Chapter 3, this volume). Here DSM represents the irregular pattern, although there are three types of DSM (Kai and Hirakawa, 1977). From this observation, it can be said that for $V_N > V_N^{**}$ WD is unstable whereas GP is stable, being the first pattern at the onset point (Figure 2.33). That is, the exchange of the structure stability is induced by noise. Visual observations for these three regions appearing between bends are as follows. For $V_N < V_N^*$, steady WD is observed for the entire region of a cell. For $V_N^{**} < V_N < V_N^{***}$, GP is formed after the fluctuation of rolls for a long time, and for $V_N^{***} < V_N$ an irregular DSM-like pattern appears aperiodically in time and space (Brand, Kai and Wakabayashi, 1985). Between V_N^* and V_N^{**}, WD fluctuates irregularly with a characteristic distance of four to eight rolls (FWD).

All slopes, even for various τ_N, converge into three characteristic voltages albeit with different slopes. A similar situation occurs for $\tau_N \gtrsim \tau_N^*$; that is, at least two bends can be observed, and these converge into the corresponding characteristic voltages, even though the slopes are negative. Obviously, the different slopes indicate that noise works differently depending upon the spatial structure (Figure 2.33). Such dependence for the coefficient a is shown in Figure 2.34. If a wide-ranging observation for V_N is carried out, the number of bends corresponding to the number of successive transitions to fully developed turbulence will be observed (Kai and Hirakawa, 1977).

Another feature worth mentioning is the fact that, although the threshold for DSM is monotonically increasing with increasing noise, the difference between the onset values for the WD and for DSM is monotonically decreasing as well as that for the GP (see also Chapter 3, this volume).

Dynamical properties

Figure 2.35 shows the noise dependence on the linear growth time τ_L around $V_c(Q)$. The divergent tendency toward $V_c(Q)$ from both sides can be observed. The tendency from the side of $V_c < V$ is sharper than that of $V < V_c$. When τ_L is

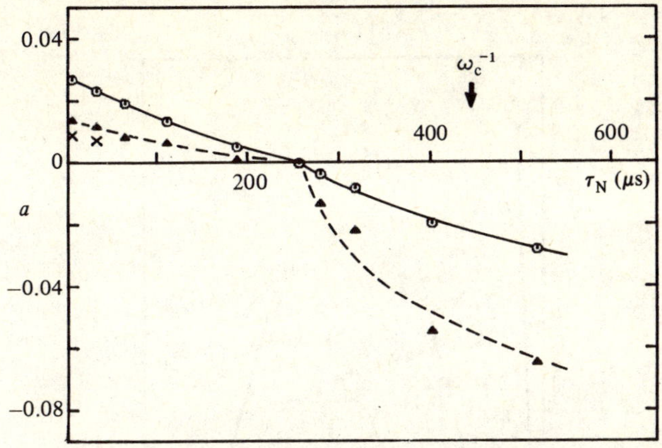

Figure 2.34. The coefficient a in (2.4.2) as a function of τ_N for each bend. At $\tau_N^* \sim 255\ \mu s$, the coefficient changes sign from positive to negative values. Symbols: ⊙, $V_N < V_N^*$; ▲, $V^* < V_N < V_N^{**}$; ×, $V_N^{**} < V_N$.

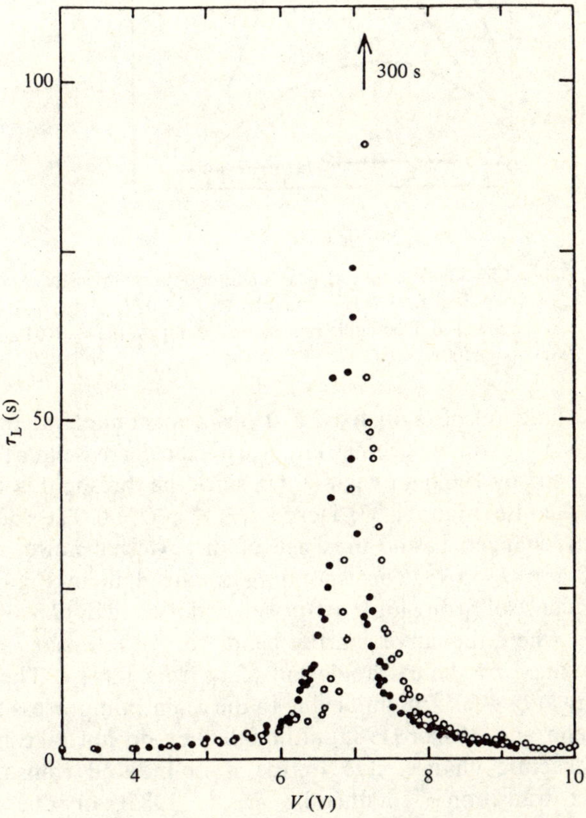

Figure 2.35. Typical example of τ_L as a function of an applied voltage ($\tau_N = 33\ \mu s$). Symbols: ●, $V_N = 1$ V; ○, $V_N = 2.8$ V.

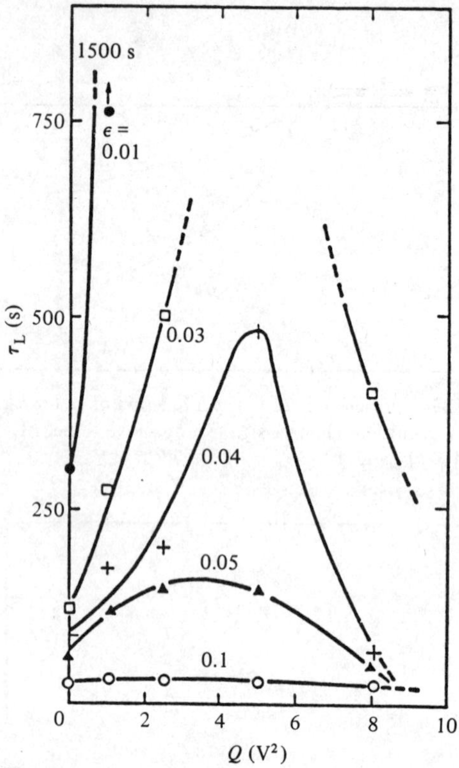

Figure 2.36. Growth time τ_L as a function of noise intensity Q ($d = 100\,\mu$m, $V_{c0} = 6.68$ V and $\tau_N = 33\,\mu$s). Symbols: ●, $\varepsilon = 0.01$; □, 0.03; +, 0.04; ▲, 0.05; ○, 0.1. The arrowed filled circle represents a point on the $\varepsilon = 0.01$ curve that fell at $\tau_L = 1500$ s.

plotted as a function of Q for fixed ε, it has a maximum value at the noise intensity $Q \sim 5\,V^2$, which becomes larger with the decrease in ε (Figure 2.36). V_c is determined by $V_c(Q)$ for each $Q \neq 0$, since the threshold is shifted by an application of noise (Figure 2.33). Here V_{c0} is V_c at $Q = 0$. The ε dependence of τ_L is slightly changed in the presence of the external noise, as shown in Figure 2.38 for $Q = 5\,V^2$, corresponding to the data in Figure 2.37. The linear dependence of τ_L on ε for $Q = 0$ is well understood by (2.3.4), but not that for $Q = 5\,V^2$ where the curve sharply bends to $\tau_L^{-1} = 0$ near $\varepsilon = 0.03$. Here $V_c(Q)$ is obtained by the extrapolation of τ_L from large ε. The anomalous increase in τ_L at $Q \sim 5\,V^2$ might be due to the re-entrant process theoretically given by Behn and Müller (1985) although they do not take into account a spatial structure change, due to the noise-induced transition for the Freedericksz transition (Horsthemke *et al.*, 1985), or due to additive thermal fluctuation effects (Mannella *et al.*, 1986).

In fact, the flow scale λ_x at $Q \sim 5\,V^2$ is slightly changed (Figure 2.38). This

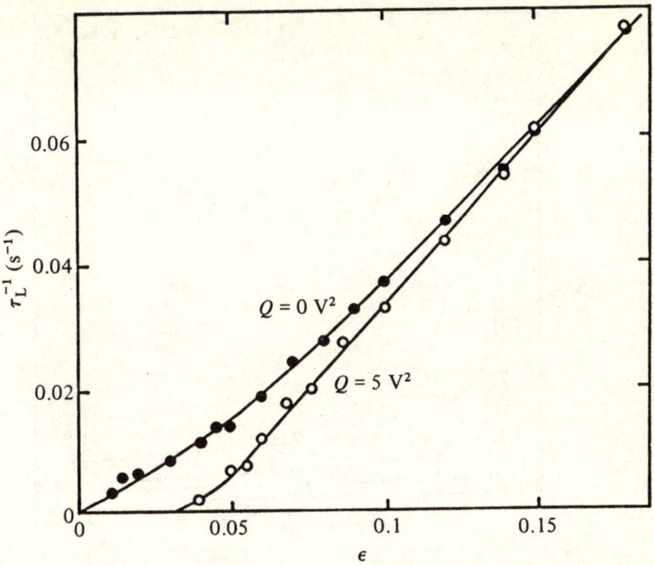

Figure 2.37. τ_L as a function of ε ($d = 100\,\mu$m, $V_{c0} = 6.68$ V).

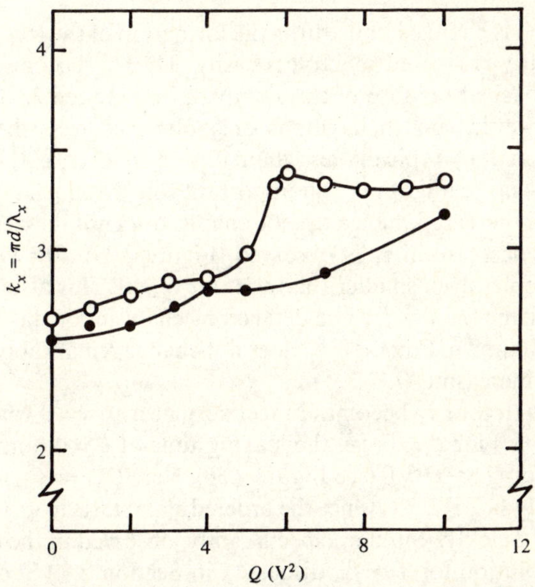

Figure 2.38. Wavenumber of WD as a function of Q. The magnitude of the jump seems to depend on Γ. Symbols: ○, $\Gamma = 10$; ●, $\Gamma = 72$.

behavior may be explained in a similar way to that described in the previous section. The superimposing of noise induces a steady flow λ_N, different from λ_0 originally expected under the deterministic field only. Consequently, the

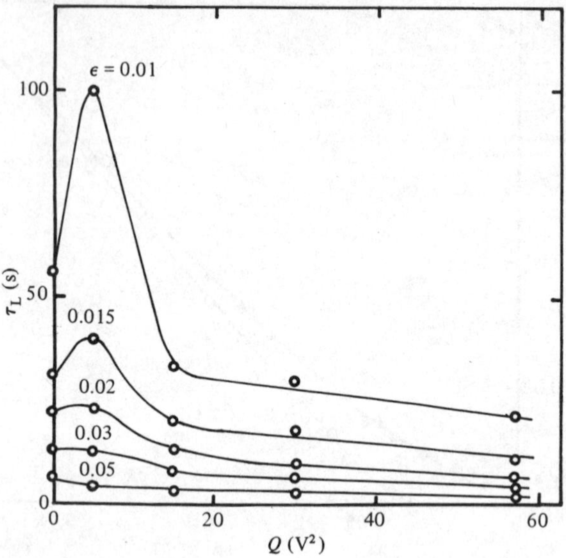

Figure 2.39. τ_L as a function of Q for the wide range ($d = 20\,\mu$m, $V_{c0} = 7.35$ V (50 Hz)).

dispersion relation is changed; it disturbs the formation of the original roll λ_0, which consequently is created much less easily. Thus τ_L becomes longer as $Q \sim 5\,V^2$ is approached because of this competition between λ_0 and λ_N (Kai *et al.*, 1985). However, such behavior is only observed near the threshold (small ε) and at small Q (typically less than $10\,V^2$). In a sense, therefore, the additive (thermal) noise might play an important role (Mannella *et al.*, 1986). When Q is further increased, such a drastic change does not happen, and only the monotonical decrease in τ_L is observed (Figure 2.39). For very large Q ($> 20\,V^2$), τ_L becomes even smaller than that for $Q = 0$. This also happens in the oscillatory regime for $f > f_c$. The detailed mechanism for this anomaly at $Q \sim 5\,V^2$ is still unknown. For $\varepsilon < 0$, τ_L does not change remarkably and seems to increase with increasing Q.

The initial growth time τ_L becomes longer very near to $\varepsilon = 0$ when at $Q > 0$ than it does at $\varepsilon = 0$ for $Q = 0$, i.e. the staying time at $\theta = 0$ (unstable state) becomes longer for $Q_* > Q > 0$ (see Figure 2.36). Here $Q_* (= V_{N*}^2)$ is a certain threshold, typically $V_{N*} \lesssim 2.5$ V. Once the ordered state starts to grow, however, it is built up very quickly. (Similar aspects can be observed in the experiment of oscillatory evolution for $f > f_c$, discussed in Section 2.5.) Such behavior even appears to be discontinuous (Feigelman and Staroselsky, 1986). These aspects are in opposition to the prediction from a homogeneous theory with the quadratic noise term. It may suggest therefore a different modification of Φ from that in MSP of the Freedericksz transition. Further investigation is needed theoretically and experimentally on this point.

We also observe that the nonlinear growth time τ_r^{-1} increases linearly with the decrease in Q. The slope of this linear relation in the GP region is sharper than that in the WD region. For the DSM mode we find that τ_r^{-1} decreases monotonically but nonlinearly as a function of the noise intensity (Kai *et al.*, 1985). There seems to be no remarkable noise dependence of the director angle θ in the steady state. For very small Q (typically $Q < 0.2\,V^2$) τ_L is not substantially changed.

2.4.5 *Summary*

Convective rolls nucleate at both lateral boundaries and grow into the stationary bulk, usually with the wavefronts parallel to the roll axis. Near $t = t_m$, both wavefronts collide at the middle of the container, and then the scale of convections starts to be modified and defects are formed to fit. An isolated defect is usually produced at the very late stage after t_m. Two types of motion of the defects can be distinguished; the climbing and the gliding motions. The climbing motion occurs for relatively small ε, compared to the gliding. The defect motion slows down at first, and then speeds up during the evolution of the system. It finally disappears, being absorbed into boundaries. In the case of a system with many defects, defects disappear by both combining with other defects and by being absorbed into the boundaries. These dynamics play an important role in the late stage of the growth, in which the new growth kinetics should be considered that correspond to the ripening process in a crystal growth.

The results obtained in MSPs are summarized as follows. An externally applied noise with correlation time $\tau_N \ll \tau_c$ can (1) delay the onset of the convection, (2) increase the threshold value for the onset, (3) make the instability more discrete, (4) suppress the spatial turbulence, and (5) realize the direct transition to turbulence when it is at a higher intensity than a certain threshold. Furthermore, different noise dependence on the growth time can be observed simultaneously following a change of flow structure, for example from WD to GP, and the structural transition is induced by an application of the external noise. In the absence of the convective structure ($\varepsilon \leqslant 0$), the influence of the external noise is rather small. But, when macroscopic structure appears (above a threshold), it becomes large. The most interesting feature of noise effects is the change in stability of the spatial modes in the pattern formation. Finally we would like to stress that the externally applied noise destabilizes the convective system at $\tau_N \gtrsim \tau_c$ but stabilizes it at $\tau_N \ll \tau_c$.

2.5 The oscillatory regime ($f > f_c$)

2.5.1 *Statistical properties of transient processes*

Evolution of the director oscillation

The temporal development of the oscillatory amplitude Y and the variance σ are shown in Figure 2.40, when the external field is increased in a steplike

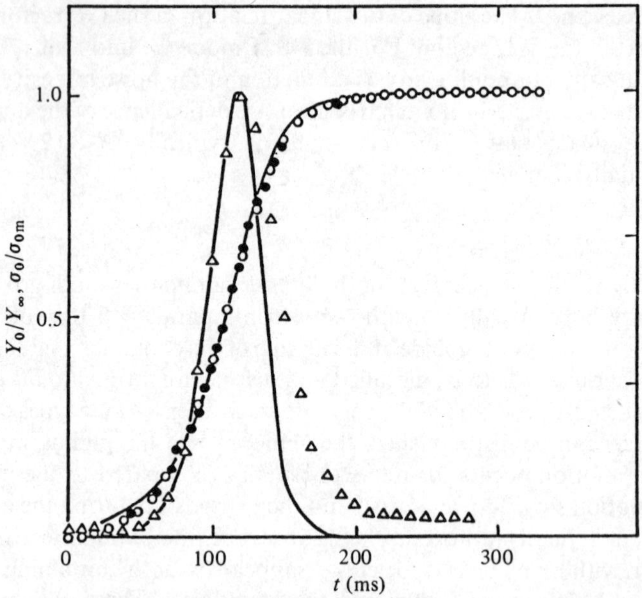

Figure 2.40. Temporal development of macroscopic order Y and variance σ at $\varepsilon = 0.098$. $\bar{Y}_0(t)$ is the mean value for a fixed t, normalized to unity at its value at $t = \infty$, $\bar{Y}_{0\infty}$; σ_0 is the variance around the $\bar{Y}_0$, normalized to unity at its maximum value, σ_{0m}. The solid lines for $\sigma_0(\triangle)$ show (2.5.1), and those for $\bar{Y}_0$ (○) depict the equation $\dot{Y} = \alpha Y - \beta Y^3 + \eta(t)$. The thickness d of the cell and the area S of electrodes are $108 \pm 5\ \mu$m and $6 \times 6\,\text{mm}^2$, respectively (Kai, Wakabayashi and Imasaki, 1986).

fashion. Symbols $\bar{Y}_0$, and σ_0 are defined in (2.3.8)–(2.3.9). The solid line for $\bar{Y}_0$ in Figure 2.40 shows the solution of the equation $\dot{Y} = \alpha Y - \beta Y^3$. $\bar{Y}_0$ and σ_0 are normalized by the steady-state value Y_∞ and the maximum value σ_m, respectively. The experimental data for $\bar{Y}_0$ fit the equation well. σ_0 shows the maximum value σ_m at $t = t_m > t_{1/2}$, where $t_{1/2}$ is defined as the time at which Y becomes half the value of Y_∞. After passing through its maximum value of σ_m, $\sigma_0(t)$ decreases towards its steady state value. The solid line for σ_0 shows the fluctuation enhanced theory for $\dot{Y} = \alpha Y - \beta Y^3 + \eta(t)$,

$$\sigma_0 \propto \left[\frac{\dot{Y}(t)}{Y(0)}\right]^2 = \frac{A\exp(2\alpha t)}{[\exp(2\alpha t) + \delta]^3}, \tag{2.5.1}$$

where A and δ are constants related to the steady state and the initial values of σ and Y (Kubo, Matsuno and Kitahara, 1973; Suzuki, 1981). The fitting of $\sigma_0(t)$ to the experimental data is done by use of the coefficients evaluated from the experimental data. Here the characteristic time t_m does not always agree with the theoretical result (Suzuki, 1981).

For large d, the relationship between $t_{1/2}$ and t_m is found to be $t_{1/2} < t_m \sim t_s$ and typically $(t_m - t_{1/2})/t_m \sim 0.06$ for $\xi = 0$, whose value increases with the noise intensity. t_s here is the time at which the derivative of $\bar{Y}_0(t)$ experimentally obtained has the maximum value. For small d, it is given as $t_m < t_{1/2} \sim t_s$ in the absence of the external noise, and as $t_{1/2} \sim t_s < t_m$ in the presence of noise. Typically $(t_m - t_{1/2})/t_m$ is in the range 0.07–0.14, with the increase in ξ from 0.05 to 0.25 (see Figure 2.41 for details). Here ξ is the ratio of the white noise intensity to the amplitude of the deterministic field. These data suggest that the potential profile depends on d and ξ. (We use ξ here as the actual noise intensity normalized by V_N/V_d, unlike $\xi(t)$ in (2.2.5). V_d is the amplitude of the deterministic field. $\eta(t)$ obviously correlates to ξ with the introduction of a certain factor. In EHD instability, however, the multiplicative term $\xi(t)Y$ is more dominant then $\eta(t)$.)

For $t < t_m$, the agreement between the theoretical curve and the experimental data of σ_0 is quite good, but not for $t \gtrsim t_m$. This deviation increases as the steady state is approached (Kai, Wakabayashi and Imasaki, 1986). When the thickness d is reduced and Γ increases, the equation $\dot{Y} = \alpha Y - \beta Y^2 + \eta(t)$ frequently seems to give a rather better fit to the experimental data than $\dot{Y} = \alpha Y - \beta Y^3 + \eta(t)$. In EHD instability, (2.2.3) is derived from bulk properties without a surface effect (anchoring). This modification of index suggests that the surface anchoring should be considered when d is very small.

When a white noise is externally superimposed on the system, the temporal evolution of σ_0 is changed. Figure 2.41 shows the result at $\varepsilon = 0.1$. Here the magnitude of Y_∞ remains the same for all ξ. White noise delays the formation of macroscopic order for the initial stage, i.e. t_m is clearly delayed. The maximum value of σ_0 also becomes larger with the increase in ξ, at least by $\xi = 0.25$. Even at steady state, a large variance ($\sigma(\infty)$) remains without going to zero in the presence of noise. A similar result has been obtained theoretically (Pasquale, Sancho, San Miguel and Tartaglia, 1986). The large σ_0 at t_m indicates large $\dot{Y}$ according to (2.5.1). This means that the evolution velocity near t_m becomes faster in the presence than in the absence of noise (Figure 2.41); that is, the slope of the potential becomes sharp. This tendency of the dynamical property seems to be in opposition to the theoretical prediction, as well as in the case for the region $f < f_c$. In the Freedericksz transition, the theory predicts that the mean first passage time (corresponding to τ_L and also t_m) and the relaxation time around $\theta \neq 0$ are reduced by application of noise. This is, however, not the case.

We should note another interesting feature of t_m. At first t_m seems to show a small maximum at $\xi \sim 0.1$. For $0.1 < \xi < 0.15$ t_m decreases slightly with the increase in ξ (see also Section 2.4.4). After such a curious dependence, t_m again increases with ξ. This may be due to competition between additive and multiplicative noise (Mannella *et al.*, 1986).

Distribution of fluctuations around the most probable path $\bar{Y}_0$

Figure 2.42 shows temporal developments of the distribution of fluctuations

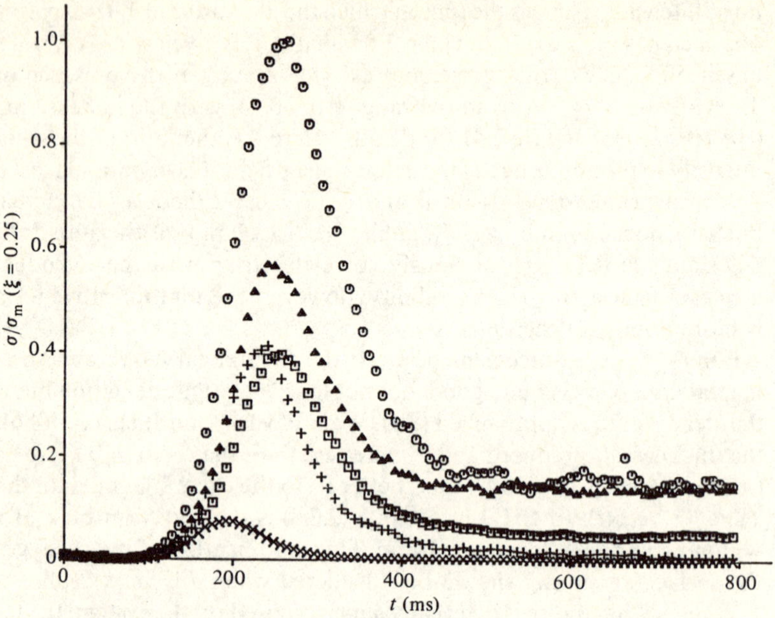

Figure 2.41. Effect of white noise intensity ξ on the variance σ. The cell used in the present study was 16 μm thick and had a plane size of $2 \times 2\,\text{mm}^2$; that is, the aspect ratio $\Gamma = 125$. The applied frequency $f = 120$ Hz, and $f_c = 60$ Hz. The threshold voltage $V_c = 40.5 \pm 0.3$ V to obtain a director oscillation. To avoid cluttering the figure, a lot of data have been eliminated for t and ξ. $\varepsilon = 0.1$. Symbols: $\times$, $\xi = 0$; $\boxdot$, 0.1; $+$, 0.15; $\blacktriangle$, 0.2; $\odot$, 0.25.

around $\bar{Y}_0$ at $\varepsilon = 0.1$ and $\xi = 0$. As seen in Figure 2.42, the distribution changes since a turn in t_m (Kai, Wakabayashi and Imasaki, 1986). We note that the distribution profiles for the early stage and for the steady state are very sharp and similar, but they are quite different from those for the transient region where the large growth rate is realized. This tendency is the same for all experimental results at small ε. In the case without external noise, the distribution has a Gaussian profile for the early stage at $t = 108$ ms (Figure 2.42), but the deviation from Gaussian profile increases with time (see Figure 2.42 and Kai, Wakabayashi and Imasaki, 1986).

The distribution near t_m becomes asymmetric with negative skewness. This tendency is especially recognized after the stage $t \sim 0.8 t_m$ where nonlinear effects of macroscopic order probably become important.

When $\xi = 0.1$, the distribution has positive skewness and is well fitted for the initial stage (for example at $t = 188$ ms; Figure 2.43) with a lognormal distribution (Aitchison and Brown, 1957; Kai, Wakabayashi and Imasaki, 1986),

$$N(Y) = (N_0 \bar{Y}_0 / Y) \exp[-(\log Y/\bar{Y}_0)^2 / 2\sigma^2], \tag{2.5.2}$$

which is often observed in statistical properties of the variable Y which has

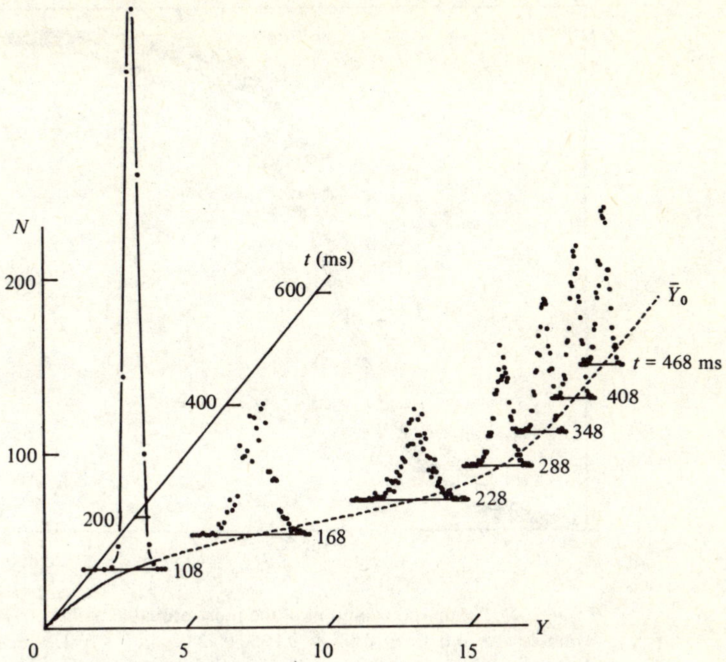

Figure 2.42. Temporal developments of distribution around $\bar{Y}_0$ at $\varepsilon = 0.1$ ($d = 16\,\mu$m, $\Gamma = 125$). $t_m = 196$ ms; $\xi = 0$.

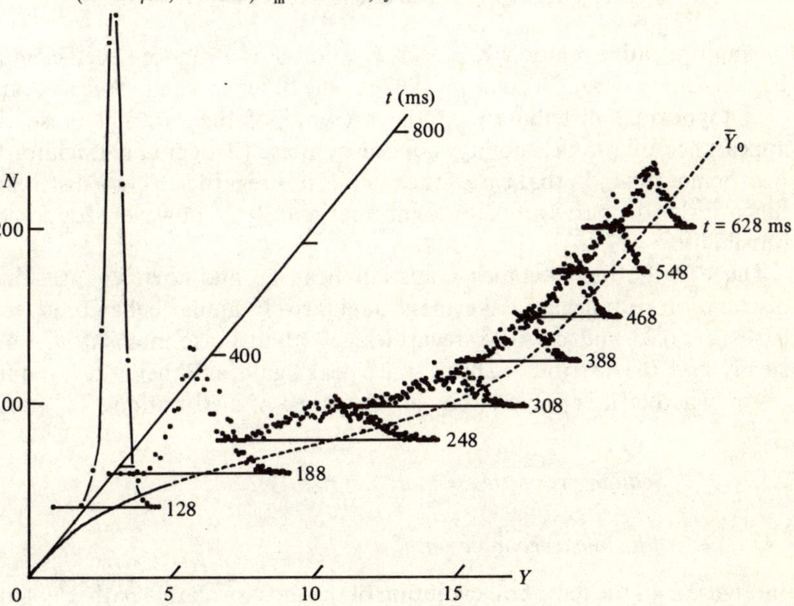

Figure 2.43. Temporal developments of distribution around $\bar{Y}_0$ at $\varepsilon = 0.1$ with $\xi = 0.1$, $t_m = 260$ ms.

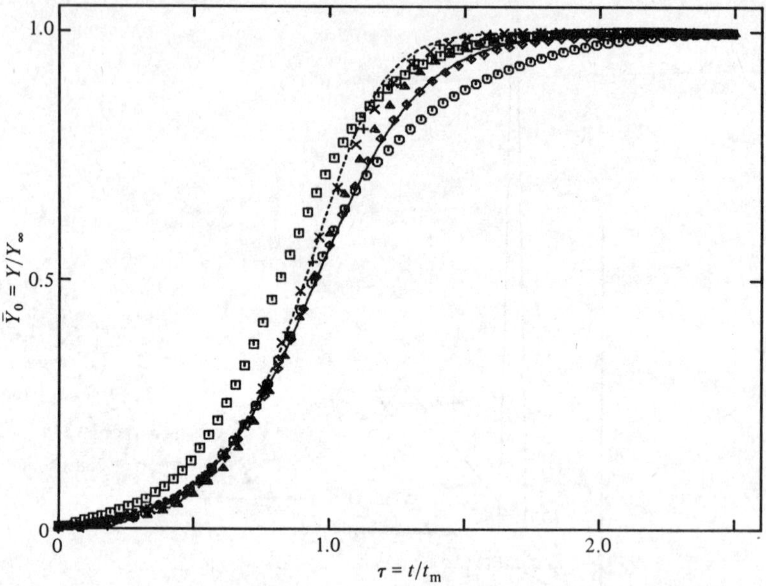

Figure 2.44. Temporal evolution of the most probable path $Y_0(\tau) = Y(\tau)/Y_\infty(\varepsilon)$. Symbols: ⊡, $\varepsilon = 0.05$; ⊙, 0.08; ◆, 0.1; ▲, 0.12; ×, 0.15; +, 0.23. The solid line shows the universal curve normalized by the onset time t_m using the equation $\dot{Y} = \alpha Y - \beta Y^2$, and the dotted line corresponds to $\dot{Y} = \alpha Y - \beta Y^3$. The best fitting is done for initial stages of all data ($d = 16\,\mu$m, $\Gamma = 125$).

the multiplicative relation $Y_{n+1} = Y_n P$, where P is a random coefficient (Kai, Higaki, Imasaki and Furukawa, 1987a). One other reason for the appearance of a lognormal distribution in the presence of the external noise is the appearance of a multistability induced by noise (Broggi and Lugiato, 1984; Fedchenia, 1984). In that case, for example, it is regarded as two distributions that overlap in part to give a lognormal one. It is, however, supposedly a possibility.

The distribution becomes Gaussian near t_m and then changes into a distribution with negative skewness: qualitatively similar behavior is seen for $0.04 < \varepsilon < 0.23$ and $\xi < 0.25$, regardless of ε and noise intensity ξ. For the steady state the distribution has a sharp peak again, and therefore it is difficult to say whether it is Gaussian or another type of distribution.

2.5.2 *Scaling properties*

The most probable path

In Figure 2.44 the temporal evolution of the most probable path $Y_0(\tau)$ for the cell with $d = 16\,\mu$m is shown for various ε, where the normalized time, τ, is given by $\tau = t/t_m$. The solid line indicates the theoretical path of $Y_0(\tau)$. The coefficients α and β are chosen so that the best fit is obtained at $0 < t < t_m$ for all ε (Kai, Tamura, Wakabayashi and Imasaki, 1987b). The data agree quite well with the evolution equation for the early stage of $Y_0(\tau)$ for $0.08 \leqslant \varepsilon \leqslant 0.23$, but not for the late stage of relatively small ε, for instance when $\varepsilon \leqslant 0.08$. In particular, the data at $\varepsilon = 0.05$ are totally different from the rest and do not fit the universal curve for all stages. Thus, although the universal curve of $Y_0(\tau)$ can certainly be recognized, the deviation from it becomes large as the external stress is closer to the onset point V_c. Similar behavior has been observed in a pressure-induced phase transition of RbI in a neutron-scattering study (Yamada, Hamaya, Axe and Shapiro, 1984), which is one example of phase transitions in nonconserving systems. In the RbI case, the evolution can be described by the equation proposed by Kolmogorov and Avrami. According to the authors of that study, such a deviation happens due to the increase in the size of the critical nucleus of a new phase near $\varepsilon = 0$; that is, due to the non-negligible size of the nucleus compared to the characteristic length determined from transport coefficients. In other words, the critical nucleus becomes too large. The present result may be such a case, when we regard each growing domain of the oscillatory structures as a nucleus.

Variance $\sigma(\tau)$

The temporal developments of the scaled variances $\sigma(\tau)$ are shown for various ε in Figure 2.45, using (2.5.1) replaced with $\tau = t/t_m$ (Suzuki, 1981). For $0 < t < t_m$, the scaling holds sufficiently for all ε, but not so clearly for the late stage. The fact that the deviation from the scaling curve increases as ε decreases can be clearly seen, similar to $Y_0(\tau)$ in Figure 2.44. This means that the distribution becomes broader than that expected from the theory. It comes from spatial inhomogeneities such as defects (these will correspond to grain boundaries in an analogous sense with a grain growth). Characteristic profiles of the distribution during the evolution can be also scaled by a Gaussian distribution for $\tau \lesssim 1.0$ and by the distribution with the fourth order coefficients for $1.0 < \tau < 1.8$ (Kai *et al.*, 1987b). The same tendencies of the scaling properties for $Y_0(\tau)$ and $\sigma(\tau)$ are observed when ξ is varied ($0 \lesssim \xi \lesssim 0.25$).

2.5.3 Summary

We summarize our results as follows. The variance σ_0 has the maximum value σ_m at $t = t_m$, and it then decreases. The temporal change of σ_0 for $t < t_m$ can be described by the noisy Landau equation. The deviation of σ_0 from it increases for the later stage, $t > t_m$. The distribution around $\bar{Y}_0$ is Gaussian for $t \ll t_m$ but becomes one with negative skewness for $t > t_m$. When the external noise is applied, the distribution for the early stage is well fitted with the lognormal one; then for $t \sim t_m$ it becomes Gaussian, and for the late stage it changes into a distribution with negative skewness (Kai, Wakabayashi and Imasaki, 1986).

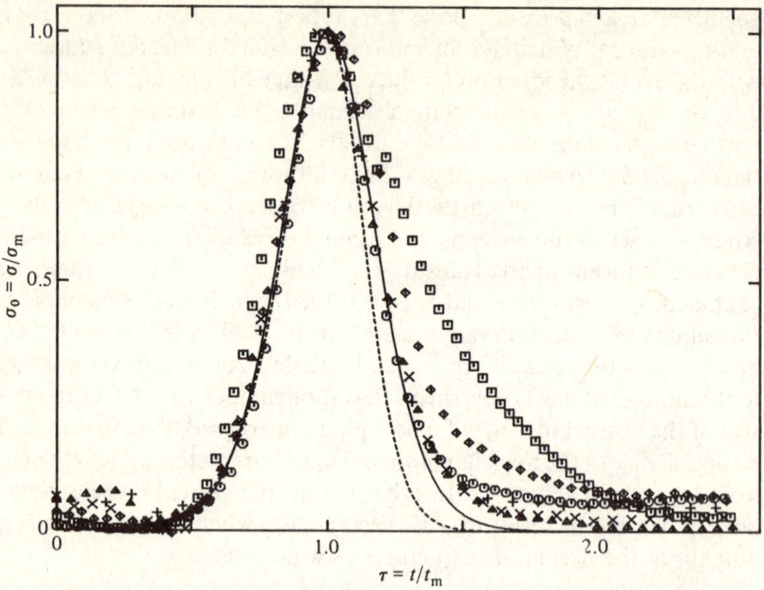

Figure 2.45. Temporal change of variance $\sigma_0(\tau) = \sigma/\sigma_m$. Symbols: ⊡, $\varepsilon = 0.05$; ⊙, 0.08; ◆, 0.1; ▲, 0.12; ×, 0.15; +, 0.23. The dotted line shows the universal curve given by (2.5.1), and the solid line corresponds to the equation $\sigma_0(\tau) = Ae^{2\alpha\tau}/(e^{\alpha\tau} + \delta)^4$, obtained from $\dot{Y} = \alpha Y - \beta Y^2 + \eta(t)$.

The application of the external noise also slows the formation process down. That is, it increases t_m, the total formation time, and σ_m. The scaling properties hold in the evolution of $Y_0(\tau)$ and in the variance $\sigma(\tau)$ for the early stage, but they do not hold for the late stage. In particular, the deviation from the scaling curve increases on approaching the instability point $\varepsilon = 0$. This perhaps comes from the growth of the spatial inhomogeneities and the increase in the nucleation size of the oscillatory domain.

Thus, the Landau equation ($\dot{Y} = \alpha Y - \beta Y^3 + \eta(t)$) is not perfect for describing the transient kinetics of systems with spatial degrees of freedom, particularly after the intermediate stage of evolution. Since we observe local properties ($\sim 10\,\mu$m in diameter) at a fixed location using a light signal, and since the fluctuations giving rise to the oscillation occur at random positions in space, the growth of the oscillation is stochastic for each observation at each burst field. The distribution at each stage obtained from the repetition therefore effectively indicates the spatial distribution of the growth of the order at that stage. The spatial inhomogeneity, which results from the spatial difference of the growth of the oscillatory order, may be responsible for the observed deviations of the variance and the shape of the distribution from predictions based on the universal properties of the Landau equation. Since spatial degrees of freedom and the effect of the externally applied noise are not taken into account in (2.5.1), a more detailed theoretical discussion is clearly required.

The behavior of σ_0 and t_m in the presence of the external noise is concerned with MSP. The externally applied noise rather strongly affects the profile of the (lognormal) distribution function during the early stages of the transition. Once the macroscopic order is formed, the external noise is relatively ineffective. The growth process of a macroscopic order for the late stage is clearly different from that for the early stage in both cases, with and without the external noise.

2.6 Conclusion

We have observed three different stages in the pattern formation process for $f > f_c$. That is, the initial linear-growth-stage, where the macroscopic order nucleates and grows without correlations; the nonlinear growth stage, where the domain boundaries of convective rolls contact together; and the final growth stage, where the defect motion plays an important role corresponding to a ripening process in a crystal growth. There are two different defect motions; the climbing and the gliding. The gliding requires larger forces to be excited than the climbing. In order to describe the kinetics of the final stage, the defect dynamics should be taken into account.

Unlike the common understanding that the externally applied noise cannot change a spatial structure, it is experimentally clear that noise can change the stability of the spatial structure in MSPs of EHD instability. The precise mechanism of this is still unknown. The externally applied noise can stabilize the system when the correlation time τ_N of the noise is much shorter than the characteristic time τ_c of the system. It can, however, destabilize when $\tau_N \gtrsim \tau_c$. The most important factor to induce drastic noise effects is the correlation time of the applied noise. No remarkable dependence can be observed when the amplitude distribution of noise is varied under fixed τ_N.

The evolution for $f > f_c$ can be also classified into three different stages. For the early stage the growth process is in good agreement with the result expected from the noisy Landau equation and the scaling properties hold everywhere except near the instability point. But it has a large deviation from the theory for the late stage, $t \gtrsim t_m$, and for the whole stage near $\varepsilon = 0$. Perhaps this comes from the fact that only a few big domains appear and grow. This may be explained in a similar way with a first order phase transition in RbI crystal (Yamada *et al.*, 1984). The large deviation near $\varepsilon = 0$ is perhaps due to the infinite correlation length. On the other hand the deviation for the late stage at relatively large ε is due to defects (spatial inhomogeneity). The distribution for the early stage in the presence of the external noise is quite different from that when external noise is absent and shows the lognormal profile. This suggests that the growth kinetics for the early stage is strongly influenced by the application of external noise. It is a generally well-known fact that the lognormal distribution comes from a multiplicative (cascade) process (Kai *et al.*, 1987a). This may be the case. It is also worth noting here that a size distribution of a crystal growth due to a second order interfacial controlled growth kinetics (SOCG) has a similar profile to the lognormal one (Hohmann and Kahlweit, 1972; Kahlweit, 1975). In addition, experimental results of crystal size-distributions are rather better fitted to the lognormal distribution than one of SOCG (Kai *et al.*, 1987a; Kai and Müller, 1985; Paramdeep *et al.*, 1983).

Finally, we have pointed out an analogy between crystal growth and dissipative structure formations. For the final stage the pattern formation might be dominated by the diffusion-like kinetics, which are realized through defect motions. Of course, it should be mentioned that there is great difference in the conserving law between the kinetics of a first order transition and of the present instability. Although the validity for such a similarity of the growth is not clear yet, the proposition of the analogous concept may be valuable for future theoretical approaches.

References

Aitchison, J. and Brown, J. A. C. 1957. *The Lognormal Distribution.* Cambridge University Press.

Arecchi, F. T. and Degiorgio, V. 1971. *Phys. Rev. A* **3**, 1108–24.

Behn, U. and Müller, R. 1985. *Phys. Lett.* **113A**, 85–8.

Brand, H. R., Kai, S. and Wakabayashi, S. 1985. *Phys. Rev. Lett.* **54**, 255–7.

Brand, H. R. and Schenzle, A. 1980. *J. Phys. Soc. Jpn.* **48**, 1382–3.

Broggi, G. and Lugiato, L. A. 1984. *Phil. Trans. R. Soc. London A* **313**, 425–8.

Croquette, V. and Pocheau, A. 1984. In *Cellular Structures in Instabilities.* (J. E. Wesfried and S. Zaleski, eds.), pp. 104–26. Lecture Notes in Physics 210. Berlin: Springer.

de Gennes, P. G. 1982. *The Physics of Liquid Crystals*, 3rd edn. Oxford: Clarendon Press.

Dubois-Violette, E., de Gennes, P. G. and Parodi, O. 1971. *J. de Phys.* **32**, 305–17.

Fedchenia, I. I. 1984. *Physica* **125A**, 577–90.

Feigelman, M. V. and Staroselsky, I. E. 1986. *Z. Phys. B* **62**, 261–6.

Fox, R. F., James, G. E. and Roy, R. 1984. *Phys. Rev. Lett.* **52**, 1778–81.

Furukawa, H. 1985. *Adv. Phys.* **34**, 703–50.

Galerne, Y., Durand, G. and Veyssie, M. 1972. *Phys. Rev. A* **6**, 484–7.

Gollub, J. P. and Benson, S. V. 1980. *J. Fluid Mech.* **100**, 449–70.

Graham, R. and Schenzle, A. 1982. *Phys. Rev. A* **26**, 1676–85.

Greenside, H. S. and Coughran, Jr. W. M. 1984. *Phys. Rev. A* **30**, 398–428.

Greenside, H. S. and Cross, M. C. 1985. *Phys. Rev. A* **31**, 2492–501.

Gunton, J. D., San Miguel, M. and Sohni, P. S. 1983. In *Phase Transitions and Critical Phenomena 8* (C. Domb and J. L. Lebowitz, eds.), p. 267. New York: Academic Press.

Helfrich, W. 1969. *J. Chem. Phys.* **51**, 4092–105.

Hijikuro, H. 1975. *Prog. Theor. Phys.* **54**, 592–4. (In this article, the nonlinear

macrodynamic equation in EHD was derived under a d.c. electric field without terms $\xi(t)\cdot Y$ and $\eta(t)$ in one-dimensional space.)
Hirakawa, K. and Kai, S. 1977. *Mol. Cryst. Liquid Crystals* **40**, 261–84.
Hohmann, H. H. and Kahlweit, M. 1972. *Ber. Bunsenges. Physik. Chem.* **76**, 933–8.
Horsthemke, W., Doering, C. R., Lefever, R. and Chi, A. S. 1985. *Phys. Rev. A* **31**, 1123–35.
Horsthemke, W. and Lefever, R. 1984. *Noise-Induced Transitions*, pp. 164–89. Berlin: Springer-Verlag.
Joets, A. and Ribbota, R. 1986a. *J. de Phys.* **47**, 595–606.
Joets, A. and Ribbota, R. 1986b. *J. de Phys.* **47**, 739–43.
Kabashima, S., Itsumi, M., Kawakubo, T. and Nagashima, T. 1975. *J. Phys. Soc. Jpn.* **39**, 1183–8.
Kabashima, S., Kogure, S., Kawakubo, T. and Okada, T. 1979. *J. Appl. Phys.* **50**, 6296–302.
Kahlweit, M. 1975. *Ad. Colloid Inter. Sci.* **5**, 1–35.
Kai, S. 1986. *Kotaibutsuri* [*Solid State Physics*] **21**, 429–40. (In Japanese.)
Kai, S., Araoka, M., Yamazaki, H. and Hirakawa, K. 1979a. *J. Phys. Soc. Jpn.* **46**, 393–409.
Kai, S., Fukunaga, H. and Brand, H. R. 1987. *J. Phys. Soc. Jpn.* **56**, 3759–62.
Kai, S., Higaki, S., Imasaki, M. and Furukawa, H. 1987a. *Phys. Rev. A* **35**, 374–9.
Kai, S. and Hirakawa, K. 1977. *Mem. Fac. Engin. Kyushu Univ.* **36**, 269–301.
Kai, S. and Hirakawa, K. 1978. *Prog. Theor. Phys. Suppl.* **64**, 212–43.
Kai, S. and Hirakawa, K. 1985. *Physics* **6**, 470–8. (In Japanese.)
Kai, T., Kai, S. and Hirakawa, K. 1977. *J. Phys. Soc. Jpn.* **43**, 717–18.
Kai, S., Kai, T., Takata, M. and Hirakawa, K. 1979b. *J. Phys. Soc. Jpn.* **47**, 1379–80.
Kai, S. and Müller, S. C. 1985. *Sci. Form* **1**, 9–39.
Kai, S., Tamura, T., Wakabayashi, S., Imasaki, M. and Brand, H. 1985. *Annual Meeting of IEEE–IAS Conference Record 85CH2207-9* pp. 1555–62.
Kai, S., Tamura, T., Wakabayashi, S. and Imasaki, M. 1987b. *Phys. Rev. A* **35**, 1438–40.
Kai, S., Wakabayashi, S. and Imasaki, M. 1986. *Phys. Rev. A* **33**, 2612–20.
Kai, S., Yamaguchi, K. and Hirakawa, K. 1975. *Jpn. J. Appl. Phys.* **14**, 1653–8.
Kawakubo, T., Yanagita, A. and Kabashima, S. 1981. *J. Phys. Soc. Jpn.* **50**, 1451–6.
Kawasaki, K. 1984a. *Ann. Phys.* **154**, 319–55.
Kawasaki, K. 1984b. *Prog. Theor. Phys. Supp.* **79**, 161–90.
Kawasaki, K. 1984c. *Prog. Theor. Phys. Supp.* **80**, 123–38.
Kawasaki, K. 1985. *Phys. Rev. A* **31**, 3880–5.
Komura, S., Osamura, K., Fujii, H. and Takeda, T. 1984. *Phys. Rev. B* **30**, 2944–7.
Kubo, R., Matsuno, K. and Kitahara, K. 1973. *J. Stat. Phys.* **9**, 51–96.
Lifshits, I. M. and Slyozov, V. V. 1961. *J. Phys. Chem. Solids* **19**, 35.
Linz, S. J. and Lucke, M. 1986. *Phys. Rev. A* **33**, 2694–703.
Mannella, R., Faetti, S., Grigolini, P., McClintock, P. V. E. and Moss, F. E. 1986. *J. Phys. A* **19**, L699–704.
Mannella, R., Moss, F. and McClintock, P. V. E. 1987. *Phys. Rev. A* **35**, 2560–6.
Masoliver, J., West, J. B. and Lindenberg, K. 1987. *Phys. Rev. A* **35**, 3086–94.

Moss, F. and Welland, G. V. 1982. *Phys. Rev. A* **25**, 3389–92.
Müller, R. and Behn, U. 1987. *Z. Phys. B.* **69**, 185–92.
Newell, A. and Whitehead, J. 1969. *J. Fluid Mech.* **38**, 279–303.
Ohta, T., Jasnow, D. and Kawasaki, K. 1982. *Phys. Rev. Lett.* **49**, 1223–6.
Paramdeep, P., Sahni, S., Srolovitz, D. J., Grest, G. S., Anderson, M. P. and Safran, S. A. 1983. *Phys. Rev. B* **28**, 2705–16.
Pasquale, F., Sancho, J. M., San Miguel, M. and Tartaglia, P. 1986. *Phys. Rev. A* **33**, 4360–6.
Pesch, W. and Kramer, L. 1986. *Z. Phys. B* **63**, 121–30.
Pieranski, P., Brochard, F. and Guyon E. 1973. *J. de Phys.* **34**, 35–48.
Pocheau, A. and Croquette, V. 1984. *J. de Phys.* **45**, 35–48.
Pomeau, Y. and Zaleski, S. 1983. *Phys. Rev. A* **27**, 2710–26.
Sagues, F. and San Miguel, M. 1985. *Phys. Rev. A* **32**, 1843–51.
Sagues, F., San Miguel, M. and Sancho, J. M. 1984. *Z. Phys. B* **55**, 269–82.
Sancho, J. M., San Miguel, M., Yamazaki, H. and Kawakubo, T. 1982. *Physica* **116A**, 560–72.
San Miguel, M. and Sagues, F. 1987. *Phys. Rev. A* **36**, 1883–93.
Schenzle, A. and Brand, H. R. 1979. *Phys. Rev. A* **20**, 1628–47.
Shiwa, Y. and Kawasaki, K. 1986. *J. Phys. A* **19**, 1387–1402.
Smith, I. W., Galerne, Y., Lagerwall, S. T., Dubois-Violette, E. and Durand, G. 1975. *J. de Phys.* **36**, (C1), 237–59.
Smythe, J., Moss, F. and McClintock, P. V. E. 1983. *Phys. Rev. Lett.* **51**, 1062–5.
Suzuki, M. 1981. *Adv. Chem. Phys.* **46**, 195–278.
Tazaki, K. and Yamaguchi, Y. 1984. *J. Phys. Soc. Jpn.* **53**, 1904–7.
Tesauro, G. and Cross, M. C. 1986. *Phys. Rev. A* **34**, 1363–79.
Toschev, S., Milchev, A. and Stoyanov, S. 1972. *J. Cryst. Growth* **13/14**, 123–7.
Tsuchiya, Y. and Horie, S. 1985. *J. Phys. Soc. Jpn.* **54**, 1–4.
Whitehead, J. A. 1983. *Phys. Fluids* **26**, 2899–904.
Wong, Y. M. and Meijer, P. H. 1982. *Phys. Rev. A* **26**, 611–16.
Yamada, Y., Hamaya, N., Axe, J. D. and Shapiro, S. M. 1984. *Phys. Rev. Lett.* **53**, 1665–8.
Yamazaki, H., Kai, S. and Hirakawa, K. 1987. *J. Phys. Soc. Jpn.* **56**, 1–4.
Zimmerman, W. and Kramer, L. 1985. *Phys. Rev. Lett.* **55**, 402–5.

3 Suppression of electrohydrodynamic instabilities by external noise

HELMUT R. BRAND

3.1 Introduction

Over the last few years the investigation of the transition to turbulence (Ahlers and Behringer, 1978; Croquette and Pocheau, 1984; Gollub and Benson, 1980; King and Swinney, 1983; Swinney and Gollub, 1981) has mainly focused on two systems: on the Rayleigh–Benard convection, which arises when a thin layer of a simple fluid is heated from below; and on the Taylor instability, for which the simple fluid in the gap between two concentric cylinders is subjected to an external torque by rotating the inner or both cylinders. In both configurations a transition from a quiescent state to spatial turbulence via some intermediate, spatially periodic patterns is observed.

A marked influence of the aspect ratio for both instabilities, as well as of the shape of the container for Rayleigh–Benard convection, is well documented (Ahlers and Behringer, 1978; Croquette and Pocheau, 1984).

No systematic study on the influence of controlled external noise on the transition to turbulence had been done prior to the experiments to be described below and very few qualitative results were known.

The big open question was: Can external noise influence or alter the transition to turbulence qualitatively in pattern-forming nonequilibrium systems, or is it a mere perturbation leaving the main features unchanged?

This was the main motivation triggering our studies on the electrohydrodynamic instability (abbreviated as EHD throughout the rest of this chapter) in nematic liquid crystals.

An early preliminary experiment on the EHD with superimposed noise had indicated (Kai, Kai, Takata and Hirakawa, 1979; subsequently confirmed by Kawakubo, Yamagita and Kabashima, 1981) that there might be a noise stabilization of the homogeneous state over the William domain pattern, a set of rolls, which is the first spatially periodic pattern in the EHD in nematics, and that this could be caused by multiplicative external noise was suggested shortly thereafter (Brand and Schenzle, 1980) and has been confirmed theoretically in detail for various models since then (Behn and Mueller, 1985; Graham and Schenzle, 1982b; Luecke and Schank, 1985).

An additional motivation was provided by the fact that some simple models for nonlinear multiplicative processes were found to be exactly solvable with respect to their static and their dynamic behavior, predicting, among other things, a linear dependence of the relaxation rates on the noise intensity (Brand and Schenzle, 1981; Schenzle and Brand, 1979a, b). So naturally the question arose as to whether a real, complicated physical system could show properties resembling those of one of the models. To achieve all these goals it was necessary to choose a system for which the strength of the external noise could be varied over a large range and for which it could be superimposed in a well controlled way on the deterministic external driving field.

The system of choice is the EHD in nematic liquid crystals, which satisfies all the requirements listed above. In a deterministic, low frequency a.c. electric field it shows a well established sequence of spatial patterns (Hirakawa and Kai, 1977; Kai and Hirakawa, 1978; S. Kai, Chapter 2, this volume). The first instability is the appearance of the Williams domains (WD) – the analogue of the simple roll pattern in convection. It is followed at higher voltages by the grid pattern (GP), which shows a spatial modulation in both directions of the plane (see Figure 2.17 of Chapter 2). At even higher driving voltages the spatially turbulent regime is reached, called the dynamic scattering mode (DSM) in the present context (Figure 2.18 of Chapter 2). Sometimes (see, e.g., Kai and Hirakawa, 1978) two sub-regimes are distinguished inside the dynamic scattering mode characterizing a change in the Fourier spectrum, but for the purpose of this chapter this is of no relevance. The EHD in nematics offers the additional advantage of being able to produce easily a large aspect ratio cell with well aligned rolls in the WD regime, and the possibility to work under conditions that are easily achieved: room temperature and atmospheric pressure.

In contrast to analogue simulations (see, e.g., Smythe, Moss and McClintock, 1983), which just implement one specific equation, the experiment described is done in a real physical system that can show many qualitatively different types of behavior in space and time as the external driving voltage is varied, i.e. the number of options available is much larger.

The external noise can be superimposed on the external driving voltage in the system considered in a well controlled way, as electric quantities can be monitored very well. This is a situation that is much more favorable than that in other systems (e.g. in superfluid ^{4}He; see, e.g., J. T. Tough, Chapter 1, this volume, and Lorenson, Griswold, Nayak and Tough, 1985), where it is frequently not even possible to visualize what is happening in space.

Here we will discuss the simplest case possible, namely the influence of spatially homogeneous external noise on the sequence of transitions. The cases of white as well as of nonwhite (or colored) noise sources will be examined, and the relative importance of additive and multiplicative noise will be elucidated. We will reiterate that there is no chance to observe experimentally the 'noise-induced long-time tails', which have attracted so much attention among theorists; this can be simply traced back to the fact that additive noise is inevitable in all real experiments as well as in analogue simulations.

It is important to note, however, that our results demonstrate in a drastic manner the impact of multiplicative noise which causes qualitative changes in both the static and the dynamic response of the system.

As the most remarkable feature we point out that external multiplicative noise can suppress all regular spatial patterns, and it induces a direct transition from a spatially homogeneous state to spatial turbulence. No theoretical treatment has ever suggested the possibility of such a phenomenon.

This chapter is organized as follows. In Section 3.2 we describe briefly the experimental set-up and present an overview of the experimental results obtained. This section is based on two recent papers: Brand, Kai and Wakabayashi (1985) and Kai *et al.* (1985). In Section 3.3 we analyze these results and relate them to the predictions of simple theoretical models to the extent to which this is possible today. The major importance played by the length of the correlation time of the noise is analyzed experimentally, and it is demonstrated that, just by changing this time-scale, a cross-over from noise-stabilization to noise-destabilization can be induced, a problem that has not been addressed properly in the theories as yet.

In the final section we give the conclusions and a perspective, in which we will also point out the importance of the study of localized noise and perturbations for the onset of turbulence in open flow systems, which are of major importance, both from a fundamental point of view and for the applications to the construction of wings for aircraft and for the bodies of ships.

3.2 Overview of the experimental results

We start with a brief description of the experimental set-up. A sample of MBBA (*N*-(p-methoxy-benzylidene)-p-butylaniline) is sandwiched between two horizontal glass plates which are covered with conducting electrodes. MBBA forms a nematic liquid crystalline phase at room temperature and under ambient pressure. In a nematic phase the molecules are, on average, aligned along a preferred direction characterized by a unit vector called the director, but they show only short range positional order (see, e.g., de Gennes, 1982). MBBA is particularly convenient for systematic experimental and theoretical studies, since all material parameters involved (static susceptibilities, such as Frank's elastic constants, and dissipative coefficients, including viscosities and electrical conductivities) have been measured in detail for this uniaxially anisotropic material.

The sample thicknesses investigated range from 20 to about 150 μm, and the lateral dimensions range from 5 to 10 mm. That is, one is always in the limit of large aspect ratios where one has a large number (30 to 500) of unit cells (electroconvective rolls in the case of the Williams domain texture).

In contrast to Rayleigh–Benard convection, where a large aspect ratio cell gives rise to equilibration times of several hours (given by the horizontal diffusion time), the time-scales involved in the same process in EHD are much shorter (seconds up to a few minutes, depending on the values for the material parameters) and thus allow for a more rapid progress in conducting the experiment.

An a.c. voltage of frequency 60 Hz and variable magnitude was applied across the sample, and the response of the system to this perturbation was monitored using a polarizing microscope. Temperature was controlled with an accuracy of ± 0.05 degrees. The frequency applied was chosen to lie well inside the low frequency conducting regime, as completely different phenomena are known to occur in the high frequency dielectric regime, where one finds a parametric oscillation of the director (see, e.g., Kai and Hirakawa, 1978, and Smith *et al.*, 1975, for a detailed discussion). The cross-over frequency between the two regimes is dependent on the conductivity of the sample, as is the threshold for the onset of the electrohydrodynamic instability.

Although electrohydrodynamic equations for nematic liquid crystals have been derived and applied to the investigation of instabilities (see Brand and Pleiner, 1987; de Gennes, 1982; Dubois-Violette, de Gennes and Parodi, 1971; Smith *et al.*, 1975) over many years, a satisfactory treatment of the consequences of space charge effects at zero or small frequencies has not been devised as yet, and thus one has to work with an a.c. frequency of more than about 5 Hz to obtain well controlled, reproducible results. Working in a frequency range which meets all the requirements discussed, one finds typically the sequence of instabilities depicted in Figures 2.17 and 2.18 of Chapter 2 and already briefly discussed in the introduction. A detailed discussion of the phase diagram with purely deterministic driving voltage can be found, e.g., in the review by Kai and Hirakawa (1978).

To investigate the influence of external noise, a noise voltage as produced by a random noise generator (NF Corp. WG-722) and amplified by a Pioneer amplifier (A470), which can be used up to 500 V, has been superimposed on the deterministic driving voltage. Both the nature (Gaussian white, Poissonian or binomial) and the correlation time τ_N of the noise could be varied. For most of our investigations we chose a value for τ_N which was much shorter than the characteristic time of the system (~ 10 ms) and thus guaranteed a clear-cut separation of time scales. In addition we will describe in this section what happens if the correlation time τ_N is increased and surpasses the characteristic time-scale of the deterministic system. The intensity of the noise $Q = \langle V(0)^2 \rangle$ was evaluated using a FFT analyzer (Iwatsu Co. SM-2100 A).

In Figure 3.1 we have plotted the resulting phase diagram in the presence of Gaussian white noise with a correlation time of 30 μs and a deterministic driving frequency of 60 Hz. The sample thickness in this experiment was 110 μm and the lateral dimensions were 8 mm each, giving rise to a square cell with an aspect ratio of about 72. Several outstanding features are noted

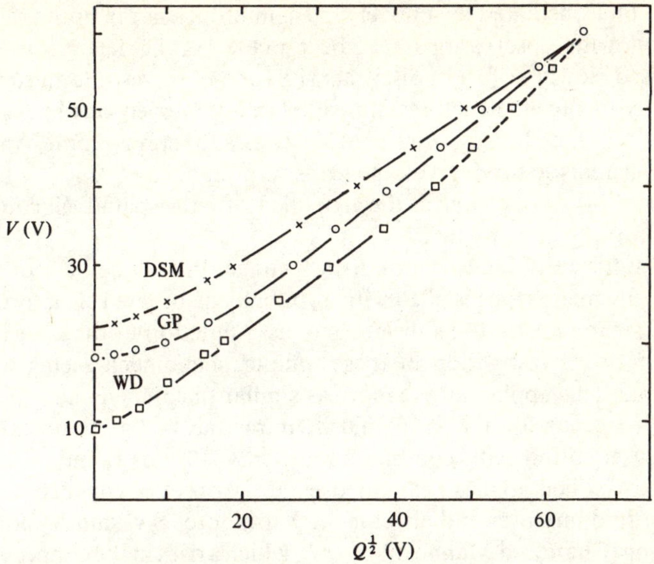

Figure 3.1. Phase diagram for Gaussian white noise with $\tau_N \simeq 30\,\mu s$. The domains of existence of the Williams domains (WD), grid pattern (GP) and dynamic scattering mode (DSM) are plotted as functions of external noise intensity.

immediately when Figure 3.1 is inspected. As the strength of the external noise is increased, the threshold value for the onset of all instabilities (WD, GP and DSM) is also increasing. That is, spatially homogeneous external noise can suppress the onset of hydrodynamic instabilities and postpone the onset of spatially periodic or irregular structures. Simultaneously it also emerges from Figure 3.1 that the range of deterministic voltages, over which the spatially periodic patterns exist, shrinks as the noise intensity is increased. In contrast to usual noise effects which are small (e.g., close to the onset of convection in the Rayleigh–Benard instability in simple fluids thermal noise is so small, Graham, 1975, that it has not been possible to detect it experimentally to this date) and typically lead to a 'smearing out' of the sharpness of the transitions (Haken, 1975, 1977), the consequences of externally superposed noise can be drastic and are of comparable or even bigger magnitude than the purely deterministic effects (e.g. the threshold voltage for Williams domains is increased by more than a factor of five in the experiments leading to Figure 3.1!).

At very high intensities of the external noise, a most exciting and novel phenomenon occurs. A direct transition from a spatially homogeneous state to spatio-temporal turbulence is observed. No stable spatially periodic structure intervenes as is always the case for lower noise intensities. This direct transition is mediated by a novel type of intermittency in both space and time. The following observations have been made. As the deterministic voltage is increased slowly at fixed noise intensity, one observes intermittent bursts of spatially turbulent structures imbedded in a homogeneous background. The location of these bursts varies irregularly as a function of time. An increase in the deterministic driving voltage leads to a higher frequency of these bursts, and their duration, as well as the area filled with the spatially incoherent spots, goes up.

For sufficiently high deterministic voltage, the whole field of view in the polarizing microscope is filled with a spatially disordered structure which also shows strong fluctuations in intensity as a function of time. Clearly a more quantitative investigation of these unique phenomena seems to be highly desirable. This applies all the more as similar phenomena have been recently found to occur for the Benjamin–Feir regime in the complex Ginzburg–Landau equation with complex coefficients (Brand, Lomdahl and Newell, 1986a, b), which applies near the oscillatory onset of convection in miscible binary fluid mixtures and also for the Kuramoto–Sivashinsky equation with damping (Chaté and Manneville, 1987), which arises in the connection of flame fronts, liquid films dropping down an inclined plane, etc. In neither of the two cases, however, has a corresponding experiment been reported so far.

We have also examined how the phase diagram changes if other types of noise sources such as Poissonian, binomial or uniform are used. One can sum up the results as follows. As long as the correlation time of the various types of noise is small compared to the characteristic time of the system, the phase diagram does not change qualitatively; only the detailed numbers for the onset values of the patterns etc. are modified. As the correlation time is increased, for example, the direct transition to turbulence appears at higher noise intensities. As τ_N becomes shorter, the slopes for the noise dependence of the onset values in the phase diagrams become steeper.

This situation changes drastically when the correlation time of the noise is increased and becomes comparable or even larger than the characteristic time of the system. In this case a destabilization takes place leading to a negative threshold shift. This is shown in Figure 3.2. As the correlation time approaches the characteristic time-scale of the system, the concept of the separation of time-scales breaks down. The 'noise' now acts cooperatively with the deterministic field to destabilize the system. As can be seen by inspection of Figure 3.2, noise with a sufficiently long correlation time can also be used to trigger the onset of the EHD without using any purely deterministic driving voltage. That is, the noise has taken over the role of one of the 'dangerous' slow modes of the system in this situation.

The underlying physical mechanism for the noise-induced stabilization for the EHD in nematic liquid crystals is quite clear. For small correlation times of the external noise the amount of space charge which must be stored to produce convection via the Carr–Helfrich effect (see, e.g., de Gennes, 1982) is reduced

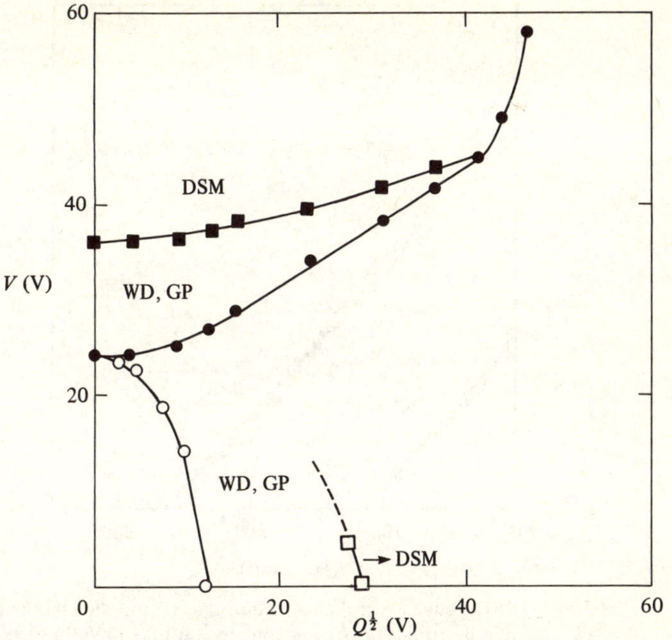

Figure 3.2. Dependence of the phase diagram on the correlation time of the noise τ_N. Symbols: ●, ■, $\tau_N \sim 30\ \mu s$; ○, □, $\tau_N \sim 1$ ms.

by the noise. Thus a higher deterministic voltage is needed to reach the same pattern in the presence of noise than without external noise.

The evaluation of the phase diagram was supplemented by dynamic observations to gain additional insight into the phenomena investigated. The nonlinear onset time τ necessary to establish a specific spatially periodic or incoherent structure after a step change in the applied voltage from a value corresponding to spatial homogeneity up to a prescribed value was measured as a function of the intensity of the external, spatially homogeneous noise. In Figure 3.3 we have plotted the inverse of this time (i.e. the relaxation rate) for deterministic voltages of $V_1 = 14$ V and $V_2 = 18$ V. As a most intriguing result we find that, for both the Williams domains and the grid pattern, a linear dependence of the relaxation rate τ^{-1} on the noise strength emerges. The slope for WD is different from that for GP. In Figure 3.4 we have added the results for a deterministic driving voltage of 25 V. Again a linear dependence of the relaxation rate on the noise strength is found for GP and WD. For the turbulent dynamic scattering mode this dependence changes and a more complicated nonlinear relation between τ^{-1} and the noise intensity arises. Nevertheless the relaxation rate is still decreasing monotonically with increasing noise intensity. A similarity between these experimental results for spatially periodic or turbulent patterns, and the properties of some simple spatially homogeneous nonlinear models, will be outlined in the next section.

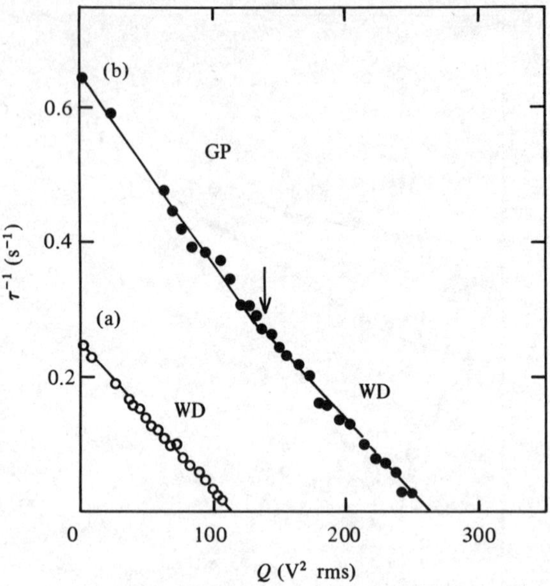

Figure 3.3. Dependence of the relaxation rate τ^{-1} on the noise intensity for two different deterministic driving voltages. (a) $V_1 = 14$ V; (b) $V_2 = 18$ V.

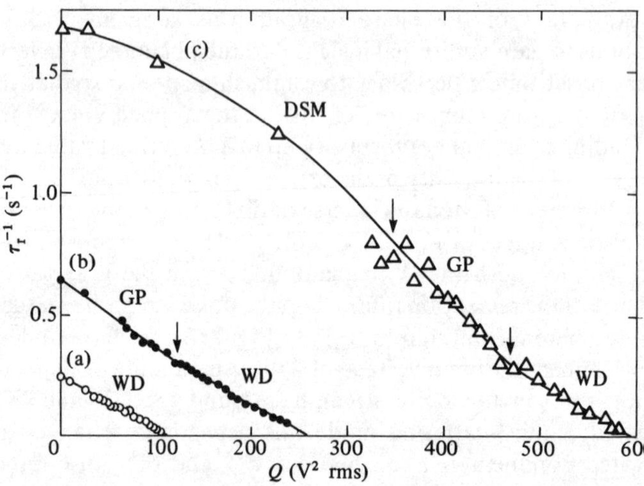

Figure 3.4. Same as Figure 3.3 with an additional curve corresponding to a voltage of $V_3 = 25$ V. We note the nonlinear dependence of τ^{-1} on the noise intensity for the turbulent DSM state. (a) $V_1 = 14$ V; (b) $V_2 = 18$ V; (c) $V_3 = 25$ V. Sample thickness $d = 100\,\mu$m.

3.3 Discussion of results and comparison with models

Initially the theoretical studies of multiplicative stochastic processes focused on the investigation of static properties such as moments and extrema of the stationary probability density in a large variety of systems, including electronic devices (Landauer, 1978; Stratonovich, 1967; Woo and Landauer, 1971), models for population dynamics and autocatalytic chemical reactions (Arnold, Horsthemke and Lefever, 1978; Horsthemke and Lefever, 1984; Horsthemke and Malek-Mansour, 1976; Kitahara, Horsthemke, Lefever and Inaba, 1982; Mori, Morita and Mashiyama, 1980), and numerous systems in quantum optics (Brand, Graham and Schenzle, 1980; Brand, Schenzle and Schroder, 1982; Schenzle and Brand, 1978, 1979a, b). A great deal of attention was paid to the fact that the number of peaks of the stationary probability density can change as a function of noise intensity (see Horsthemke and Lefever, 1984, for a review), and an analogy to phase transitions in systems close to thermodynamic equilibrium was suggested. This analogy was critically examined (Brand, 1985) based on the fact that neither moments nor relaxation rates show any special behavior in the vicinity of the point where the number of peaks in the stationary probability density changes (Brand, Schenzle and Schroder, 1982).

Concerning the dynamic behavior of nonlinear multiplicative stochastic processes, it became apparent that very few models could be solved exactly analytically (Brand and Schenzle, 1981; Graham and Schenzle, 1982a; Schenzle and Brand, 1979a, b).

The most physical model which turned out to be exactly solvable is of the form

$$\dot{w} = dw - bw^3 + w\xi. \tag{3.3.1}$$

It arises naturally in optical systems and in coupled nonlinear oscillators. The eigenvalue spectrum of the Fokker–Planck equation associated with (3.3.1) for Gaussian white noise ($\langle \xi \rangle = 0$, $\langle \xi \xi_\tau \rangle = Q\delta(\tau)$) was found to have a discrete as well as a continuous part. The same holds for two state noise (Ishii and Kitahara, 1982).

Most remarkably, it turned out that the relaxation rates associated with the discrete part of the eigenvalue spectrum decrease linearly with increasing noise intensity Q

$$\lambda_m = 2m(d - mQ), \quad (d/Q \geqslant 2m). \tag{3.3.2}$$

This predicted behavior coincides with what had been seen in early experiments by Kabashima, Kogure, Kawakubo and Okada (1979) on coupled nonlinear oscillators in electronic circuits.

That multiplicative noise can lead to a stabilization of the system has been shown by Graham and Schenzle (1982 b), who generalized (3.3.1) incorporating an inertia term. They found in their model a linear increase of the threshold with increasing noise intensity. This result was generalized by Luecke and

Schank (1985) to the case where one has a modulation by either noise or a periodic forcing.

As we have seen in the preceding section, the experiments on the EHD in nematics with superimposed external noise show a stabilization effect (for small correlation times of the noise) as well as a linear dependence of the relaxation rates on the noise intensity Q for the Williams domains and the grid pattern. That a different slope is obtained in Figures 3.3 and 3.4 for WD versus GP can be traced back to the fact that a different spatial pattern is involved. So far nobody has attempted to discuss the properties of (3.3.1) including slow spatial modulations in the plane (x–y) of the nematic layer. As a generalization of (3.3.1) we suggest

$$\frac{1}{\gamma}\ddot{w} + \dot{w} = \tilde{d}w - bw^3 + w\xi + \zeta + D_1\frac{\partial^2 w}{\partial x^2} + D_2\frac{\partial^2 w}{\partial y^2}. \qquad (3.3.3)$$

In writing down (3.3.3) we have included – aside from the spatial derivative terms – a source of additive noise ζ and the inertia of the macroscopic variable w. As will be discussed in the next section, additive noise is inevitable for both real experiments and analogue simulations. To solve (3.3.3) clearly represents a formidable task, but it seems that all contributions in (3.3.3) are necessary to account for the experimental observations. Nevertheless, it is rather remarkable that the exact analytic solution of the (very simple) model (3.3.1) for Gaussian white noise gives the qualitatively correct dependence of the relaxation rates on the noise intensity. This implies that the linear decrease with increasing noise intensity is a rather robust property which might also hold – at least approximately – for (3.3.3).

As one goes to the turbulent DSM state the situation becomes more complicated and, probably, the possibility to write down an amplitude equation of the type of (3.3.3) is lost. Accordingly we find in the experiments described in the preceding section a nonlinear dependence of the nonlinear relaxation rate on the noise intensity in the DSM state.

It seems that it would be highly desirable to investigate the properties of (3.3.3) in more detail theoretically, even without the presence of the spatial derivative terms. An open question is, e.g., the dependence of the onset value on the noise intensity, and it seems to us that – although a noise-induced stabilization has been observed – it is important to keep the additive noise and an inertia term to obtain the nonlinear dependence on the noise intensity of the threshold values for WD and GP.

3.4 Conclusions and perspective

In the bulk part of this chapter we have focused on the description of our experiments on the influence of external noise on the phase diagram and the nonlinear dynamic behavior of the EHD in nematic liquid crystals. Other systems studied experimentally under the influence of external noise include an early investigation of autocatalytic chemical reactions (de Kepper and Horsthemke, 1978), the study of coupled nonlinear oscillators (Kabashima *et al.*, 1979) and of superfluid ^{4}He in restricted geometries (Griswold and Tough, 1987; Lorenson *et al.*, 1985; and see J. T. Tough, Chapter 1, this volume, for a detailed description). For the case of the influence of noise on the EHD in nematics a large body of work has been accumulated in both the dielectric and the conducting regime by Kai and collaborators (Kai *et al.*, 1979; Kai, Tamura, Wakabayashi and Imasaki, 1987; Kai *et al.*, 1985; Kai, Wakabayashi and Imasaki, 1986), which is reviewed in Chapter 2 by Kai. As a last example for the importance of multiplicative noise we mention the dye laser, the statistical properties of which have been investigated experimentally by Kaminishi, Roy, Short and Mandel (1981) and which have been studied theoretically in detail (Fox, James and Roy, 1984; Graham, Hoehnerbach and Schenzle, 1982). Many properties of multiplicative stochastic processes have been examined using analogue and digital simulations (including, e.g., Faetti *et al.*, 1983; Mannella *et al.*, 1986; Smythe, Moss and McClintock, 1983; see also Chapters 7–9 in this volume).

We would like to stress, however, that only for the EHD in nematics has the influence of external noise on a spatially periodic or turbulent pattern been investigated systematically including a visualization of the changes in the spatial structures.

The possibility of a very exciting phenomenon, namely of noise-induced long-time tails in time-dependent moments and correlation functions in the range of parameter values, for which (3.3.1) leads to a continuous spectrum, has attracted a considerable amount of attention in the theoretical community (Brenig and Banai, 1982; Graham and Schenzle, 1982a; Suzuki, 1981; Suzuki, Takesue and Sasagawa, 1982). Unfortunately this prediction will be very hard to test experimentally, since a small admixture of additive noise in (3.3.1), which is always present in experiments, will wipe out the long-time tails. The associated eigenvalue spectrum loses its continuous part in this case (Schenzle and Brand, 1979a) and as a consequence all time dependent moments and correlation functions show a purely exponential decay in the long-time limit (Brand, 1984). This analysis has been confirmed recently in analogue simulations (Mannella *et al.*, 1986).

The study of the influence of spatially homogeneous external noise on pattern forming hydrodynamic instabilities as described in this chapter might turn out to be a first step towards the understanding of the onset of turbulence in open flow systems such as shear layers, the Dean and the Goertler instability, etc. For these systems it is known that localized noise and/or localized perturbations upstream can cause turbulent behavior further downstream. And this resembles the 'noise-induced' structures found by Deissler (1985), when he studied the complex Ginzburg–Landau equation with complex coefficients for travelling waves under the influence of localized weak additive noise. Recently the same type of behavior has been found by

H. R. Brand and R. J. Deissler for this equation in the presence of localized multiplicative noise (corresponding to velocity fluctuations) and a small constant driving force. These studies possibly provide a new gateway of attack for the understanding of the onset of turbulence in open flow systems, which are extremely important for the applications in aeronautics and ship-building.

Acknowledgements

It is a pleasure to thank Shoichi Kai for our stimulating collaboration over several years.

The experiments described were carried out while the present author was visiting Kyozi Kawasaki and Hajime Mori at the Department of Physics, Kyushu University, Fukuoka, 812, Japan, as a fellow of the Japan Society for the Promotion of Science.

Support by the Deutsche Forschungsgemeinschaft is gratefully acknowledged.

References

Ahlers, G. and Behringer, R. P. 1978. *Phys. Rev. Lett.* **41**, 948.
Arnold, L., Horsthemke, W. and Lefever, R. 1978. *Z. Phys. B* **29**, 367.
Behn, U. and Mueller, R. 1985. *Phys. Lett. A* **113**, 85.
Brand, H. R. 1984. *Prog. Theor. Phys.* **72**, 1255.
Brand, H. R. 1985. *Phys. Rev. Lett.* **54**, 605.
Brand, H., Graham, R. and Schenzle, A. 1980. *Opt. Comm.* **32**, 359.
Brand, H. R., Kai, S. and Wakabayashi, S. 1985. *Phys. Rev. Lett.* **54**, 555.
Brand, H. R., Lomdahl, P. S. and Newell, A. C. 1986a. *Phys. Lett. A* **118**, 67.
Brand, H. R., Lomdahl, P. S. and Newell, A. C. 1986b. *Physica D–Nonlinear Phenomena* **D23**, 245.
Brand, H. R. and Pleiner, H. 1987. *Phys. Rev. A* **35**, 3122.
Brand, H. and Schenzle, A. 1980. *J. Phys. Soc. Jpn.* **48**, 1382.
Brand, H. and Schenzle, A. 1981. *Phys. Lett. A* **81**, 321.
Brand, H., Schenzle, A. and Schroder, G. 1982. *Phys. Rev. A* **25**, 2324.
Brenig, L. and Banai, N. 1982. *Physica* **D5**, 208.
Chaté, H. and Manneville, P. 1987. *Phys. Rev. Lett.* **58**, 112.
Croquette, V. and Pocheau, A. 1984. In *Cellular Structures in Instabilities.* (J. E. Wesfreid and S. Zaleski, eds.), pp. 104–26. Lecture Notes in Physics. NY: Springer.
de Gennes P. G. 1982. *The Physics of Liquid Crystals*, 3rd edn. Oxford: Clarendon Press.
Deissler, R. J. 1985. *J. Stat. Phys.* **40**, 371.
de Kepper, P. and Horsthemke, W. 1978. *C. R. Acad. Sci.* **C287**, 251.
Dubois-Violette, E., de Gennes, P. G. and Parodi, O. 1971. *J. Phys.* **32**, 305.
Faetti, S., Festa, C., Fronzoni, L., Grigolini, P., Marchesoni, F. and Palleschi, V. 1983. *Phys. Lett.* **A99**, 25.
Fox, R. F., James, G. E. and Roy, R. 1984. *Phys. Rev. Lett.* **52**, 1778.
Gollub, J. P. and Benson, S. V. 1980. *J. Fluid Mech.* **100**, 499.

Graham, R. 1975. In *Fluctuations, Instabilities and Phase Transitions* (T. Riste, ed.), pp. 215–79. NY: Plenum.
Graham, R., Hoehnerbach, M. and Schenzle, A. 1982. *Phys. Rev. Lett.* **48**, 1396.
Graham, R. and Schenzle, A. 1982a. *Phys. Rev. A* **25**, 1731.
Graham, R. and Schenzle, A. 1982b. *Phys. Rev. A* **26**, 1676.
Griswold, D. and Tough, J. T. 1987. *Phys. Rev. A* **36**, 1360.
Haken, H. 1975. *Rev. Mod. Phys.* **47**, 65.
Haken, H. 1977. *Synergetics – An Introduction.* NY: Springer.
Hirakawa, K. and Kai, S. 1977. *Mol. Cryst. Liq. Cryst.* **40**, 261.
Horsthemke, W. and Lefever, R. 1984. *Noise Induced Transitions.* Berlin: Springer.
Horsthemke, W. and Malek-Mansour, M. 1976. *Z. Phys.* **B24**, 357.
Ishii, K. and Kitahara, K. 1982. *Prog. Theor. Phys.* **68**, 665.
Kabashima, S., Kogure, S., Kawakubo, T. and Okada, T. 1979. *J. Appl. Phys.* **50**, 6296.
Kai, S. and Hirakawa, K. 1978. *Prog. Theor. Phys. Suppl.* **64**, 212.
Kai, S., Kai, T., Takata, M. and Hirakawa, K. 1979. *J. Phys. Soc. Jpn.* **47**, 1379.
Kai, S., Tamura, T., Wakabayashi, S. and Imasaki, M. 1987. *Phys. Rev. A* **35**, 1438.
Kai, S., Tamura, T., Wakabayashi, S., Imasaki, M. and Brand, H. R. 1985. *IEEE–IAS Conference Records* **85CH**, 1555.
Kai, S., Wakabayashi, S. and Imasaki, M. 1986. *Phys. Rev. A* **33**, 2612.
Kaminishi, K., Roy, R., Short, R. and Mandel, L. 1981. *Phys. Rev. A* **24**, 370.
Kawakubo, T., Yamagita, A. and Kabashima, S. 1981. *J. Phys. Soc. Jpn.* **50**, 1451.
King, G. P. and Swinney, H. L. 1983. *Phys, Rev. A* **27**, 1240.
Kitahara, K., Horsthemke, W., Lefever, R. and Inaba, Y. 1982. *Prog. Theor. Phys.* **68**, 665.
Landauer, R. 1978. *Phys. Today* **31**, 23.
Lorenson, C. P., Griswold, D., Nayak, V. U. and Tough J. T. 1985. *Phys. Rev. Lett.* **55**, 1494.
Luecke, M. and Schank, F. 1985. *Phys. Rev. Lett.* **54**, 1465.
Mannella, R., Faetti, S., Grigolini, P., McClintock, P. V. E. and Moss, F. E. 1986. *J. Phys.* **A19**, L699.
Mori, H., Morita, T. and Mashiyama, K. T. 1980. *Prog. Theor. Phys.* **63**, 1865.
Schenzle, A. and Brand, H. 1978. *Opt. Comm.* **27**, 485.
Schenzle, A. and Brand, H. 1979a. *Phys. Rev. A* **20**, 1628.
Schenzle, A. and Brand, H. 1979b. *Phys. Lett. A* **69**, 313.
Smith, I. W., Galerne, Y., Lagerwall, S. T., Dubois-Violette, E. and Durand, G. 1975. *J. Phys. Colloq.* **C36**, 1–237.
Smythe, J., Moss, F. and McClintock, P. V. E. 1983. *Phys. Rev. Lett.* **51**, 1062.
Stratonovich, R. L. 1967. *Topics in the Theory of Random Noise*, vol. II. NY: Gordon and Breach.
Suzuki, M. 1981. *Adv. Chem. Phys.* **46**, 195.
Suzuki, M., Takesue, S. and Sasagawa, F. 1982. *Prog. Theor. Phys.* **68**, 98.
Swinney, H. L. and Gollub, J. P. (eds.) 1981. *Hydrodynamic Instabilities.* NY: Springer.
Woo, J. W. F. and Landauer, R. 1971. *IEEE J. Quant. Electron.* **7**, 435.

4 Colored noise in dye laser fluctuations

R. ROY, A. W. YU and S. ZHU

4.1 Introduction

The investigation of the coherence properties of light sources has traditionally involved the study of correlation functions of different orders of the electric field (Born and Wolf, 1975; Mandel and Wolf, 1965). Field and intensity correlation functions have been measured for laser light (for a review, see Lax and Zwanziger, 1973) and have been compared with predictions obtained from quantum laser theory (Haken, 1970, 1981; Lax, 1969; Louisell, 1973; Sargent, Scully and Lamb, 1974). This was, in fact, one of the major confirmations of the correctness of the basic formalism of quantum laser theory. Most of the experiments that probed the coherence properties of lasers were, however, carried out on a very limited variety of lasing media. Interest was also centered on the regime of operation near threshold, where the light from the laser is extremely weak; hence photoelectron counting and correlation techniques were developed for the measurement of coherence properties of the laser light.

While these early experiments were performed with the main intent of testing the basic theory, the large variety of lasers now available, and their ever more demanding applications, make it necessary to develop new methods for the measurement and analysis of laser fluctuations. These techniques should be applicable to the measurement of noise in high power lasers; at the same time, in order to delineate the limits of usage, they must be sensitive enough to quantitatively determine the quantum mechanical sources of noise which may lie buried in external noise that is many orders of magnitude larger. The identification of the noise sources and their time scales is of importance to the development of noise reduction techniques. Further, if we wish to create novel states of the electromagnetic field, e.g. to permit precision measurements not presently possible, it is necessary to control and modify the fluctuations of the laser phase and/or amplitude. These are some of the motivations for the development of novel techniques of laser noise measurement and analysis that are easily applicable to a variety of lasers, operating over a wide range of powers.

One of the central assumptions in laser noise theory until recent times was that of the delta-correlated nature of the noise sources. This arose from the status of the theory of stochastic processes. Solutions to stochastic equations could be conveniently obtained when the noise source was white, and, indeed, this was a physically justifiable assumption in many cases. In the He–Ne laser, for example, the time scale of quantum fluctuations is much shorter than the time scale for the growth or decay of the macroscopic electric field. The latter is largely determined by the cavity decay rate. In the past few years it has become clear that there may exist dominant sources of noise which cannot be treated as delta-correlated. Thus the theory of lasers with colored noise sources, i.e. those with finite time scales, has been recognized recently as being of particular importance in practice.

Noise sources in lasers are often separated into intrinsic quantum noise and external (often referred to as 'pump') noise. We will follow this convention in this chapter; at the same time it should be pointed out that some of the external noise sources are often quite unavoidable, though the extent to which they are present can be controlled to some degree. For example, the hydrodynamic fluctuations in the jet of a dye laser are a source of external noise. They depend on the precise nature of the nozzle, the solvent used, the velocity of flow, the operating temperature and the circulation system used. These fluctuations are always present, though their strength can be reduced by optimizing the operating conditions. The time scale of the noise may be comparable to or longer than that for growth or decay of the total laser field, and hence cannot be treated as delta-correlated.

Noise sources are also often classified as additive and multiplicative. Transformations of the equations can often be used to convert one kind of noise to the other, but it is frequently true that additive noise is intrinsic while multiplicative noise is external in character. We will make it clear, as the occasion arises, what the origin and nature of the noise sources are, to avoid any confusion in the terminology.

In this chapter we will discuss the role of colored noise in laser fluctuations and how the magnitude and time scale of such noise sources can be measured. We will restrict ourselves to the dye laser, on which most of the measurements have been made. Further, we will restrict our treatment to the single mode laser, since this is the system for which comparisons between theory and experiment are most conveniently made. It is also an example of practical importance, since the tunable single mode dye laser is extensively used in spectroscopy. In Section 4.2 we review studies of dye lasers in steady state operation. The variance of the intensity fluctuations and correlation functions of the laser intensity fluctuations about the steady state average will be described and compared with the predictions of current theories. Section 4.3 deals with the transient dynamics and fluctuations of the laser. A recently developed technique based on the first passage time (FPT) concept for the measurement and analysis of laser noise will be described. The determination of parameters for the laser noise sources will be addressed. In Section 4.4 we will report on recent experiments that examine the consistency of the parameters determined by correlation measurements, FPT measurements, and power spectral densities of the argon and dye lasers. We will conclude the chapter with a discussion of the results and a survey of future directions for research.

4.2 Fluctuations about the steady state

A laser in steady state operation will exhibit intensity fluctuations about the average intensity; the extent of these fluctuations depends on the operating parameters and nature of the laser medium. An important statistical quantity that characterizes these fluctuations is the normalized variance, $\langle(\Delta I)^2\rangle/\langle I\rangle^2$, where $\Delta I = (I(t) - \langle I\rangle)$, and $\langle I\rangle$ is the steady state average intensity. This quantity may be measured by photon counting. The probability distribution $p(n, T)$ for counting n photoelectrons in a time interval T is determined experimentally and the normalized variance above can be calculated. To obtain information about the dynamics of the excursions of the intensity about the average, the correlation functions of the intensity may be measured. The normalized correlation of the intensity fluctuations $\lambda(\tau)$ is defined as $\langle\Delta I(t)\Delta I(t+\tau)\rangle/\langle I\rangle^2$; it is easily seen that its value for zero time delay ($\tau = 0$) is $\langle(\Delta I)^2\rangle/\langle I\rangle^2$. The function $\lambda(\tau)$ is often measured by a digital correlator to which are input the pulses from two photomultiplier tubes on which the laser light being investigated is incident.

The experiments of Abate, Kimble and Mandel (1976) were the first to indicate that the photon statistics of the dye laser were quite different from those predicted by conventional laser theory. They compared the results of their experiments to the predictions of an augmented semiclassical laser theory based on the scaled Langevin equation for the complex laser field E of the form (see the appendix to this chapter for a brief review of the semiclassical laser equations with noise)

$$\dot{E} = aE - |E|^2 E + q(t), \tag{4.2.1}$$

where $q(t)$ is a complex noise source that represents the effect of spontaneous emission. It is assumed to be delta-correlated,

$$\langle q^*(t)q(t')\rangle = 2P\delta(t-t') \quad (i,j = 1,2), \tag{4.2.2}$$

which is certainly justified if the spontaneous emission time scale is short compared to the time scales for growth or decay of the macroscopic electric field, and its strength $P = 2$ in the appropriately scaled time and electric field units. 'a' is the pump parameter for the laser, and depends on the pump rate, cavity decay rate and molecular decay rates. This Langevin equation is easily converted to the corresponding Fokker–Planck equation. The steady state distribution of the laser intensity and its first two moments are to a good

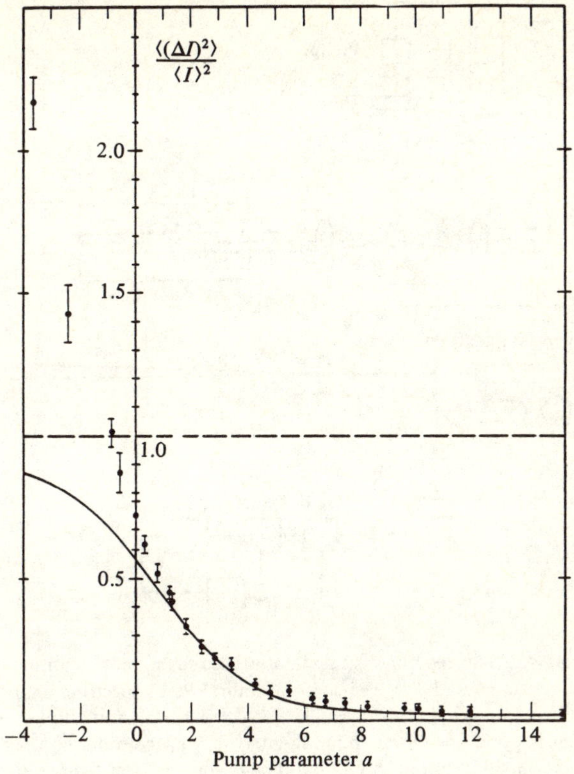

Figure 4.1. These measurements of the normalized intensity fluctuations by Abate, Kimble and Mandel (1976) first demonstrated the departure of the dye laser from normal laser behavior; the solid line shows the expected results for a laser with only additive noise.

approximation given by

$$P_s(I) = N\mathrm{e}^{-(I-a)^2/4} \tag{4.2.3}$$

$$\langle I \rangle = a + \frac{2\mathrm{e}^{-a^2/4}}{\pi^{1/2}[1 + \mathrm{erf}(a/2)]} \tag{4.2.4}$$

$$\langle (\Delta I)^2 \rangle = 2 - a(\langle I \rangle - a) - (\langle I \rangle - a)^2. \tag{4.2.5}$$

These measurements did not span a very large range of operation of the laser, but even within this range deviations from the predictions of the model above for the variance of the intensity fluctuations and the intensity correlation function were noticed (Figure 4.1).

Theoretical predictions that deviations could occur had just been made by Schaefer and Willis (1976) and by Dembinski, Kossakowski and their colleagues (Baczynski, Kossakowski and Marzalek, 1977; Dembinski and

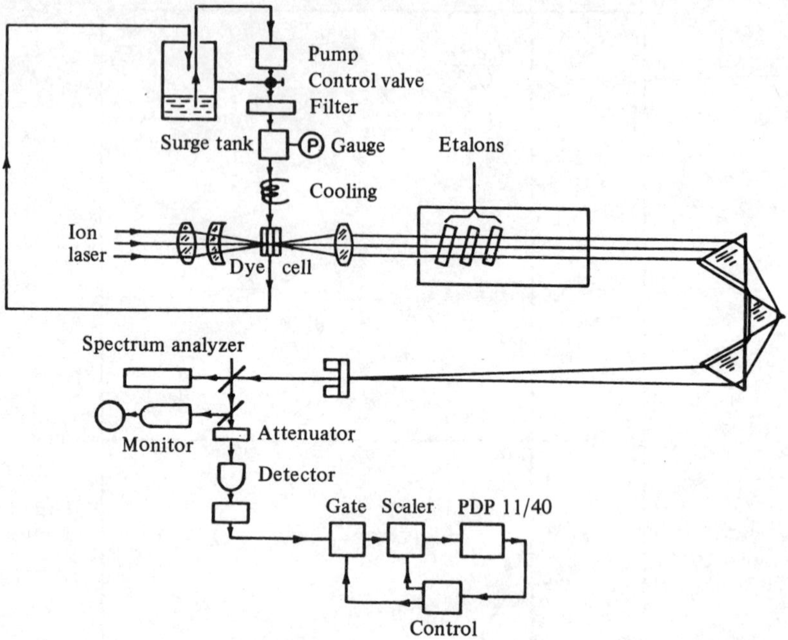

Figure 4.2. The experimental apparatus used for photon counting experiments (Kaminishi *et al.*, 1981). The confocal Fabry–Perot spectrum analyzer monitors the mode structure of the laser. The detector is a photomultiplier tube. The scaler is gated open for a counting interval T, of the order of a microsecond, short compared to the correlation time of the dye laser light. A flowing dye cell was used in the dye laser.

Kossakowski, 1976a, b), based on considerations of dye molecule triplet state interactions with radiation. This was not the conclusion reached by the experimenters, though it was not ruled out by them. It appeared possible even then that fluctuations of the pump argon ion laser could be at least partly responsible for the deviations observed. There existed no theory at that time that could incorporate the effect of pump laser fluctuations in a comprehensive fashion. The effect of triplet states was examined in further detail (Roy, 1979a, b; Roy and Mandel, 1977), and it was not until extensive measurements were performed on single mode dye lasers that measured the change in intensity fluctuations with dye flow velocity (Short, Roy and Mandel, 1980) and over a wide range of operating parameters (Kaminishi, Roy, Short and Mandel, 1981) (Figure 4.2 shows the experimental arrangement) that attention began to be focussed on the role of 'pump fluctuations', a term that includes both fluctuations in the pump laser and those in the active dye medium. The normalized variance of the intensity fluctuations, $\langle(\Delta I)^2\rangle/\langle I\rangle^2$, showed a marked dependence on the velocity of dye flow, and measurements near threshold revealed that this quantity could reach values as high as several hundred. This came as a great surprise, since the maximum value of

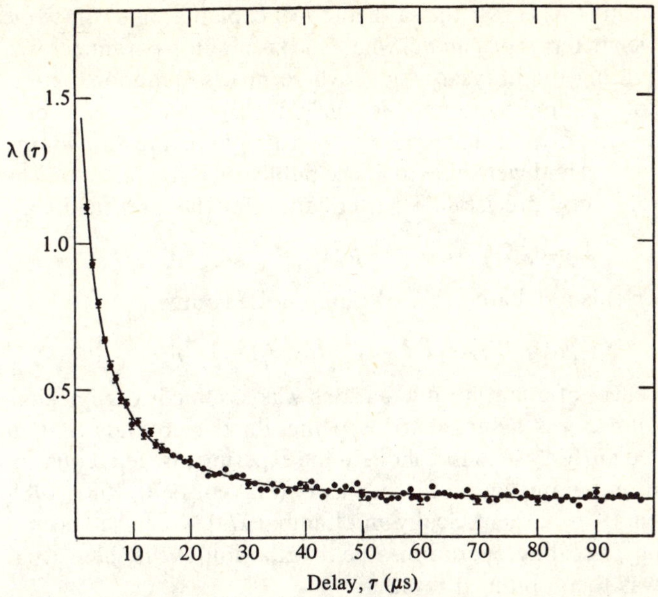

Figure 4.3. A typical correlation function for the intensity of the dye laser. The solid line is an empirical fit of the data with three exponentials, $c_1 e^{-\gamma_1 \tau} + c_2 e^{-\gamma_2 \tau} + c_3 e^{-\gamma_3 \tau}$. It is found that the coefficients do not decrease rapidly, behavior that is contrary to the predictions of conventional laser theory.

$\langle(\Delta I)^2\rangle/\langle I\rangle^2$, predicted by conventional laser theory and verified by experiments on He–Ne lasers, was unity, and, even if triplet states did play a significant role, such extremely large values were not expected (Roy, 1979a, b; Roy and Mandel, 1977). Such large values of $\langle(\Delta I)^2\rangle/\langle I\rangle^2$ could only occur if there were anomalously large fluctuations of the laser intensity, far larger than those of a thermal light source or of a laser operated well below threshold. In Kaminishi *et al.* (1981) a simple explanation was presented in terms of pump fluctuations which momentarily drove the laser from an ambient operating point below threshold to above threshold. This would essentially turn the laser on and off and induce enormous fluctuations of the laser output.

Another departure from conventional behavior presented by the dye laser was the nature of its correlation functions (Figure 4.3). Measurements of the intensity autocorrelation function $\lambda(\tau)$ demonstrated that instead of being closely approximated by a single exponential, as was the case for a He–Ne laser, the function needed at least three different exponentials for a fit. A new theory for the dynamics of the dye laser intensity excursions about the steady state average was clearly required to explain this behavior.

At the time when these measurements on the dye laser were being performed, several theorists had realized the importance of multiplicative stochastic processes in physics (Fox, 1972, 1978; Horsthemke and Malek-Mansour, 1976; Schenzle and Brand, 1979). Thus, it was within months of the publication of

Kaminishi *et al.* (1981) that a theoretical explanation of those measurements was given in terms of a model where the laser pump parameter was taken to be a fluctuating quantity, $a_0 + p(t)$, where $p(t)$ is a random pump noise term (Graham, Hohnerbach and Schenzle, 1982). In a remarkable calculation, Graham and his colleagues obtained an analytic expression for the correlation function $\lambda(\tau)$ and were able to fit the published data in Kaminishi *et al.* (1981) with very good precision. Their equation for the laser field was

$$\dot{E} = a_0 E - AE|E|^2 + p(t)E, \tag{4.2.6}$$

where $p(t)$ is a delta-correlated pump noise source, with

$$\langle p_i(t) p_j(t') \rangle = P' \delta_{ij} \delta(t - t') \quad (i, j = 1, 2). \tag{4.2.7}$$

The effect of quantum fluctuations was assumed to be negligible, and the pump noise was assumed to be white, on the grounds that the laser was operated fairly close to threshold in the experiments. This equation retains the saturation parameter A of laser theory (Haken, 1970, 1981, 1985; Lax, 1969; Louisell, 1973; Sargent, Scully and Lamb, 1974), which had been eliminated in (4.2.1) by a scaling operation. The distribution of the intensity in the steady state was then obtained to be

$$P_s(I) = N I^{[(a_0/2P') - 1]} e^{-AI/2P'} \tag{4.2.8}$$

with

$$\lambda(0) = \langle (\Delta I)^2 \rangle / \langle I \rangle^2 = 2P/a_0 \tag{4.2.9}$$

and

$$\lambda(\tau) = \frac{e^{-\alpha^2 P' \tau}}{16\alpha^2 \Gamma(\alpha)} \times \int_{-\infty}^{+\infty} \frac{dx(\alpha^2 + x^2) x \sinh(\pi x) |\Gamma[(\alpha + ix)/2]|^2 e^{-x^2 P' \tau}}{\cosh(\pi x) - \cos(\pi \alpha)}, \tag{4.2.10}$$

where $\alpha = (a_0/2P')$ and $\Gamma(\alpha)$ is the gamma function.

An analysis of additional measurements of the laser intensity autocorrelation functions revealed shortly that the situation was perhaps more complicated (Short, Mandel and Roy, 1982). Though the theory of Graham, Hohnerbach and Schenzle (1982) did indeed fit the one correlation function, it failed to fit further data taken at different operating points of the laser. A suggestion was made in Short, Mandel and Roy (1982) that the pump noise should not be considered delta-correlated, and a characteristic time scale should be assigned. Dixit and Sahni (1983) did just this, and demonstrated the importance of colored noise in the dye laser in their seminal paper; though the equations were not amenable to analytic treatment, they were able to use computer simulations, following the algorithm introduced by Sancho (Sancho, San Miguel, Katz and Gunton, 1982), to calculate the correlation functions of the laser intensity. They were able to fit three sets of experimental data, taken at different operating points of the laser, with the same set of noise parameters, varying only the operating point of the laser. A separate equation was written to describe the colored pump noise, which was assumed to be exponentially correlated, following Ornstein and Uhlenbeck (Uhlenbeck and Ornstein, 1954). The pump noise equation was of the form

$$\mathrm{d}p(t)/\mathrm{d}t = -\gamma p(t) + \gamma q'(t) \tag{4.2.11}$$

with

$$\langle q'_i(t)q'_j(t')\rangle = P'\delta_{ij}\delta(t-t') \ \ (i, j = 1, 2) \tag{4.2.12}$$

which implies

$$\langle p_i(t)p_j(t')\rangle = (P'\gamma/2)\mathrm{e}^{-\gamma|t-t'|}. \tag{4.2.13}$$

The role of colored noise, as opposed to white noise, was thus recognized as being important to explain the observed behavior of the dye laser. The value of γ estimated by these authors was $\simeq 10^6\,\mathrm{s}^{-1}$. A large number of papers have appeared since then that develop various approximation schemes to treat the problem of colored multiplicative noise in the dye laser (Hernandez-Machado, San Miguel and Katz, 1985, 1986; Jung and Hänggi, 1987; Jung and Risken, 1984, 1985, 1986; Lindenberg, West and Cortes, 1984; Marchesoni, 1986; Schenzle, 1986; Schenzle and Graham, 1983).

It should be noted, however, that the approach to the theory is quite phenomenological in the form stated above. Further, the effect of quantum noise has been completely neglected. A subtraction procedure was also used to compare the measured correlation functions to theory (Dixit and Sahni, 1983; Graham, Hohnerbach and Schenzle, 1982). To address these issues, rederivation of the laser theory is necessary that begins from a microscopic model. Such a theory has been developed (Fox, James and Roy, 1984a, b); a four level model of laser operation was employed, with a stochastic, classical pump field and a quantized laser field. In the final comparison of the experimental data to the theory, the quantum noise was once again neglected, but no subtraction procedure was used.

A thorough investigation of the photon statistics of the single mode dye laser operated near and below threshold was carried out by Lett (Lett, 1986; Lett, Short and Mandel, 1984). Their measurements of $\lambda(0)$ displayed a prominent peak when plotted versus the mean intensity $\langle I\rangle$. Computer simulations of the colored noise driven laser equations were used to obtain a fit to the data. Recently, we have used an ansatz due to Hänggi (Hänggi, Mroczkowski, Moss and McClintock, 1985) to examine the role of colored noise in these measurements (Fox and Roy, 1987). It was found that the data could not be fit by white noise; it was necessary to use colored noise (Figure 4.4). No computer simulations are required for these calculations, though a simple numerical iterative procedure is used to implement the ansatz. The ansatz replaces the white noise correlation strength by an effective strength that depends on the correlation time $\tau_c(\tau_c = 1/\gamma)$. Quantum fluctuations are included in the laser

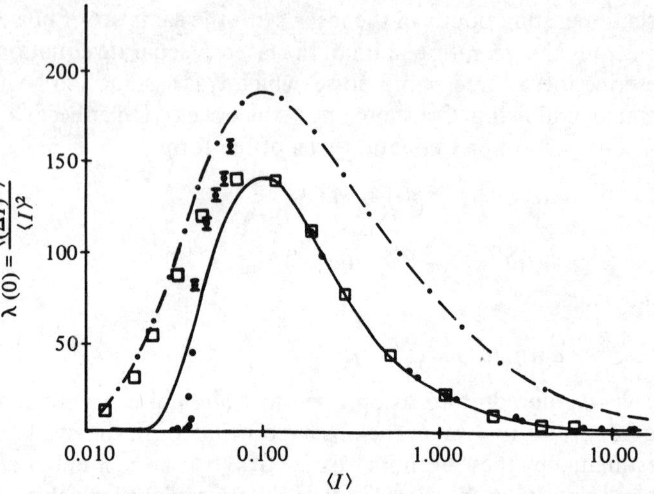

Figure 4.4. The experimental measurements, ⧲ are those of Lett, Short and Mandel (1984). The solid line is a fit obtained by Monte Carlo simulations. The dot-dashed line is a fit using (4.2.15) (white noise). The application of the Hänggi ansatz (Hänggi *et al.*, 1985) results in the fit shown by the square boxes.

model, since they are important near and below threshold, and cannot be neglected. The equation for the laser field is now

$$\dot{E} = a_0 E - A|E|^2 E + p(t)E + q(t) \tag{4.2.14}$$

which, together with (4.2.2), (4.2.11), (4.2.12) and (4.2.13) describes the dye laser with quantum noise and colored pump noise. The steady state distribution of the laser intensity is, for white pump noise ($\gamma \to \infty$),

$$P_s(I) = N(P + P'I)^{\xi} \exp(-AI/P'), \tag{4.2.15}$$

where

$$\xi = (a_0 - P')/P' + AP/P'^2. \tag{4.2.16}$$

The Hänggi ansatz replaces P' with an effective noise strength

$$\bar{P} = P'/(1 + 2A\langle I\rangle/\gamma), \tag{4.2.17}$$

which now introduces the time scale of the pump noise. A fit to the experimental data can be obtained by a numerical procedure. The time scale of the pump fluctuations determined from this data of Lett, Short and Mandel (1984) is found to be of the order of 1 ms, but later estimates by Lett (1986) appear to be much shorter, $\simeq 10^{-5}$ s. The strength of the pump noise is $\simeq 1.2 \times 10^4\,\text{s}^{-1}$, but the quantum noise strength cannot be determined from these measurements because of the unknown scale factors used in the presentation of the data.

A recent study by San Miguel and colleagues has examined the detailed nature of the intensity correlation functions of the single mode dye laser (San Miguel, Pesquera, Rodriguez and Hernandez-Machado, 1987). They show that the initial slope of the correlation function depends on the spontaneous noise strength, and that if only multiplicative noise were present, the slope would be zero. Experimental measurements by Lett have indeed shown a flattening of the initial part of the correlation function (Lett, 1986). It would be of interest to compare the theoretical predictions with precise measurements of the initial slope.

The nature of the phase transition analogy for the dye laser threshold behavior, which was the origin of interest in the dye laser (Baczynski, Kossakowski and Marzalek, 1977; Dembinski and Kossakowski, 1976a, b; Roy, 1979a, b; Roy and Mandel, 1977; Schaefer and Willis, 1976), has also been studied recently by Lett (Lett, Gage and Chyba, 1987). The phase transition at threshold for conventional lasers is considered to be a second order transition (Haken, 1970, 1981, 1985; Lax, 1969; Louisell, 1973; Sargent, Scully and Lamb, 1974). The results of Lett, Gage and Chyba (1987) indicate that even though the triplet states probably do not play a prominent role, the pump fluctuations may be responsible for initiating a first order transition at threshold. The suggestion is made that the colored nature of the multiplicative noise may in fact be responsible for the first order transition; the order of the phase transition is dependent on the correlation time of the multiplicative noise.

Is deterministic chaos present in single mode dye lasers? It is present in multimode dye lasers (Atmanspacher and Scheingraber, 1986), but there is no evidence to indicate its presence in single mode systems. Chyba and colleagues (Chyba *et al.*, 1986, 1987) have examined the question of deterministic chaos arising from the turbulent flow of dye through a dye cell. There is recent evidence (Chyba *et al.*, 1987) to believe that the dye flow is turbulent, and that chaotic behavior may thus be imposed on the argon ion laser pump beam. It appears that replacing the dye cell with a dye jet removes the chaotic nature of the transmitted argon laser beam. However, the results are dependent strongly on the sampling times used for the transient digitizer. Further investigations are clearly necessary to obtain a thorough understanding of this question.

4.3 Fluctuations in the transient dynamics

Though a great deal of information can be efficiently extracted from the steady state operation of the dye laser, it is necessary to operate the laser near and below threshold to obtain any information regarding the quantum (additive) noise, since it manifests itself only in this regime when the laser is operated in the steady state. When the laser is operated above threshold, the correlation times of the laser become very short, and do not fall within the range of the digital correlators normally used for these measurements. Further, photon counting and correlation experiments are quite delicate and have to be performed with great care. There is, then, a strong motivation to develop new measurement techniques that overcome some of the difficulties mentioned above. With these aims in mind, we have carried out experiments on the transient dynamics of the dye laser.

Arecchi and colleagues (Arecchi and DeGiorgio, 1971; Arecchi, DeGiorgio and Querzola, 1967) and Meltzer and Mandel (1970, 1971) performed the first detailed measurements of the transient dynamics of Q-switched He–Ne lasers. They measured the time evolution of the photon counting distribution for the laser light and its moments. Satisfactory agreement with conventional laser theory was obtained. Photomultiplier tubes and photon counting techniques were used in their measurements. Today, there are extremely fast photodiodes commonly available, and these can be profitably used if the laser is pumped far above threshold. Instead of the conventional photon counting techniques, we have developed easily implemented analog measurements for the quantitative analysis of laser fluctuations based on the concept of first passage times. Our measurements are, so far, the only statistical analysis of dye laser dynamics that examine the growth of the laser intensity from spontaneous emission noise to steady state. These experiments enable us to measure the quantum noise and pump noise strengths and the time scale of the pump noise.

The first passage time (FPT) concept (Darling and Siegert, 1953; Gardiner, 1983; Kramers, 1940; Landauer and Swanson, 1961; Montroll and Shuler, 1958; Risken, 1984; Schuss, 1980; Stratonovich, 1963; Van Kampen, 1982; Weiss, 1977) is easily translated to practical measurements as follows. The time taken by the laser intensity to grow from spontaneous emission to a given reference value of the intensity (the first passage time) is measured repeatedly by Q-switching the laser, and a histogram is formed of these first passage times. The quantum noise determines the initial growth of the laser intensity, while the external noise strongly influences the fluctuations of the intensity as the laser approaches steady state. The characteristics of these noise sources may be discerned from a statistical analysis of the first passage time distributions.

The experimental arrangement (Roy, Yu and Zhu, 1985; Zhu, Yu and Roy, 1986) is shown in Figure 4.5. The laser is Q-switched abruptly from far below threshold by an acousto-optic modulator inserted into the laser cavity. It is adjusted at the Bragg angle; the loss is $\simeq 70\%$ and then decreased to the static insertion loss of $\simeq 2\%$ by a pulse. The laser intensity is allowed to grow and reach steady state and remain there for several hundred microseconds. It is then turned off and the whole process is repeated several hundred thousand times. The rise time of the acousto-optic modulator is dependent on the diameter of the laser beam, and can be as short as $\simeq 20$ ns for a beam diameter of $\simeq 100\,\mu$m. The photodiode rise time is $\simeq 1$ ns. The laser is forced into single mode operation by a Pellin–Broca prism and coated etalons within the cavity. A Faraday rotator and compensator plate maintain unidirectional operation of the ring laser. A confocal Fabry–Perot interferometer is used to monitor the mode structure of the laser during the experiment. At the time that the pulse turns off the acousto-optic modulator

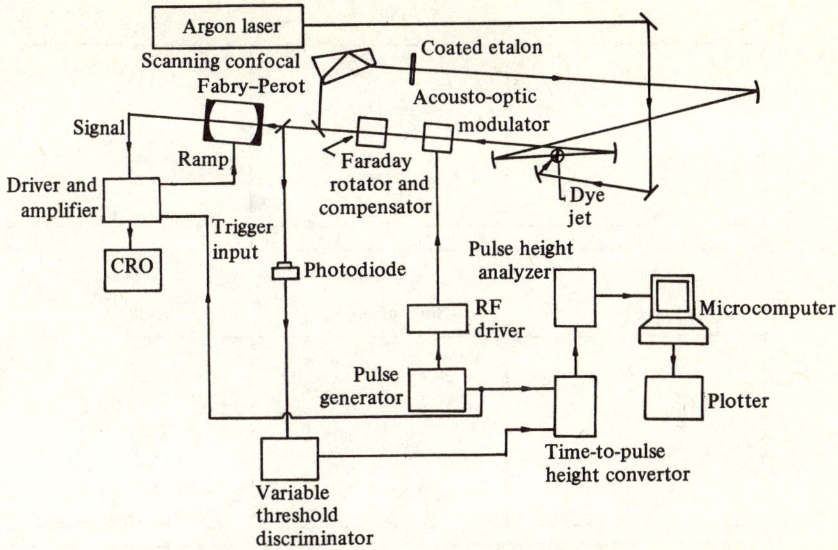

Figure 4.5. The experimental arrangement for the measurement of first passage time distributions. The acousto-optic modulator is used to Q-switch the single mode ring dye laser. The Faraday rotator maintains unidirectional operation. A dye jet is used in this laser.

and thus turns on the laser, a trigger pulse is sent to the start input of the time-to-amplitude convertor (TAC). When the photodiode output voltage reaches a certain preset value, the variable threshold discriminator generates a pulse to the stop input of the TAC. The output pulse from the TAC is proportional in amplitude to the delay between the start and stop pulses. A pulse height analyzer arranges these pulses in a histogram, which is the first passage time distribution that we set out to measure. The data from the pulse height analyzer is stored in a microcomputer, and analyzed statistically.

Haake, Haus and Glauber (1981) have examined the FPT distributions for (4.2.1), and, based on their theory, we have derived detailed expressions for the mean, variance and skewness of the distributions (Roy, Yu and Zhu, 1985; Zhu, Yu and Roy, 1986). These are given below:
the mean,

$$\langle t\rangle_{\mathrm{H}} \simeq (1/2a_0)[C + \ln(a_0^2/PA) + \ln(I_0/(1-I_0))]; \qquad (4.3.1)$$

the variance,

$$\langle(\Delta t)^2\rangle_{\mathrm{H}} \simeq (\pi^2/24a_0^2); \qquad (4.3.2)$$

and the skewness,

$$\frac{\langle(t-\langle t\rangle_{\mathrm{H}})^3\rangle_{\mathrm{H}}}{\langle(\Delta t)^2\rangle_{\mathrm{H}}^{3/2}} \simeq 2(\sqrt{6}/\pi)^3 \sum_{\mathscr{L}=1}^{\infty} (1/\mathscr{L}^3) \simeq 1.14; \qquad (4.3.3)$$

where $C = 0.5772\ldots$ is Euler's constant and $I_0 \equiv AI_{\mathrm{ref}}/a_0$ is the reference

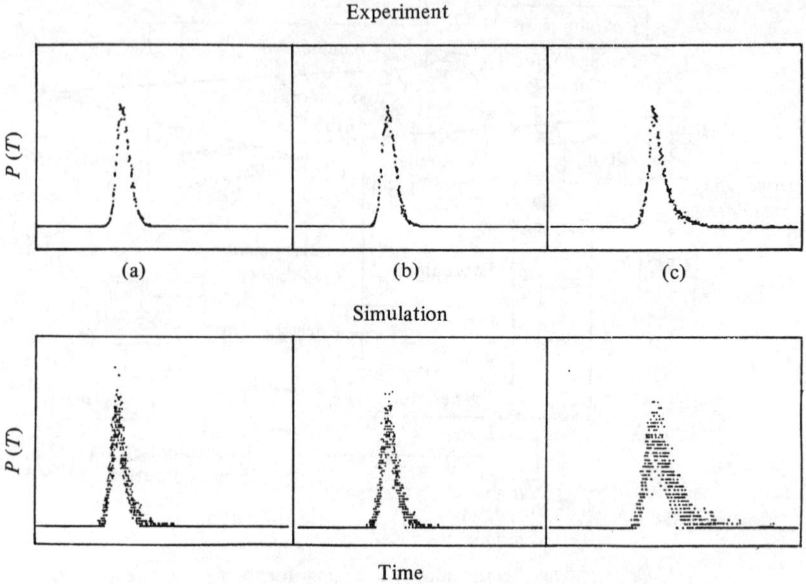

Figure 4.6. Comparison of measured passage time distributions with the results of Monte Carlo simulations of a single mode laser model with quantum fluctuations and pump noise. (a) $I_{ref}/I_{ss} = 0.24$; (b) $I_{ref}/I_{ss} = 0.66$; (c) $I_{ref} I_{ss} = 0.99$. The laser parameters are kept constant for all the computations. Twenty-two such distributions were measured at this operating point of the laser. The laser parameters are $a_0 = 0.7 \times 10^6\,s^{-1}$, $A = 0.11 \times 10^6\,s^{-1}$, $P = 0.004\,s^{-1}$, $P' = 10^4\,s^{-1}$, $\gamma = 2 \times 10^6\,s^{-1}$. The total time scale for each graph is 37.3 μs. The laser was operated at about 6% above threshold.

intensity normalized by the semiclassical steady state intensity $I_{ss} = a_0/A$.

It is evident from these expressions that the delay after which the intensity develops to an appreciable fraction of the steady state value depends on the quantum noise strength, and also on the parameter a_0. However, the variance of the FPT distributions depends entirely on a_0, which contradicts the experimental results shown in Figure 4.6, where the width of the distributions clearly increases as the value of I_0 approaches unity.

Thus (4.2.1) does not lead to predictions that correspond to experiment, and (4.2.6), though adequate for the description of fluctuations of the laser about the steady state and fairly far above threshold, does not contain the quantum noise that is necessary if the laser oscillation is to build up from scratch. It is then necessary to use (4.2.14) for the statistical analysis of the FPT distributions. Several authors have examined the theoretical problem of calculating FPT distributions (Arecchi and Politi, 1980; Arecchi, Politi and Ulivi, 1982; de Pasquale, Tartaglia and Tombesi, 1979, 1981; Fox, 1986a; Gordon and Aslaksen, 1970; Suzuki, 1977; Young and Singh, 1985), and some analytic results are now available that directly address the results of our experiments (de Pasquale, Sancho, San Miguel and Tartaglia, 1986a,b; Fox, 1986b; E. Peacock-Lopez and K. Lindenberg, 1986, private communication). In Roy, Yu and Zhu (1975) and Zhu, Yu and Roy (1986) these approximations have been compared with one another. We have also obtained expressions for the mean and variance of the first passage time distributions when the quantum noise and multiplicative pump noise are included, but have not yet taken account of the colored nature of the pump noise in our analytic approximations. Our analysis is based on the following Fokker–Planck equation for the probability function $Q(x, t)$ (Roy, Yu and Zhu, 1985; Zhu, Yu and Roy, 1986)

$$\partial Q/\partial t = -(\partial/\partial x)[(a_0 x - Ax^3 + (P/2x) + (P'x/2))Q] + \tfrac{1}{2}(\partial^2/\partial x^2)[(P + P'x^2)Q] \qquad (4.3.4)$$

where x is the amplitude of the complex electric field E. If the laser field starts initially from $x = 0$ and arrives at the reference value $x = I_{\text{ref}}^{1/2}$, the mean and the variance of the FPT distribution are given by

$$\langle t \rangle = \int_0^{I_{\text{ref}}^{1/2}} \mathrm{d}x/V(x) \int_0^x \mathrm{d}y(V(y)/D(y)) \qquad (4.3.5)$$

and

$$\langle (\Delta t)^2 \rangle = 4 \int_0^{I_{\text{ref}}^{1/2}} \mathrm{d}x/V(x) \int_0^x \mathrm{d}y/V(y) \int_0^y \mathrm{d}\xi(V(\xi)/D(\xi)) \times \int_0^\xi \mathrm{d}\eta(V(\eta)/D(\eta)), \qquad (4.3.6)$$

where

$$V(x) = \mathrm{e}^{U(x)}, \quad U(x) = \int^x \mathrm{d}\eta(F(\eta)/D(\eta)), \qquad (4.3.7)$$

$$F(x) = a_0 x - Ax^3 + (P/2x) + (P'x/2) \qquad (4.3.8)$$

and

$$D(x) = (P + P'x^2)/2. \qquad (4.3.9)$$

These integrals may be evaluated approximately. We find that the mean first passage time is still given by (4.3.1), but that the variance is now strongly dependent on the strength of the multiplicative noise when I_0 approaches unity:

$$\langle (\Delta t)^2 \rangle \simeq (\pi^2/24a_0^2) + (P'/2a_0^3)[(1.5 - I_0)/(1 - I_0)^2 + \ln(I_0/(1 - I_0)) - 1.5 + \ln(4 \times 10^6) + \ln(a_0/A)]. \qquad (4.3.10)$$

To make a detailed comparison of the data with the predictions of (4.2.14), we still need to resort to numerical simulations on the computer. This is done

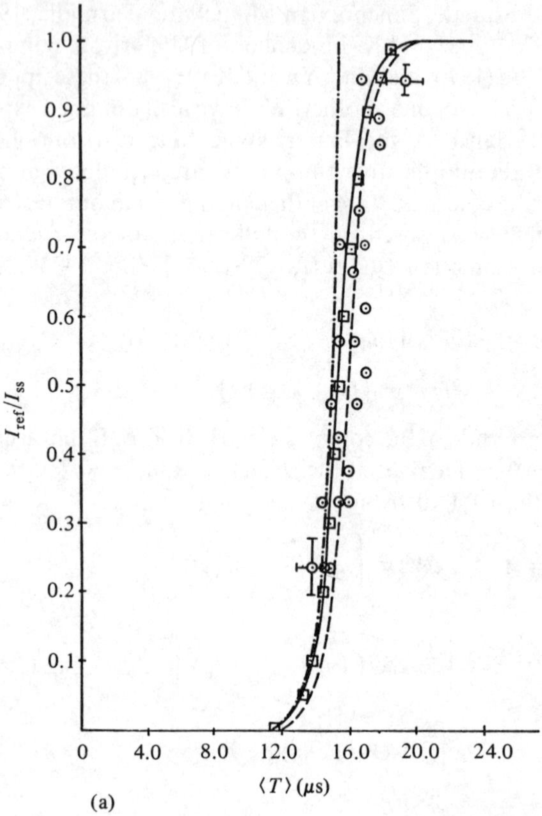

Figure 4.7. (a) Mean first passage times; (b) variance of the distributions; and (c) skewness of the distributions of which three were shown in Figure 4.6. Symbols: ⊙, experimental measurements; ⊡, numerical simulations; ———, from (4.3.1)–(4.3.3); ·—·—·, from the theory of de Pasquale *et al.* (1986b); ----, from Fox (1986b); ······, from (4.3.10).

by Monte Carlo techniques, the algorithm followed being that of Sancho *et al.* (1982). The colored nature of the pump noise can then be incorporated without difficulty, and FPT distributions obtained from step by step integration of the stochastic differential equations. The distributions shown in the lower half of Figure 4.6 were obtained in this manner. The laser was operated at about 6% above threshold for these experiments. It is seen that the shape of the measured distributions can be accurately reproduced by the simulations. The somewhat greater fuzziness of the simulated distributions arises from the fact that 10000 trajectories have been used for the simulation, whereas several hundred thousand trajectories were measured in the experiments. A comparison of the mean, variance and skewness of the passage time distributions with the theoretical results given above as well as the results of de Pasquale *et al.* (1986a,b) and Fox (1986b) are shown in Figures 4.7(a), (b) and (c). Also shown

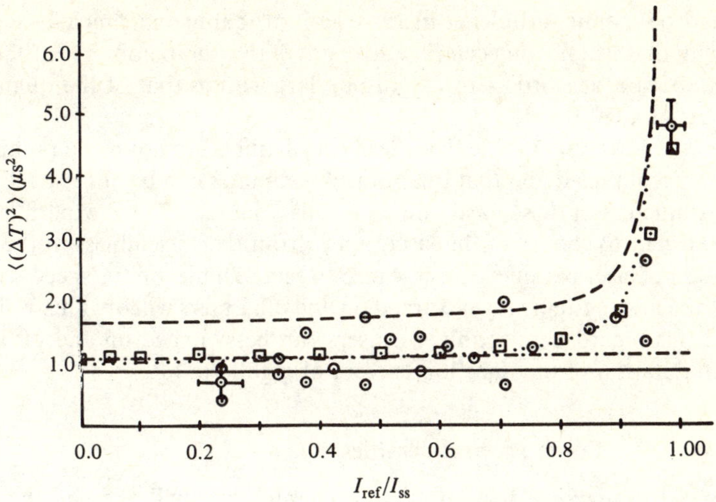

(b)

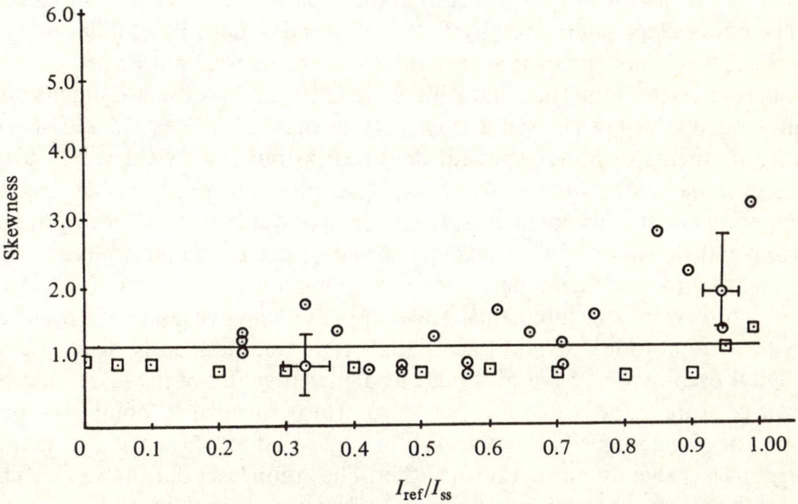

(c)

in these figures are the results of numerical simulations. The parameters which describe the noise sources for the laser can be obtained in this way, by fitting the distributions and the moments of the distributions, and are given in the figure caption. Results of measurements on the laser pumped at about 20% above threshold are given in Roy, Yu and Zhu (1985) and Zhu, Yu and Roy (1986) and are very similar to the ones shown here. In this laser, operated with a dye jet, the time scale of the pump fluctuations is given by $\gamma \simeq 2 \times 10^6\,s^{-1}$, which is somewhat faster than that estimated by Lett (Lett, 1986; Lett, Short and Mandel, 1984), but this is not unexpected, since the nature of the hydrodynamic turbulence in a dye jet is probably on a much faster time scale than that in a flowing cell. The strength of the pump noise is $\simeq 10^4\,s^{-1}$, which is about seven orders of magnitude larger than that of the quantum noise strength, $0.004\,s^{-1}$.

It is quite remarkable that the effects of quantum noise and pump noise can be disentangled and that quantitative estimates can be obtained for the noise parameters of these vastly different noise sources. These experiments on the transient dynamics of the laser, apart from their usefulness in the determination of laser parameters, serve also to set the limits on the speed with which a given laser system may be turned on and off. Lasers will obviously differ widely in this regard; semiconductor lasers can be switched on and off by a direct modulation of the injection current at gigahertz rates.

4.4 Power spectral densities

In the remaining section of this chapter, we will present the results of measurements of power spectra that we have carried out in our laboratory to test the consistency of the determination of parameters by different methods. The parameters determined by Lett and Mandel (Lett, 1986) differ from our values; this is not surprising since the experiments were performed on rather different laser geometries and differed also in that we use a dye jet while a flowing dye cell was used by Lett. It is thus of interest to see whether measurements of power spectral densities on our laser yield values for the noise parameters similar to those determined from the laser transient experiments. It is also of interest to compare power spectra of the argon pump laser and dye laser, and to investigate the source of the pump fluctuations.

In order to investigate the origin of the pump noise, we have performed a few straightforward measurements. The pump laser and dye laser were monitored with two photodiodes, and their signals were simultaneously displayed on a digital oscilloscope. Several thousand digitized values of the laser intensities can be stored, and a fast Fourier transform performed to obtain the power spectra of the intensity fluctuations. In Figure 4.8 we show the dye laser and argon laser signals, one below the other. The argon laser output was stabilized and the beam was monitored before it had passed through the dye jet. Two sets of traces are shown, corresponding to time scales of $5\,\mu s$ and $100\,\mu s$ per division. The dye laser output was $\simeq 15\,mW$. The dye laser output fluctuations are seen to follow the pump laser fluctuations very closely in both these sets of traces. In Figure 4.9 we show the power spectra of the two laser outputs. The lower spectra are for the range 0–6.22 MHz; the upper ones are expanded versions of the initial portions of the lower graphs. The dye laser spectrum levels out at about 1 MHz, but the argon laser spectrum shows broad peaks at $\simeq 1.2$ MHz and $\simeq 2.4$ MHz, and at higher frequencies. It is clear from this spectrum that the Ornstein–Uhlenbeck pump noise can at best be a lumped approximation to the actual spectrum. We can at best estimate an effective

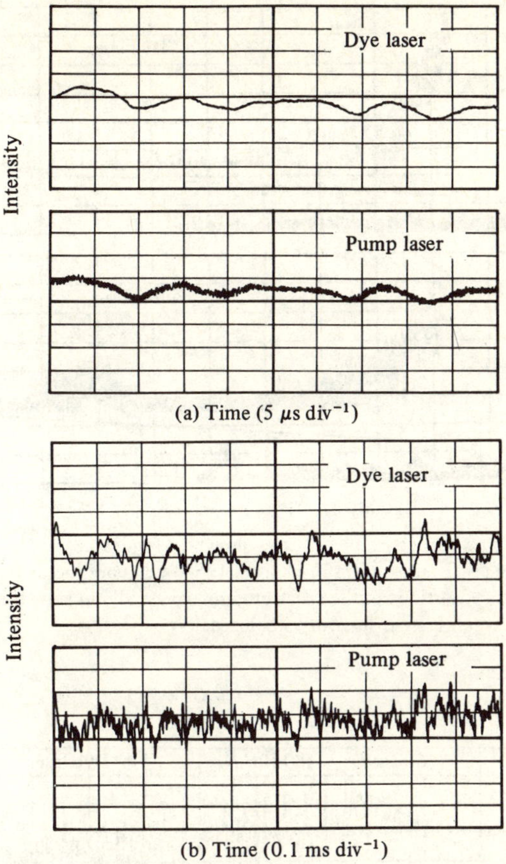

Figure 4.8. Pump laser and dye laser intensity fluctuations, displayed at two different time scales. The vertical scales are different and only show relative fluctuations. The dye laser output was $\simeq 15$ mW single mode, while the argon laser output was $\simeq 2.0$ W at 514.5 nm, with the intensity stabilized. Only a very small fraction of the light was incident on the photodiodes. The argon laser beam was sampled before it was incident on the dye jet. There is a clear correspondence of the pump beam noise with the fluctuations of the dye output.

decay rate, γ_{eff}, for the pump noise, defined as $\Sigma c_i \gamma_i / \Sigma c_i$ where γ_i are the decay rates for individual contributions. Such an estimate gives a $\gamma_{\text{eff}} \simeq 1.8$ MHz; this is in good (perhaps fortuitously close!) correspondence with the time scale determined for the pump noise from the FPT measurements, $\gamma \simeq 2 \times 10^6\ \text{s}^{-1}$. Detailed spectra of the low frequency regime show no components at 60 Hz or its harmonics when the pump laser is intensity stabilized. If, however, the argon laser is operated under current control, when it has far larger fluctuations, particularly at 60 Hz and harmonic frequencies, this is evident in the dye laser output as well. This is demonstrated in the power spectra

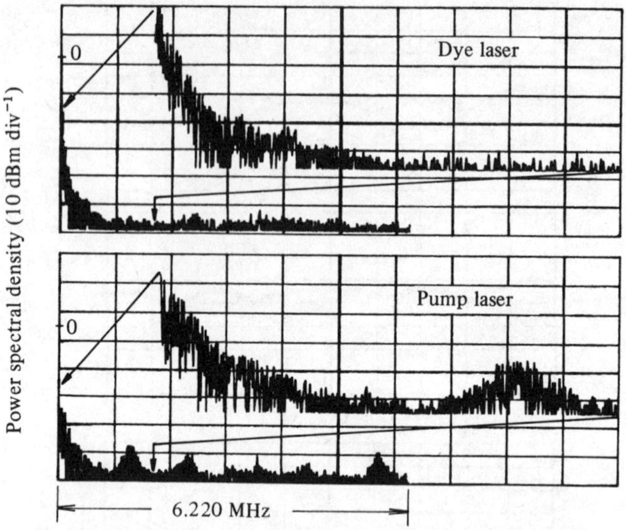

Figure 4.9. Power spectra of the traces shown in Figure 4.8(b). The spectra are obtained by fast Fourier transforms from 12500 points of the digitized data. Low frequency portions of the spectrum are expanded and shown in greater detail. The argon laser is intensity stabilized.

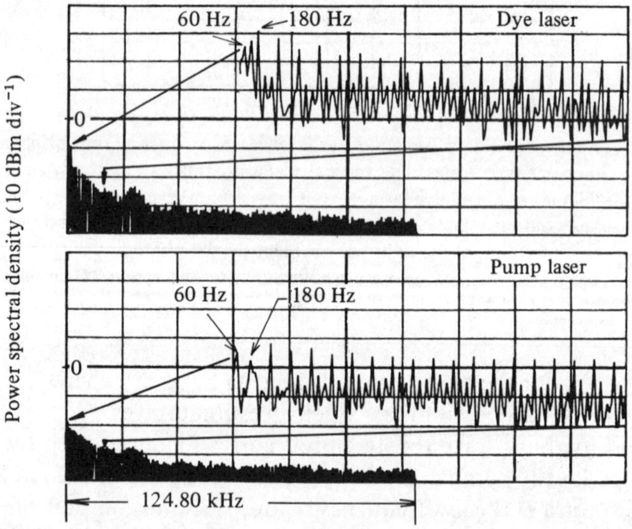

Figure 4.10. Power spectra of the argon and dye lasers when the pump laser is not intensity stabilized. Peaks at 60 Hz as well as at a number of higher harmonics are seen in the pump laser spectrum. Corresponding peaks are also seen in the dye laser spectrum.

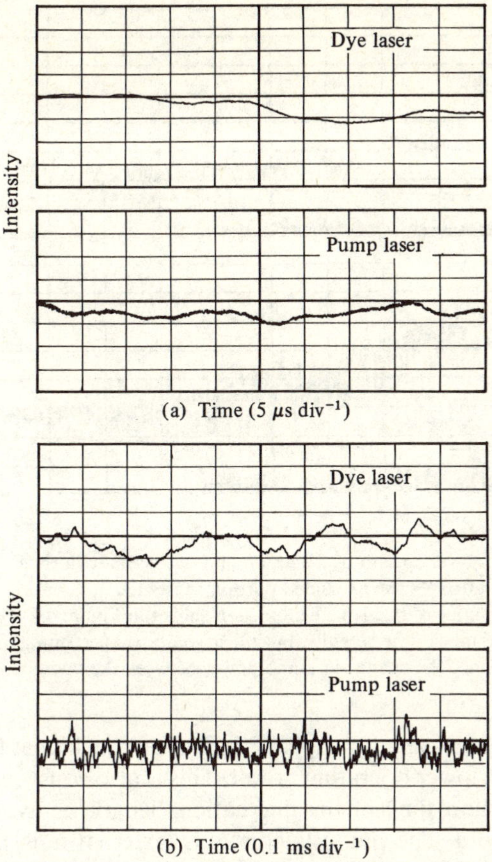

Figure 4.11. Pump and dye laser relative intensity fluctuations at two different time scales, with the dye laser output at $\simeq 0.1$ mW. The close correspondence seen in Figure 4.8 between the pump noise and dye laser fluctuations is almost completely lost.

(Figure 4.10), where the peaks at 60 Hz and several higher harmonic components present in the argon laser are shown that have corresponding contributions in the dye laser spectrum.

The situation is somewhat different when the dye laser is operated closer to threshold, with an output of $\simeq 0.1$ mW. For the two sets of traces shown in Figure 4.11, the correspondence between the pump and dye laser output is much fainter than before. From the power spectra in Figure 4.12, it is seen that the dye laser noise is much more concentrated at the low frequency end than for operation at the higher power. The argon laser spectrum is very similar to that shown in Figure 4.10. The change in its power output is very small, and we would expect the spectra to be almost identical. It is still a source of noise at

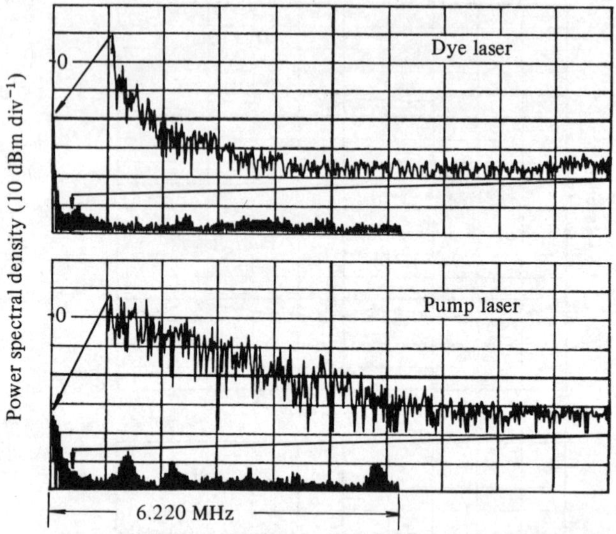

Figure 4.12. Power spectra of the argon and dye lasers with the dye laser output at $\simeq 1.0$ mW. The expanded portion shows the first 450 kHz, which is about one-quarter of the expanded portion shown in Figure 4.9. The dye laser noise falls off much more rapidly, but the argon laser spectrum appears unchanged, since the change in pump power between the two operating points is small.

faster time scales, but the dye laser appears to filter out the higher frequencies more effectively at the lower operating power. This is also consistent with the idea that the correlation time of the fluctuations lengthens as the laser is operated near threshold. The pump fluctuations (external noise) serve as a probe to clearly reveal the critical slowing down of the laser near threshold. A fuller analysis of this behavior, together with a simple analytic approximation for the propagation characteristics of the noise from pump to secondary lasers, has been recently published (Yu, Agrawal and Roy, 1987).

It is interesting to note that the quantum noise strength is most easily measured, perhaps, by the FPT distributions, which also serve to determine the pump noise characteristics. The measurements of the time scale and magnitude of the pump noise can be most easily obtained, and in the greatest detail, from the power spectra measured above. Both the techniques are in reasonable agreement. Digital correlation measurements are most valuable when the laser is operated close to threshold. Each technique has its own strengths, and can be used to advantage in a given situation.

4.5 Conclusions

We may thus conclude that both the dye jet and the argon laser contribute pump noise to the dye laser output. Clearly, one must account for both in a realistic model of the laser. A model which considers the pump noise to be Ornstein–Uhlenbeck in nature, with a single time scale is not a very accurate one. However, it may still reproduce the results of experiments in an overall sense with reasonable accuracy.

There remain many other points which should be considered if a detailed model of the laser is required.

(i) The transverse mode structure of the laser has not been considered in any of the theoretical work mentioned so far. This may be an important aspect of the laser to include in future models. Indeed, the question of the influence of transverse mode structure on the spontaneous emission component of laser light has been a matter of some concern in the realm of semiconductor lasers.

(ii) The presence of more than one mode, even if they are considerably smaller (by a factor of 100 or so) than the dominant one, is unavoidable in the transient growth period of the laser. These modes may contribute to fluctuation phenomena yet unaccounted for in theoretical models.

(iii) The question of the effect of pump noise on laser linewidth is an important one that is at present under investigation in our laboratory.

(iv) Multimode lasers, with many tens or hundreds of oscillating modes, have been analysed from the point of view of deterministic chaos, but the influence of quantum and pump noise on these systems has not been considered in much detail.

(v) A basic assumption in the investigations reported here is the validity of the adiabatic elimination of the polarization and the population inversion. The extremely fast (sub-picosecond) decay times of the molecular polarization certainly justify the elimination of the polarization. The decay time for the population inversion (several nanoseconds) is faster than the cavity decay time ($\simeq 0.1\ \mu$s), and this would appear to justify the adiabatic elimination of this quantity, and this has been assumed to be the case in this chapter. This assumption is probably justified for the steady state operation of the laser. It should probably be scrutinized more carefully for the application to the analysis of the transient dynamics.

(vi) The equations for the laser field considered here are third order equations. This is valid for operation up to about 20% above threshold. If the laser is operated at higher excitations, the effects of saturation may need to be included to a fuller extent. This has been considered in Roy, Yu and Zhu (1985) and Zhu, Yu and Roy (1986) where computer calculations have been used to examine the extent of difference between the third order theory and one that contains the full saturation behavior of the laser.

Despite these many areas that remain to be explored, it is clear that some progress has been made in the study of pump noise and quantum noise on laser fluctuations. New techniques have been developed for the study and characterization of these noise sources that are particularly suitable for lasers pumped far above threshold, and that are able to disentangle quantum and pump noise contributions in different regimes of operation of the dye laser. Statistical measurements will play a particularly crucial role as the applications of lasers grow more demanding, for the limits of precision measurements and optical information processing methods depend on our ability to recognize and control sources of noise in laser systems.

4.6 Appendix

In this appendix we will briefly review the semiclassical equations for a laser, augmented by the contributions of different noise sources. The reduction of these equations to the forms used in this chapter, (4.2.1), (4.2.6), and (4.2.14) is discussed. The origin of the noise sources and their inclusion in the laser model is discussed. The treatment follows that of Haken (Haken, 1981, 1985; Sargent, Scully and Lamb, 1974) and Lett (1986). The equations for the electric field, polarization and population inversion of the active medium are obtained from the molecule, radiation and interaction Hamiltonians, initially in their quantum form. Interaction of the molecules and the radiation with reservoirs is included to introduce decay rates of the field, polarization and inversion. Associated with these dissipative processes are fluctuating forces. These quantum equations are reduced to the semiclassical level by factorization and replacement of the operators by their classical expectation values. The equations obtained in this manner are

$$\dot{b} = (-\mathrm{i}\omega - \kappa)b - \mathrm{i}\sum g\alpha_\mu + F(t), \tag{4.6.1}$$

$$\dot{\alpha}_\mu = (-\mathrm{i}\omega_0 - \gamma)\alpha_\mu + \mathrm{i}g^* b\sigma_\mu + \Gamma_\mu(t), \tag{4.6.2}$$

$$\dot{\sigma}_\mu = \gamma_{11}(\sigma_0 - \sigma_\mu) + 2\mathrm{i}(g\alpha_\mu b^* - \mathrm{cc}) + \Gamma'_\mu(t). \tag{4.6.3}$$

In these equations, b is a dimensionless complex amplitude of the electric field of the laser and ω is its angular frequency; κ is the decay rate for the field due to losses such as the cavity transmission; g is a coupling constant, proportional to the dipole matrix element ζ:

$$g(x) = -\mathrm{i}[\omega/(2\hbar\varepsilon_0 V)]^{1/2}\zeta \mathrm{e}^{\mathrm{i}kx}, \tag{4.6.4}$$

V is the volume occupied by the cavity mode, and k is the propagation number. α_μ is the complex polarization of the molecule, in dimensionless units. The index μ refers to a given molecule. $F(t)$ represents noise contributions to the cavity mode due to fluctuations of the cavity length, mirror vibrations, etc. The resonance frequency ω_0 of the molecules (we will assume that $\omega = \omega_0$) and decay rate γ of the molecular polarization are the parameters introduced in the second equation. The term $\mathrm{i}g^* b\sigma_\mu$ describes the connection of α_μ with the population inversion σ_μ. $\Gamma_\mu(t)$ is the fluctuating force connected with the dissipative decay of the molecular polarization. In the third equation, the rate of change of the population inversion σ_μ is given. σ_0 is the equilibrium inversion due to pumping and incoherent decay processes if no laser field is present. γ_{11} is the decay rate for the population towards equilibrium. $\Gamma'_\mu(t)$ is the fluctuating force associated with the dissipation γ_{11} and cc denotes the 'complex conjugate'. We note that there may be several separate sources for the random force terms, with different time scales. We will represent all of these by a single term. Further, some of the sources of noise cannot be considered delta-correlated; for these reasons we will not attempt to apply the fluctuation–dissipation theorem.

The fast dependence on time may now be removed, with the assumption of resonance, if we let

$$b(t) = E(t)\,\mathrm{e}^{-\mathrm{i}\omega t}, \tag{4.6.5}$$

$$\alpha_\mu(t) = p_\mu(t)\,\mathrm{e}^{-\mathrm{i}\omega t}, \tag{4.6.6}$$

$$F(t) = f(t)\,\mathrm{e}^{-\mathrm{i}\omega t}, \tag{4.6.7}$$

$$\Gamma_\mu(t) = \bar{\Gamma}_\mu(t)\,\mathrm{e}^{-\mathrm{i}\omega t}, \tag{4.6.8}$$

and

$$\Gamma'_\mu(t) = \bar{\Gamma}'_\mu(t)\,\mathrm{e}^{-\omega t}. \tag{4.6.9}$$

If we make these substitutions and sum over all the molecules to obtain the macroscopic polarization P and inversion D ($P = \sum p_\mu$ and $D = \sum \sigma_\mu$), we obtain

$$\mathrm{i}[(\partial/\partial t) + \kappa]E = gP + \mathrm{i}f(t) \tag{4.6.10}$$

$$\mathrm{i}[(\partial/\partial t) + \gamma]P = -Dg^*E + \mathrm{i}r(t) \tag{4.6.11}$$

$$\mathrm{i}[(\partial/\partial t) + \gamma_{11}]D = \mathrm{i}\sigma\gamma_{11} - 2[gPE^* - \mathrm{cc}] + \mathrm{i}s(t), \tag{4.6.12}$$

where $\sigma = N\sigma_0$, $r(t) = \sum \bar{\Gamma}_\mu(t)$ and $s(t) = \sum \bar{\Gamma}'_\mu(t)$.

We now summarize the relevant information regarding the noise sources. In a dye laser system, $f(t)$ arises from effective cavity length fluctuations, mirror vibrations, and dust. The time scale of these noise sources could range from slow (milliseconds or more for acoustic and mechanical noise) to medium fast (tenth of a microsecond for dye jet turbulence). The fluctuations in the polarization, $r(t)$, could be due to collisions and interactions with solvent molecules. These are on a very fast, subpicosecond, time scale. Somewhat longer time scale (nanoseconds) fluctuations in the polarization are due to spontaneous emission events. Fluctuations in the population inversion originate from spontaneous emission and non-radiative decay processes in the dye. Another source is the noise due to pump laser fluctuations. Dye concentration fluctuations could also contribute to $s(t)$. The time scales for these processes could range from milliseconds to nanoseconds.

Arecchi, Lippi, Puccioni and Tredicce (1984) have classified lasers based on whether adiabatic elimination of the polarization and inversion can be performed on the laser equations.

Class A lasers: $\gamma, \gamma_{11} \gg \kappa$. In this case the equations for P and D may be solved for the steady state conditions and P and D can be adiabatically eliminated. Many gas lasers and the dye laser are examples of this category.

Class B lasers: $\gamma \gg \gamma_{11}, \kappa$. The polarization can be adiabatically eliminated, but not the population inversion. Examples are carbon dioxide and solid state lasers.

Class C lasers: γ, γ_{11} and κ are all comparable. All three equations have to be retained. Examples of this class are some far-infrared lasers. Deterministic chaos has been observed in single mode lasers of this type.

Let us now proceed to outline the adiabatic elimination of the P and D variables. If we solve (4.6.11) for P in the steady state,

$$P = (1/\gamma)[\mathrm{i}Dg^* E + r(t)]. \tag{4.6.13}$$

Substitution into the field equation gives

$$[(\partial/\partial t) + \kappa]E = (1/\gamma)|g|^2 DE + q(t), \tag{4.6.14}$$

where

$$q(t) = f(t) - (\mathrm{i}g/\gamma)r(t). \tag{4.6.15}$$

The population inversion equation is now

$$\left[\frac{\partial}{\partial t} + \gamma_{11} + \frac{4|g|^2|E|^2}{\gamma}\right]D = \gamma_{11}\sigma + s'(t), \tag{4.6.16}$$

where $s'(t) = s(t) + 4\,\mathrm{Im}[\mathrm{i}gEr(t)/\gamma]$. We will neglect the second term here, since it will usually turn out to be much smaller than the first one. Equations (4.6.14) and (4.6.16) describe a Class B laser with noise terms. If the population inversion can also be adiabatically eliminated, we get from solving (4.6.16) in the steady state

$$D = [\sigma + \{s(t)/\gamma_{11}\}]/[1 + \{4|g|^2|E|^2/(\gamma\gamma_{11})\}]. \tag{4.6.17}$$

The equation for the electric field is then

$$[(\partial/\partial t) + \kappa]E = \frac{|g|^2 E(\sigma + s(t)/\gamma_{11})}{\gamma\left[1 + \dfrac{4|g|^2|E|^2}{\gamma\gamma_{11}}\right]} + q(t). \tag{4.6.18}$$

When the laser intensity is not too high, we may assume

$$[4|g|^2|E|^2/(\gamma\gamma_{11})] \ll 1. \tag{4.6.19}$$

The denominator in (4.6.18) can be expanded to obtain

$$\begin{aligned}\dot{E} = &-\kappa E + (|g|^2\sigma E/\gamma) - [4|g|^4 E|E|^2\sigma/(\gamma^2\gamma_{11})] \\ &+ [E|g|^2 s(t)/(\gamma\gamma_{11})] - [4|g|^4 s(t)E|E|^2/(\gamma^2\gamma_{11}^2)] + q(t).\end{aligned} \tag{4.6.20}$$

This equation can be written as

$$\dot{E} = a_0 E - AE|E|^2 + p(t)[E - 4|g|^2|E|^2 E/(\gamma\gamma_{11})] + q(t), \tag{4.6.21}$$

where we have defined

$$a_0 = F_1 - \kappa = [(|g|^2\sigma/\gamma) - \kappa] \tag{4.6.22}$$

$$F_1 = |g|^2\sigma/\gamma \tag{4.6.23}$$

$$A = [4|g|^4\sigma/(\gamma^2\gamma_{11})] \tag{4.6.24}$$

$$p(t) = |g|^2 s(t)/(\gamma\gamma_{11}). \tag{4.6.25}$$

This is the basic form of the electric field equation for the laser, as given in (4.2.14). There are both additive and multiplicative noise terms in this Langevin equation. Equation (4.2.6) is obtained by omission of the additive noise term. The field is expressed in dimensionless units, as stated at the outset. In Roy, Yu and Zhu (1985) and Zhu, Yu and Roy (1986) we have used the dimensionless field defined by

$$E = \tfrac{1}{2}[\zeta^2/(\hbar^2\gamma_a\gamma_b)]^{1/2}\mathbf{E}, \tag{4.6.26}$$

where ζ is the dipole moment of the transition, γ_a and γ_b are the lifetimes of the upper and lower lasing levels. The strengths P and P' of the correlation functions of $q(t)$ and $p(t)$ in (4.2.2), (4.2.7) and (4.2.13) are expressed directly in s^{-1}. The pump noise term corresponding to the saturation term is often neglected since it is usually small compared to $p(t)E$ (Roy, Yu and Zhu, 1985).

The field equation (4.2.1) is obtained as follows. Let

$$\varepsilon = (2A/P)^{1/4} E, \tag{4.6.27}$$

and

$$\bar{t} = (AP/2)^{1/2} t \tag{4.6.28}$$

be a dimensionless time. With these definitions, one obtains, if the fluctuating saturation term is neglected,

$$\dot{\varepsilon} = (\bar{a}\varepsilon - \varepsilon|\varepsilon|^2) + \bar{p}(\bar{t})\varepsilon + \bar{q}(\bar{t}). \tag{4.6.29}$$

All quantities with the dimensions of s^{-1} are scaled by the factor $(AP/2)^{-1/2}$. Lett (1986) has taken 12.6 ms to correspond to one unit of dimensionless time in his experiments on single mode lasers. To simplify the notation, we have written this last equation as in (4.2.1),

$$\dot{E} = aE - AE|E|^2 + q(t), \tag{4.6.30}$$

where the pump noise term was left out.

Acknowledgements

We thank G. Vemuri and G. Gray for assistance with numerical analysis of data and experimental measurements. The work at Georgia Tech. was supported by a grant from the US Department of Energy, Office of Basic Energy Sciences, Chemical Sciences Division. It is a pleasure to thank Ron

Fox, Glenn James, P. Lett, L. Mandel and R. Short for their contributions at various stages of this work. The recognition of pump noise as an important source of laser intensity fluctuations originated from Leonard Mandel's physical insight. Special thanks are due to him for his guidance of the earlier stages of the work reviewed here. We also thank F. T. Arecchi, P. Hänggi, P. Jung, K. Lindenberg, E. Peacock-Lopez, H. Risken, M. San Miguel and S. Singh for sending us preprints and reprints of their work and helpful discussions.

References

Abate, J. A., Kimble, H. J. and Mandel, L. 1976. *Phys. Rev. A* **14**, 788.

Arecchi, F. T. and DeGiorgio, V. 1971. *Phys. Rev. A* **3**, 1108.

Arecchi, F. T., DeGiorgio, V. and Querzola, B. 1967. *Phys. Rev. Lett.* **19**, 1168.

Arecchi, F. T., Lippi, G., Puccioni, L. and Tredicce, J. 1984. *Opt. Comm.* **51**, 308.

Arecchi, F. T. and Politi, A. 1980. *Phys. Rev. Lett.* **45**, 1219.

Arecchi, F. T., Politi, A. and Ulivi, L. 1982. *Nuovo Cimento* **71**, 119.

Atmanspacher, H. and Scheingraber, H. 1986. *Phys. Rev. A* **34**, 253.

Baczynski, A., Kossakowski, A. and Marzalek, T. 1977. *Z. Phys.* **26**, 93.

Born, M. and Wolf, E. 1975. *Principles of Optics*, 5th edn. Oxford: Pergamon Press.

Chyba, T., Christian, W., Gage, E., Lett, P. and Mandel, L. 1986. In *Optical Instabilities* (R. W. Boyd, M. G. Raymer and L. M. Narducci, eds.), pp. 253–5.

Chyba, T. H., Gage, E. C., Ghosh, R., Lett, P., Mandel, L. and McMackin, I. 1987. *Opt. Lett.* **12**, 422.

Darling, D. A. and Siegert, A. J. F. 1953. *Ann. Math. Stat.* **24**, 624.

Dembinski, S. T. and Kossakowski, A. 1976a. *Z. Phys. B* **24**, 141.

Dembinski, S. T. and Kossakowski, A. 1976b. *Z. Phys. B* **24**, 207.

de Pasquale, F., Sancho, J. M., San Miguel, M. and Tartaglia, P. 1986a. *Phys. Rev. A* **33**, 4360.

de Pasquale, F., Sancho, J. M., San Miguel, M. and Tartaglia, P. 1986b. *Phys. Rev. Lett.* **56**, 2473.

de Pasquale, F., Tartaglia, P. and Tombesi, P. 1979. *Physica* **99A**, 581.

de Pasquale, F., Tartaglia, P. and Tombesi, P. 1981. *Z. Phys. B* **43**, 353.

Dixit, S. and Sahni, P. 1983. *Phys. Rev. Lett.* **50**, 1273.

Fox, R. F. 1972. *J. Math. Phys.* **13**, 1196.

Fox, R. F. 1978. *Phys. Rep.* **48**, 178.

Fox, R. F. 1986a. *Phys. Rev. A* **33**, 467.

Fox, R. F. 1986b. *Phys. Rev. A* **34**, 3405.

Fox, R. F., James, G. E. and Roy, R. 1984a. *Phys. Rev. Lett.* **52**, 1778.

Fox, R. F., James, G. E. and Roy, R. 1984b. *Phys. Rev. A* **30**, 2482.

Fox, R. F. and Roy, R. 1987. *Phys. Rev. A* **35**, 1838.

Gardiner, C. W. 1983. *Handbook of Stochastic Methods.* Berlin: Springer-Verlag.

Gordon, J. P. and Aslaksen, E. W. 1970. *IEEE J. Quantum Electron.* **QE-6**, 428.

Graham, R., Hohnerbach, M. and Schenzle, A. 1982. *Phys. Rev. Lett.* **48**, 1396.

Haake, F., Haus, J. W. and Glauber, R. J. 1981. *Phys. Rev. A* **23**, 3255.

Haken, H. 1970. In *Encyclopedia of Physics* (S. Flugge, ed.), vol. XXV/2c. Berlin: Springer.
Haken, H. 1981. *Light*, vol. I. Amsterdam: North-Holland.
Haken, H. 1985. *Light*, vol. II. Amsterdam: North-Holland.
Hänggi, P., Mroczkowski, T., Moss F. and McClintock, P. V. E. 1985. *Phys. Rev. A* **32**, 695.
Hernandez-Machado, A., San Miguel, M. and Katz, S. 1985. *Phys. Rev. A* **31**, 2362.
Hernandez-Machado, A., San Miguel, M. and Katz, S. 1986. In *Recent Developments in Nonequilibrium Thermodynamics: Fluids and Related Topics* (J. Casas–Vazquez, D. Jou and J. M. Rubi, eds.), pp. 368–71. Berlin: Springer.
Horsthemke, W. and Malek-Mansour, M. 1976. *Z. Phys. B* **24**, 307.
Jung, P. and Hänggi, P. 1987. Preprint.
Jung, P. and Risken, H. 1984. *Phys. Lett.* **103A**, 38.
Jung, P. and Risken, H. 1985. *Z. Phys. B* **59**, 469.
Jung, P. and Risken, H. 1986. In *Optical Instabilities* (R. W. Boyd, M. G. Raymer and L. M. Narducci, eds.), pp. 361–3. Cambridge University Press.
Kaminishi, K., Roy, R., Short, R. and Mandel, L. 1981. *Phys. Rev. A* **24**, 370.
Kramers, H. A. 1940. *Physica (Utrecht)* **7**, 284.
Landauer, R. and Swanson, J. A. 1961. *Phys. Rev.* **121**, 1668.
Lax, M. 1969. In *Brandeis Lectures* (M. Chretien, E. P. Gross and S. Deser, eds.), vol. II, pp. 271–478. New York: Gordon and Breach.
Lax, M. and Zwanziger, M. 1973. *Phys. Rev. A* **7**, 750.
Lett, P. 1986. PhD thesis, University of Rochester.
Lett, P., Gage, E. C. and Chyba, T. H. 1987. *Phys. Rev. A* **35**, 746.
Lett, P., Short, R. and Mandel, L. 1984. *Phys. Rev. Lett.* **52**, 341.
Lindenberg, K., West, B. J. and Cortes, E. 1984. *Appl. Phys. Lett.* **44**, 175.
Louisell, W. H. 1973. *Quantum Statistical Properties of Radiation.* New York: J. Wiley.
Mandel, L. and Wolf, E. 1965. *Rev. Mod. Phys.* **37**, 231.
Marchesoni, F. 1986. *J. Appl. Phys.* **59**, 666.
Meltzer, D. and Mandel, L. 1970. *Phys. Rev. Lett.* **25**, 1151.
Meltzer, D. and Mandel, L. 1971. *Phys. Rev. A* **3**, 1763.
Montroll, E. W. and Shuler, K. 1958. *Adv. Chem. Phys.* **1**, 361.
Risken, H. 1984. *The Fokker–Planck Equation.* Berlin: Springer-Verlag.
Roy, R. 1979a. *Phys. Rev. A* **20**, 2093.
Roy, R. 1979b. *Opt. Comm.* **30**, 90.
Roy, R. and Mandel, L. 1977. *Opt. Comm.* **23**, 306.
Roy, R., Yu, A. W. and Zhu, S. 1985. *Phys. Rev. Lett.* **55**, 2794.
Sancho, J. M., San Miguel, M., Katz, S. and Gunton, J. D. 1982. *Phys. Rev. A* **26**, 1589.
San Miguel, M., Pesquera, L., Rodriguez, M. A. and Hernandez-Machado, A. 1987. *Phys. Rev. A* **35**, 208.
Sargent, III, M., Scully, M. O. and Lamb, Jr., W. E. 1974. *Laser Physics.* Reading, MA: Addison-Wesley.
Schaefer, R. B. and Willis, C. R. 1976. *Phys. Rev. A* **13**, 1874.
Schenzle, A. 1986. In *Optical Instabilities* (R. W. Boyd, M. G. Raymer and

L. M. Narducci, eds.), pp. 376–9. Cambridge University Press.
Schenzle, A. and Brand, H. 1979. *Phys. Rev. A* **20**, 1628.
Schenzle, A. and Graham, R. 1983. *Phys. Lett.* **98A**, 319.
Schuss, Z. 1980. *Theory and Applications of Stochastic Differential Equations*, New York: J. Wiley.
Short, R., Mandel, L. and Roy, R. 1982. *Phys. Rev. Lett.* **49**, 647.
Short, R., Roy, R. and Mandel, L. 1980. *App. Phys. Lett.* **37**, 973.
Stratonovich, R. L. 1963. *Topics in the Theory of Random Noise.* New York: Gordon and Breach.
Suzuki, M. 1977. *J. Stat. Phys.* **16**, 477.
Uhlenbeck, G. E. and Ornstein, L. S. 1954. In *Selected Papers in Noise and Stochastic Processes* (N. Wax, ed.), pp. 93–111. New York: Dover.
Van Kampen, N. G. 1982. *Stochastic Processes in Physics and Chemistry.* Amsterdam: North-Holland.
Weiss, G. H. 1977. In *Stochastic Processes in Chemical Physics* (I. Oppenheim, K. Shuler and G. H. Weiss, eds.), pp. 361–78. Cambridge, MA: MIT Press.
Young, M. R. and Singh, S. 1985. *Phys. Rev. A* **31**, 888.
Yu, A. W., Agrawal, G. P. and Roy, R. 1987. *Opt. Lett.* **12**, 806.
Zhu, S., Yu, A. W. and Roy, R. 1986. *Phys. Rev. A* **34**, 4333.

5 Noisy dynamics in optically bistable systems

E. ARIMONDO, D. HENNEQUIN and
P. GLORIEUX

5.1 Introduction

Lasers are part of a large class of non-linear systems which covers a great deal of scientific domains, including physics, chemistry, biology, sociology, etc. All these non-linear systems have common properties and present universal features which become more and more extensive as new studies are performed.

In recent years, the interest in these non-linear systems has considerably increased. In particular, it is now well established that the role of fluctuations in non-equilibrium systems is far more relevant than in systems at thermodynamical equilibrium, and the study of their influence on stationary and transient states has become very important.

Among the different aspects presented by the non-linearity in lasers, two of them will be particularly studied here:

(i) The first one is the optical bistability (OB). An optical system is bistable if it has two output states I for the same value of the excitation A over some range of excitation values (see Gibbs, 1985, for a recent review). A typical response curve of such a system is given in Figure 5.1. Here, the system is defined as bistable between $A_{\downarrow}$ and $A_{\uparrow}$, where the effectively occupied state depends on the history of the input parameter (see the caption to Figure 5.1). If OB is a common phenomenon in different fields, the OB transient phenomena have far more general features. If the A parameter is abruptly switched on from a value below the up-switching value $A_{\uparrow}$ to a value slightly above, the laser intensity switches rapidly on after a delay which depends on the laser operating conditions (Arimondo, Gabbanini, Menchi and Zambon, 1986). Such an evolution, characterized by a long lethargic stage of slow time evolution, followed by a fast transition to another stable state (Figure 5.2), is encountered in a large variety of systems operating near a phase transition. It has been studied, for instance, in explosive chemical reactions (Baras, Nicolis, Malek Mansour and Turner, 1983; Francowitz, Malek Mansour and Nicolis, 1984; Francowitz and Nicolis, 1983), and in laser physics this type of behavior has appeared independently of OB, as in the build-up of laser action (Arecchi and De

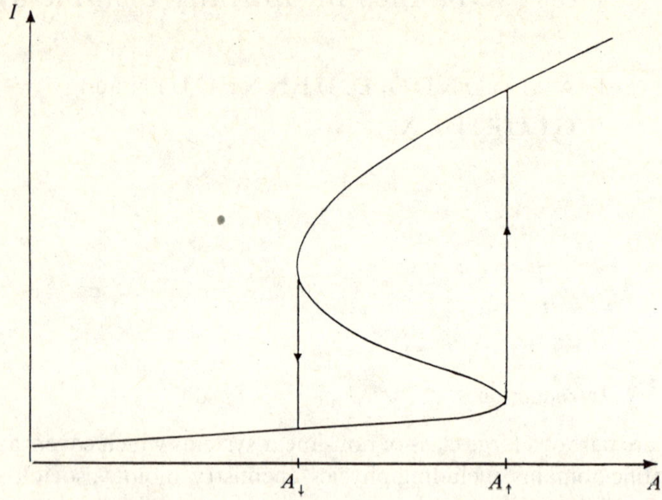

Figure 5.1. Principle of optical bistability. Increasing the *A* parameter, the system remains on the lower state up to $A_\uparrow$ where it switches to the upper state. Decreasing the *A* parameter, the system remains on the upper state down to $A_\downarrow$ where it switches to the lower state. The system is bistable between $A_\uparrow$ and $A_\downarrow$.

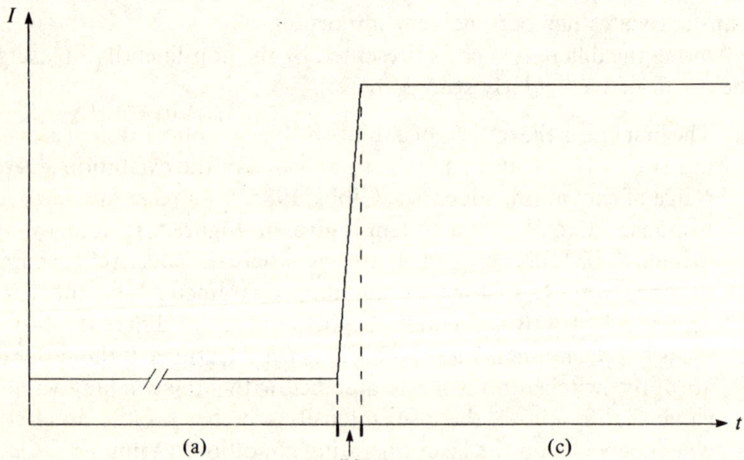

Figure 5.2. Schematic representation of 'explosive' behavior. The time evolution presents a long lethargic stage (a), followed by a fast transition (b) to the final stable state (c).

Giorgio, 1971; Glorieux and Dangoisse, 1987; Lefebvre, Bootz, Dangoisse and Glorieux, 1983; Wascat, Dangoisse, Glorieux and Lefebvre, 1983), or in lethargic gain (Chung, Lee and DeTemple, 1981). Experimental studies concerning the influence of noise on OB are presented here.

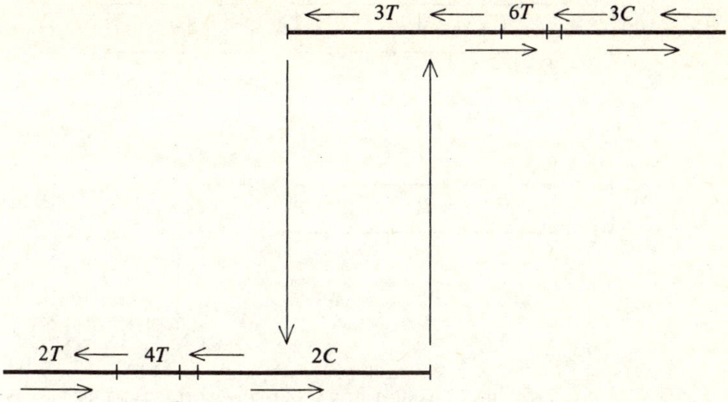

Figure 5.3. Generalized bistability (GB) principle. GB is produced by the coexistence of two attractors. Here, the first attractor is a chaotic one, coming from a period-doubling cascade T, $2T$, $4T$, $8T$,..., where T is an eigenperiod of the system. This chaotic attractor is called '$2C$' because its maximum spectral component corresponds to the period $2T$. The second attractor is a $3T$ periodic attractor, which leads itself to another chaotic attractor after a series of period doubling. Increasing the control parameter, the system remains on the $2C$ attractor until it switches to the $3T$ one. Decreasing the control parameter, the system remains on the $3T$ attractor until it becomes unstable, and then switches on the $2C$ one.

(ii) The second part of this chapter may be considered as an extension of the first one. In OB, the transition occurs between two continuous wave (cw) states, i.e. two point attractors. Some systems present generalized bistability (GB), where the transition occurs between two complex attractors, evolving from a periodic to a chaotic state following the now well-known period doubling route to chaos (Figure 5.3) (Arecchi, Meucci, Puccioni and Tredicce, 1982; Midavaine, Dangoisse and Glorieux, 1985). The term 'chaotic' refers to a behavior which does not show any regularity, but which remains governed by deterministic rules. Such attractors have been studied in physics (see, e.g., Libchaber and Maurer, 1982, for turbulence), chemistry (e.g., Roux, Rossi, Bachelart and Vidal, 1980), biology (e.g., Glass, Guevara and Shrier, 1983). In laser physics, such attractors have been observed also in systems which do not present GB, such as, for example, hybrid bistable systems (Chrostowski, Vallee and Delisle, 1983; Hopf, Kaplan, Gibbs and Shoemaker, 1982), passive all-optical devices (Harrison, Firth, Emshary and Al-Saidi, 1983; Nakatsuka *et al.*, 1983), or lasers (Gioggia and Abraham, 1983). So, before studying the influence of fluctuations on GB, it will be necessary to examine the influence of fluctuations on one specific attractor, and those results are presented here. Influence of noise on interaction between several attractors is at the present time in a stage of very fast progress (Arecchi, Badii and Politi, 1984a, b, 1985).

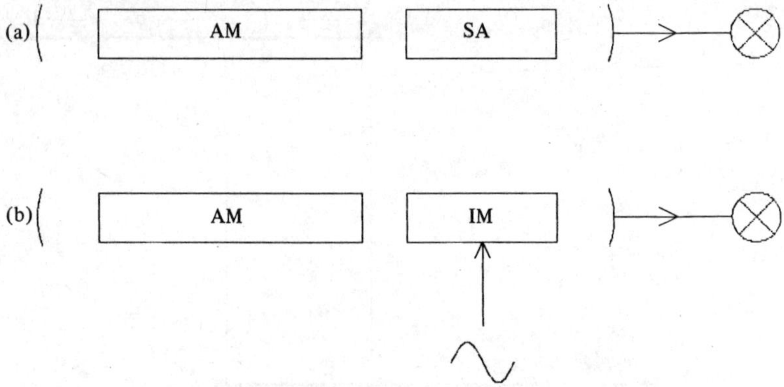

Figure 5.4. Principle of the LSA and of the LIM. The LSA, (a), is constituted by an amplifier medium (AM) and a saturable absorber (SA) inside the laser cavity. The LIM, (b), is composed of an amplifier medium (AM) and the modulator (IM) inside the laser cavity.

The study of the influence of noise on OB or chaos requires systems to be properly adapted. In particular, the time scales of the systems have to be adapted to the existing noise generators and the laser parameters have to be well-defined and easily controllable. An optical bistable device responding to these requirements is the laser with saturable absorber (LSA), which is composed of amplifier and absorber media placed inside an optical cavity (Figure 5.4a). If the absorber reaches saturation at a laser intensity lower than that required by the amplifier, the LSA presents OB (Arimondo *et al.*, 1986). However, period doubling and chaos have never been observed in an LSA, but a quite similar system, the laser with internal modulation (LIM), composed of an amplifier medium and a modulator inside an optical cavity (Figure 5.4b), presents this evolution to chaos as different parameters, such as the modulation amplitude or the laser frequency, are varied (Arecchi *et al.*, 1982; Dangoisse, Glorieux and Midavaine, 1986; Midavaine, Dangoisse and Glorieux, 1985).

The LSA and the LIM present other important advantages, that we have made good use of by employing, in the first case, a CO_2 LSA containing SF_6 as the saturable absorber, and, in the second case a CO_2 laser with an internal elasto-optic modulator.

The CO_2 LSA is simple, inexpensive to operate and has well-defined laser parameters. The existing theories permit a good description of the system (Arimondo, Bootz, Glorieux and Menchi, 1985; Arimondo, Casagrande, Lugiato and Glorieux, 1983; Tachikawa, Tanii, Kajita and Shimizu, 1986). The LSA studies gain much interest with the practical applications of the optical bistable devices in the domain of optical computers (Gibbs, 1985).

The laser with an internal elasto-optic modulator is characterized by a very weak level of internal noise, easily accessible experimental parameters, and time scales particularly well adapted to an experimental study. A very simple model permits us to describe all the phenomenology of the system (Dangoisse, Glorieux and Hennequin, 1987). Finally, applications in the telecommunication domain increase the interest of these studies.

This chapter is composed of two parts. The first one deals with the phenomena induced by noise in OB in general, and more particularly in the LSA. The second part is devoted to reviewing the influence of noise on the period doubling scenario, and in particular in the LIM. In each part, after a short introduction, we describe the experimental set-up and results of our groups, and a comparison with experimental results obtained by other groups. Finally, the simulations carried out with models describing the experimental systems studied in the first sections are presented, and the results of these calculations are compared with the experimental ones.

5.2 Noise in optical bistable devices

5.2.1 Introduction

Some distinctions have to be made between different optical bistable devices. The first one concerns the system itself: *passive* (non-laser) OB and *active* (laser) OB. Although the two fields have evolved independently, there are great similarities between them: both are described by coupled Bloch–Maxwell equations with boundary conditions. Another criterion of classification is to separate the *all-optical* devices from the *hybrid* devices. In all-optical devices, the non-linearity arises from a direct interaction of the light with matter. In hybrid devices, non-linearity arises from an electrical feed-back of the output signal to the driving of some element placed inside the cavity. Examples of practical achievement of these different devices are shown in Figure 5.5. In a passive all-optical device, the beam, coming, e.g., from a laser, goes through a non-linear element, e.g. a saturable absorber or semi-conductor (Figure 5.5a). The set-up of the passive hybrid device is the same, except that there is a feed-back, for instance to an intracavity elasto-optic or acousto-optic crystal (Figure 5.5b). In an active bistable device, the non-linear element is inside the laser cavity (Figure 5.5c) and may be the active medium, a saturable absorber or any other non-linear optical device.

The influence of noise in the optical bistable devices may be observed either in the transient or in the cw regime, monitoring different parameters which characterize the system and are experimentally accessible, such as output intensity, laser frequency, polarization, laser mode, etc. In most experiments, the mean intensity and the threshold intensities have been monitored. Bonifacio and Lugiato have predicted that the steady-state intensity distribution in the bistability region has a two-peaked character (Bonifacio and Lugiato, 1978). The experimental observation of this double-peaked distribution at steady state with internal fluctuations is difficult, owing to the very long lifetime of each state (McCall *et al.*, 1982). In the transient, the so-called

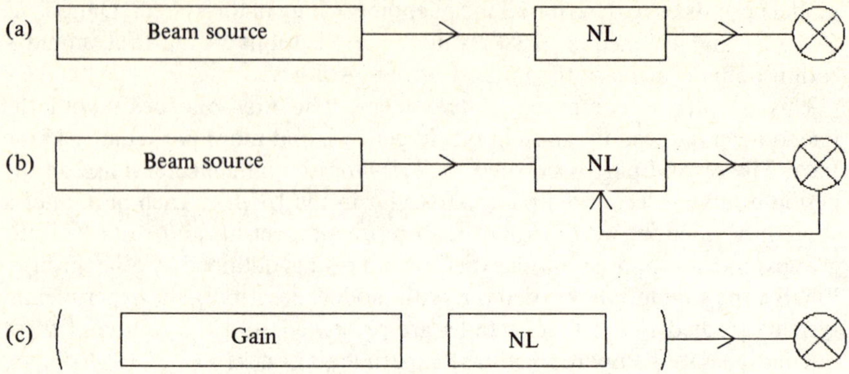

Figure 5.5. Examples of optical bistable devices. In a passive optical device, (a), the beam goes through the non-linear element (NL), where bistability arises by a direct light–matter interaction. In a passive hybrid device, (b), a feed-back on the non-linear element produces bistability. In active devices, (c), the non-linear element is inside the cavity.

transient noise-induced optical bistability, or transient bimodality (TB), appears in systems where the response to a given excitation consists of a lethargic delay τ_D after which the system abruptly switches to another state. In these conditions, the probability distribution, which is one-peaked in the deterministic case, develops a second peak during a sizeable interval of the time evolution. Baras *et al.* (1983) first predicted TB in the case of explosive chemical reactions, and they pointed out that the same phenomenon should arise whenever the evolution of the system involves a long induction period followed by an abrupt switching to a final stable state. Broggi and Lugiato (1984) proposed OB as a candidate for the study of this effect. TB has effectively been observed in all-optical bistable passive (Lange, Mitschke, Deserno and Mlynek, 1985) or active (Arimondo, Dangoisse and Fronzoni, 1987) devices, and we review in the following paragraph the present status of experimental results.

5.2.2 *Experimental set-up: the CO_2 laser with saturable absorber*

The optical bistable device used is a CO_2 laser with SF_6 as a saturable absorber which has been described in detail elsewhere (Arimondo, Dangoisse, and Fronzoni, 1987; Arimondo and Menchi, 1985) and is schematically presented in Figure 5.6. The laser is composed of a CO_2 discharge amplifier and an intracavity cell containing SF_6 gas at 15 mTorr pressure typically. This system exhibits bistability cycles as a function of various parameters such as the pressures of the amplifier or the absorber, the cavity laser length or the discharge current. The experimental control parameter is chosen in order to permit a simple interpretation of the experimental results. Consequently, it should modify only one of the laser control parameters. In the LSA, these

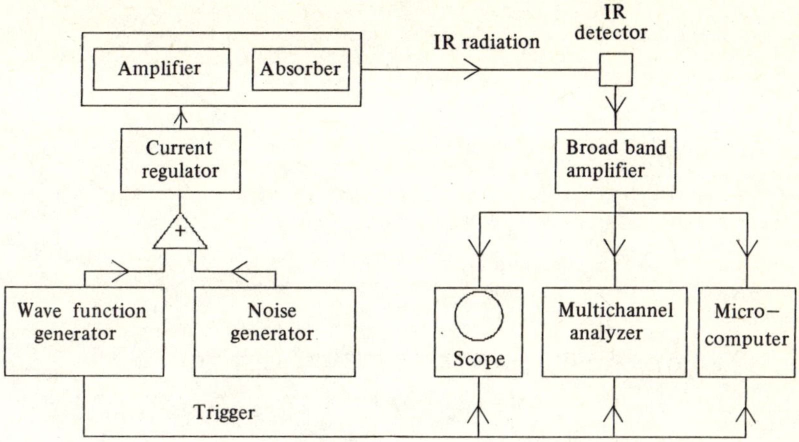

Figure 5.6. Diagram of the LSA experimental arrangement.

parameters are the following ones: the ratio, a, of the infrared saturation intensity in the amplifier to that in the absorber, which must be larger than unity to observe OB; the ratio A $(\bar{A})$ between the amplifier gain (the absorber loss) and the laser cavity losses. The most adapted experimental parameter is the discharge current i, which modifies the A parameter only. So, the bistable cycle is obtained by scanning the A parameter acting on the laser current regulator (Figure 5.7).

In our experiments, the behavior of the system is characterized by two parameters: the mean output intensity and the threshold intensity. The mean output intensity I, measured in reduced units of the saturation intensity value, in the absence of intracavity saturable absorber, is given by $I = 0$ for $A < 1$ and $I = A - 1$ for $A > 1$ (Figure 5.7a). In the presence of the saturable absorber, the mean output intensity presents the optically bistable response between the off state and a non-zero solution I_+ (Figure 5.7b). The influence of fluctuations on these two parameters has been studied by adding external noise to the system. The criteria for the choice of the noise parameter are essentially the same as those governing the choice of the control parameter; noise, therefore, is added to the discharge current, using a pseudo-random noise generator described elsewhere (Arimondo, Dangoisse and Fronzoni, 1987). The produced noise has a bandwidth of 80 kHz. It may be considered as white because it has a frequency spectrum which is broad as compared to the relaxation rates of the amplifier and absorber populations, which are about 10 kHz and 1 kHz, respectively. The noise rms amplitude $\sigma = \langle \delta A^2 \rangle^{1/2}$ has been derived by measuring the current noise and using the experimental relation connecting the control parameter A to the amplifier discharge current.

In the experiments showing the steady-state bimodality, a commercial noise generator producing a gaussian noise with adjustable spectral width was used. A microprocessor system sampled the laser output intensity during a time that

(a)

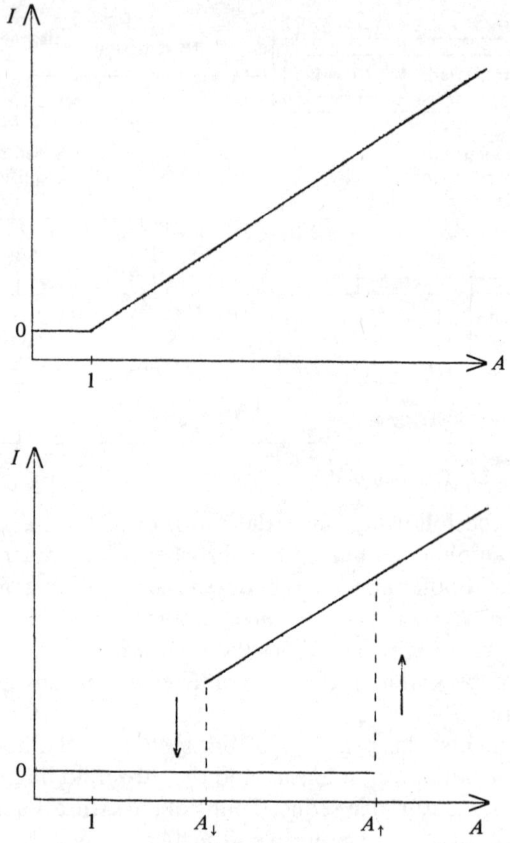

(b)

Figure 5.7. Output intensity I of a laser versus the control parameter A, in (a) without saturable absorber, and in (b) with saturable absorber.

is very long compared to the lifetime of the states and then plotted the probability distribution of the output intensity.

Another relevant characteristic of a bistable system is the switching time, i.e. the time required to undergo the transition from one branch of the hysteresis cycle to the other one. In the LSA experiments mentioned above, the laser current, whence the laser control parameter, A, is abruptly increased by a step generator to a value A_f just above the up-switching value $A_{\uparrow}$. In this case, the output laser intensity displays the slowing-down behavior shown in Figure 5.14(a). In the first lethargic step, the output intensity remains zero. After a delay time τ_D, there is a rapid switching to the upper branch. This delay time is not constant, and depends on the final value A_f of the control parameter. But even for a given value A_f, fluctuations of τ_D are observed. If the A_f value becomes very close to the up-switching value $A_{\uparrow}$, the probability distribution of the switching time becomes broader and asymmetric, and at the same time the mean value $\langle \tau_D \rangle$ becomes longer (Figure 5.8). In the LSA

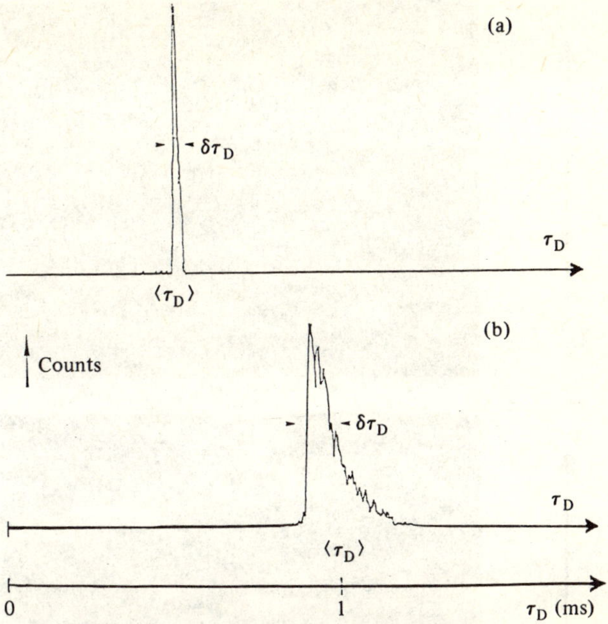

Figure 5.8. Distribution of the switching-up delay τ_D for a step function increase to a final value A_f much above the threshold $A_\uparrow$ in (a), and close to $A_\uparrow$ in (b). CO_2 laser operating on 10P(16) line with 12 mTorr pressure SF_6 gas absorber.

experiments, the $\langle \tau_D \rangle$ value varies between 100 μs and a few milliseconds, a time scale particularly convenient for detailed studies. To measure the fluctuations in the response time, the time τ_D between the discharge current change and the first passage of the laser output power to a certain percentage (typically 70%) of its final value is determined through an electronic counter. Histograms of the τ_D time distribution are obtained by repeating the switching process many times. By slightly modifying this set-up, the probability distribution of the laser output power may be plotted for different times τ around the $\langle \tau_D \rangle$ value.

In the next section, we describe the results acquired by this set-up, and those obtained by other groups, with other systems.

5.2.3 Noise-induced effects

Noise-induced change of the mean output and threshold intensities

The noise on the control parameter modifies the bistability cycle as shown in Figure 5.9. An important experimental result is the observation that a noise with mean value equal to zero applied externally modifies the mean level $\langle I \rangle$ of

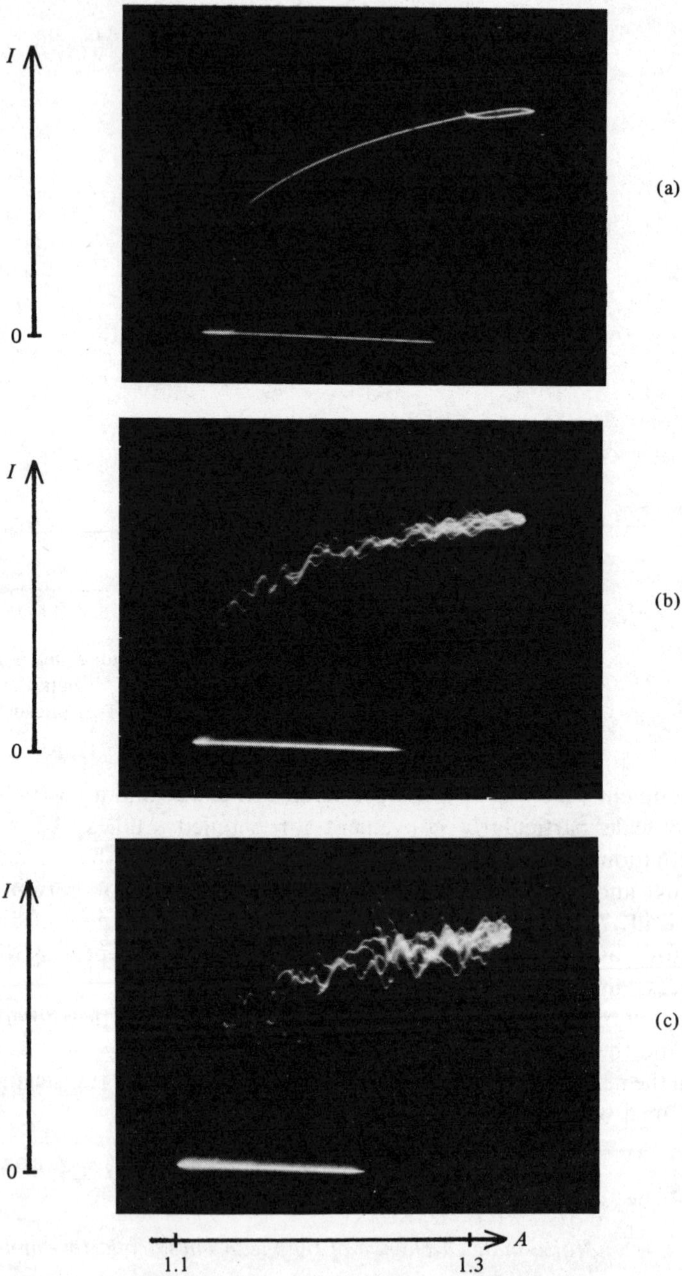

Figure 5.9. Bistability cycles of the LSA output intensity I in arbitrary units versus the control parameter A. (a) Without external noise. (b) and (c) When a noise with rms amplitude $\sigma = 0.15$ and 0.3, respectively, was added to the control parameter. The LSA system is the same as in Figure 5.7.

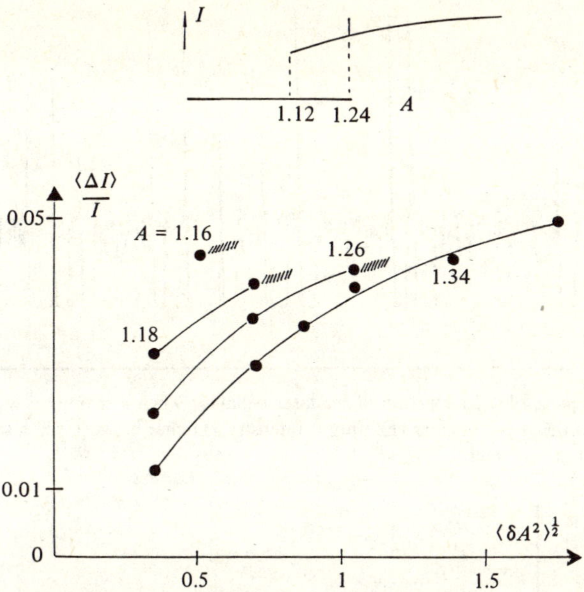

Figure 5.10. Relative modification in the average power level produced by the external noise as a function of the noise level measured by the peak-to-peak value, and the operation point of the laser. The dashed region indicates that, at that noise level, the laser fluctuates between the off and on states. The inset indicates the A-limiting values of the bistability cycle.

the laser output power. The average value of the output power I has been measured as a function of the noise level. The LSA experimental results are presented in Figure 5.10. It is clear from this diagram that the application of noise increases the average output power of the laser. Larger modifications are obtained near the limiting values of the bistable cycle.

It can also be seen in Figure 5.9 that the down-switching and up-switching points are shifted to lower values. However, the $A_{\uparrow}$ shift is more important than the $A_{\downarrow}$ one. Thus, the width of the cycle decreases in presence of noise.

High noise levels: cw bimodality

At large noise levels, the laser fluctuates between the off and on states, and the probability distribution of the output power becomes two-peaked. A time sequence corresponding to this behavior is shown in Figure 5.11. Figure 5.12 presents two series of histograms for the distribution of the output intensity I for two different values of the noise bandwidth. The two-peaked structure centered around the zero value and the upper value of the bistability cycle appears very clearly. This bimodality has been observed for the first time by McCall *et al.* (1982, 1985) in a hybrid bistable optical device consisting of an electro-optic crystal between crossed polarizers. The transmitted laser beam was strongly attenuated so that the shot noise of the detector was the

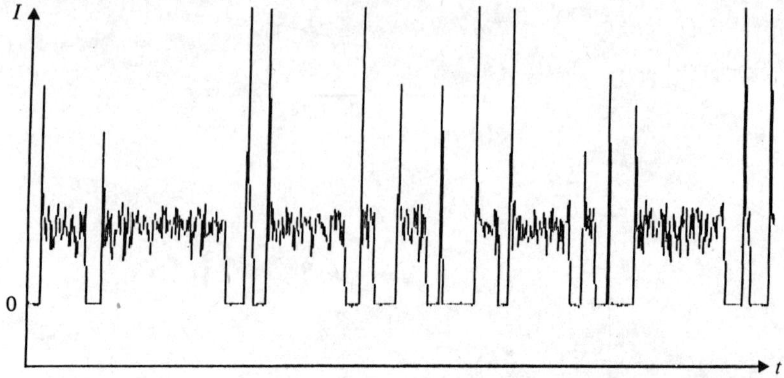

Figure 5.11. Illustration of the laser behavior when a strong noise is applied to the *A* parameter. The output intensity switches between the laser-off and the laser-on states.

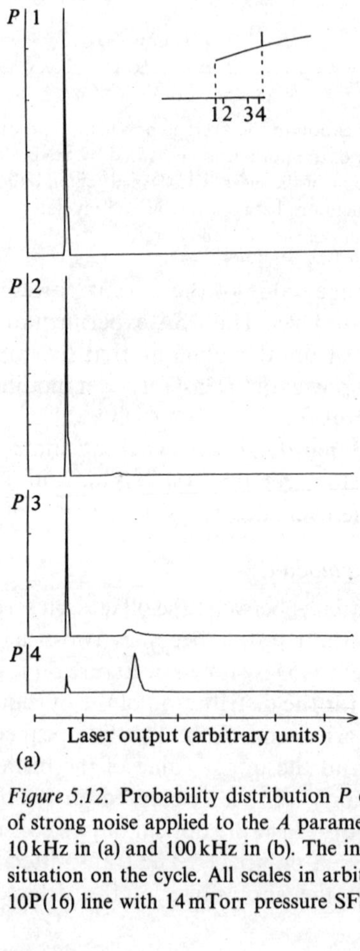

(a)

Figure 5.12. Probability distribution *P* of the output laser power *I* in the case of strong noise applied to the *A* parameter. Gaussian noise with a half-width of 10 kHz in (a) and 100 kHz in (b). The inset indicates the position of each situation on the cycle. All scales in arbitrary units. CO_2 laser operating on 10P(16) line with 14 mTorr pressure SF_6 gas absorber.

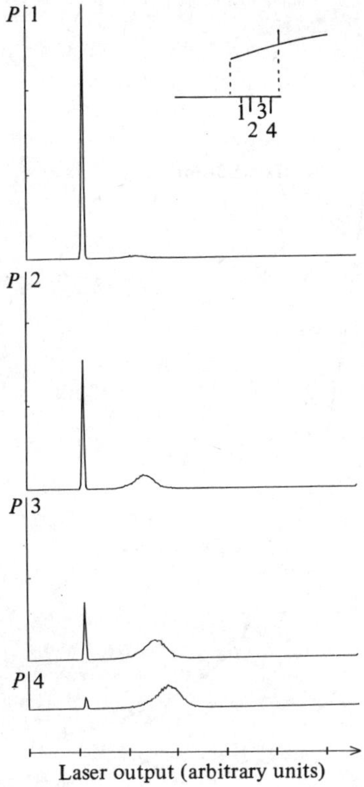

(b)

dominant noise source. Data were taken in two modes. The first one corresponds to the treatment used to derive the data presented in Figure 5.12 and leads to similar histograms. In the second mode, the system is forced into the upper or lower state by an external pulse, and the time to return to the original state is recorded. Figure 5.13 reports the results for that return time versus the control parameter value.

Statistics of delay times

In discussing the transient features, experimental results regarding the duration of the lethargic stage will be presented first. Figure 5.14 presents two records of the LSA transient response obtained in the same conditions of laser operation, with and without external noise, respectively. It appears that the duration of the lethargic stage decreases in the presence of noise. A similar result has been found by Lange *et al.* using a passive optical-bistable device, consisting of sodium vapor in an argon atmosphere placed inside a confocal Fabry–Perot resonator (Lange *et al.*, 1985; Mitschke, Deserno, Mlynek and Lange, 1985; and Chapter 6 of this volume). A cw dye laser served to resonantly excite the sodium D1 line, and the control parameter was the input light power.

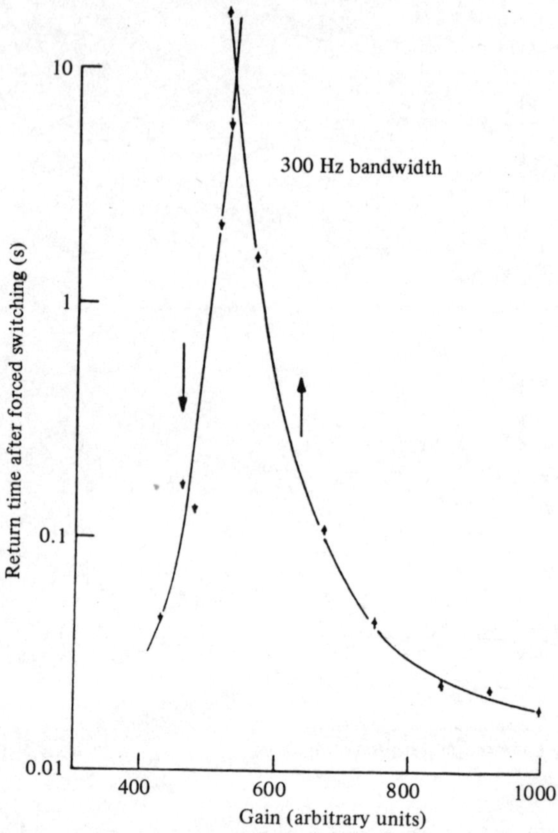

Figure 5.13. Return time after forced switching to the lower state (↓ points) and to the upper state (↑ points) as a function of feed-back gain which is the control parameter in the experiments. Switch-up and switch-down gains were $\simeq 1000$ and $\simeq 400$. From McCall *et al.* (1982).

As the aforementioned authors have also pointed out, the fact that the value τ_D for the lethargic time is reduced by increasing noise is an important result, since, owing to the unavoidable presence of noise in the apparatus, it implies that the deterministic delay τ_D^d cannot be obtained experimentally. This fact has to be kept in mind when predictions based on deterministic models are checked.

Lange *et al.* have also measured the distribution of τ_D. Typical histograms obtained are presented in Figure 6.12(a) in Chapter 6 of this volume, for different external noise levels, and without added noise. Besides the decrease in the mean value with increasing noise, described above, it may be remarked that (i) there is a broad asymmetric distribution of τ_D, and (ii) the distribution of τ_D can become narrower with increasing noise levels.

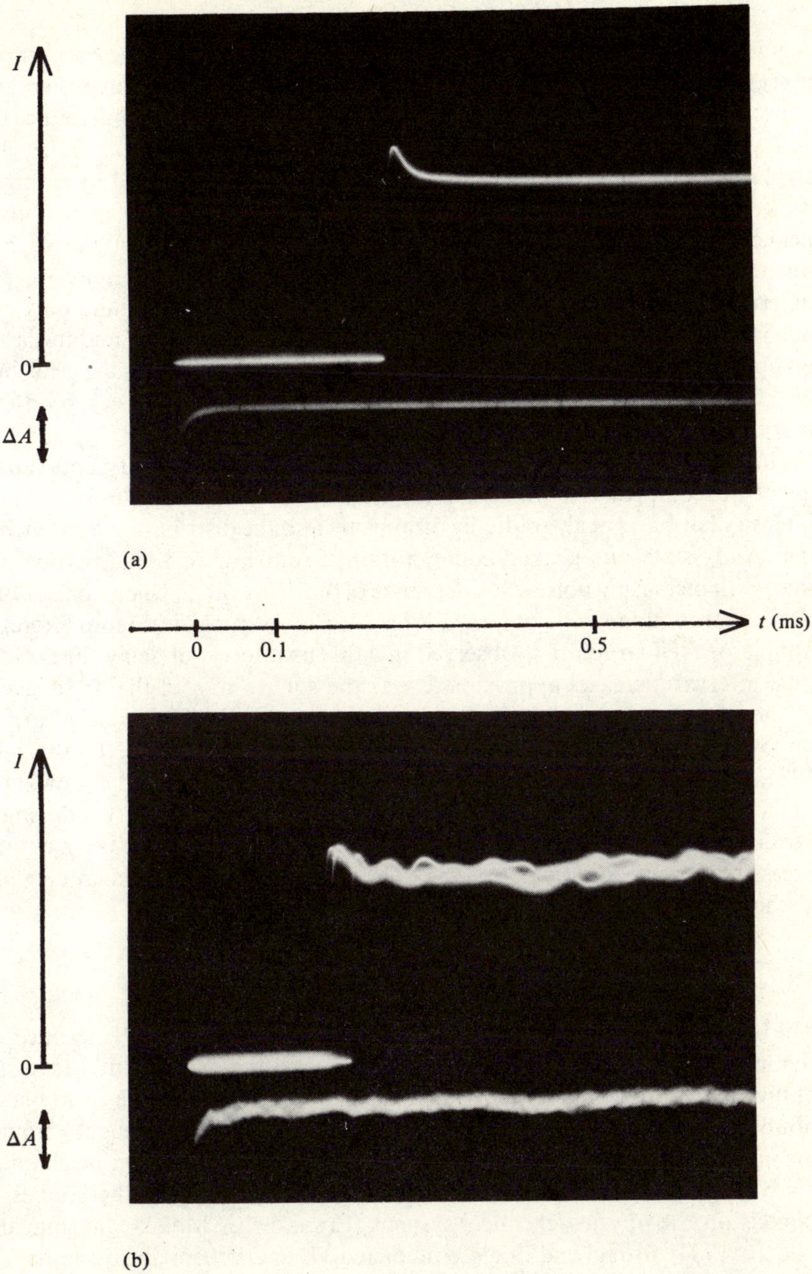

Figure 5.14. Switching-up delays in the LSA for a pulse ΔA in the A control parameter, without noise in (a) and adding external noise with rms amplitude $\sigma = 0.15$ in (b). From Arimondo, Dangoisse and Fronzoni (1987).

Transient bimodality

If, under the same physical conditions as those for Figure 5.13, the laser output power I is monitored at a time τ in an interval around the $\langle \tau_D \rangle$ value, then, as a consequence of the fluctuations, the laser output power may present either the zero value or the final power value. Thus, the probability distribution $P(I, \tau)$ of the laser field amplitude, that in the steady state is represented by a single-peaked distribution, in a sizeable time interval around $\langle \tau_D \rangle$ is a double-peaked distribution. Results concerning this TB are shown in Figure 5.15 at different delay times τ and different noise levels. Without noise, in the deterministic case, the distribution $P(I, \tau)$ should consist of one and only one delta-function peak, describing the zero output power for $\tau < \tau_D^d$, and the laser-on operation for $\tau > \tau_D^d$. The presence of noise obviously broadens the peak, but, also, for a considerable range of intermediate τ values around $\langle \tau_D \rangle$, produces a double-peaked distribution.

The distribution is initially one-peaked at zero laser intensity and subsequently becomes double-peaked. In the last stage of the time evolution, the zero intensity left-hand peak gradually disappears and the distribution approaches the steady-state one-peaked configuration. From Figure 5.15, it is obvious that, with increasing noise level, the range of the delay time τ where the double-peaked structure may be observed is larger. For example, in Figure 5.15(c), a double-peaked structure is observed in a 200 μs interval of delay times. This time interval increases approximately as the square root of the noise level.

The first experimental evidence of the phenomenon of TB has been carried out by Lange *et al.* on the optical-bistable device described above (Lange *et al.*, 1985; Mitschke *et al.*, 1985). These experiments were started following a suggestion by Broggi and Lugiato of the use of TB as a procedure producing 'a novel kind of OB', such that the double-peaked distribution is experimentally accessible (Broggi and Lugiato, 1984). The next section will consider that work in the context of the theoretical developments.

5.2.4 *Comparison with theories and numerical simulations*

The interest in the influence of noise on optical-bistable devices is testified by the many theoretical studies on this subject, and the experimental investigations have been generally motivated by theoretical results. The steady-state double-peaked distribution was predicted by Bonifacio and Lugiato from a mean field model of the optical bistability and a fully quantum mechanical analysis (Bonifacio and Lugiato, 1978). Likewise, following the studies of Nicolis on the physico-chemical systems (Frankowitz, Malek-Mansour and Nicolis, 1984), Broggi and Lugiato predicted TB in OB resolving by numerical integration the Fokker–Planck equation (Broggi and Lugiato, 1984; Broggi, Lugiato and Colombo, 1985). Since this work, the studies point to the multiplicative noise (Lugiato, Colombo, Broggi and Horowicz, 1986), the influence of the correlation time of the noise (Moore, 1985, 1986), and more

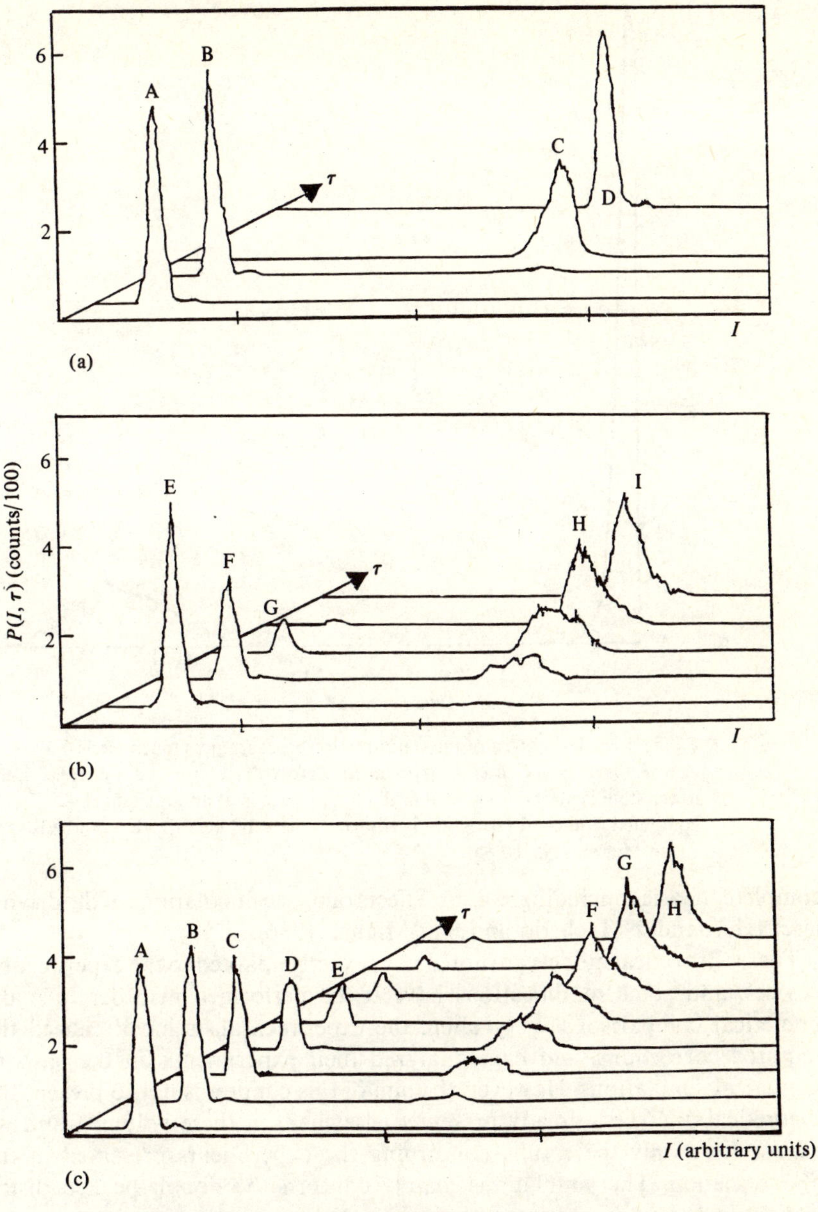

Figure 5.15. Time evolution of the probability distribution $P(I, \tau)$ at different delay times τ and on modifying the noise level. (a) Without external noise. (b) and (c) With external noise having rms of 0.06 and 0.25, respectively. In (a), the curves A, B, C and D correspond to a delay time of 800 μs, 980 μs, 1080 μs and 1400 μs, respectively. In (b) and (c) τ was increased from A to H from 720 μs to 1000 μs in 40 μs steps. I is measured in arbitrary units. From Arimondo, Dangoisse, and Fronzoni (1987).

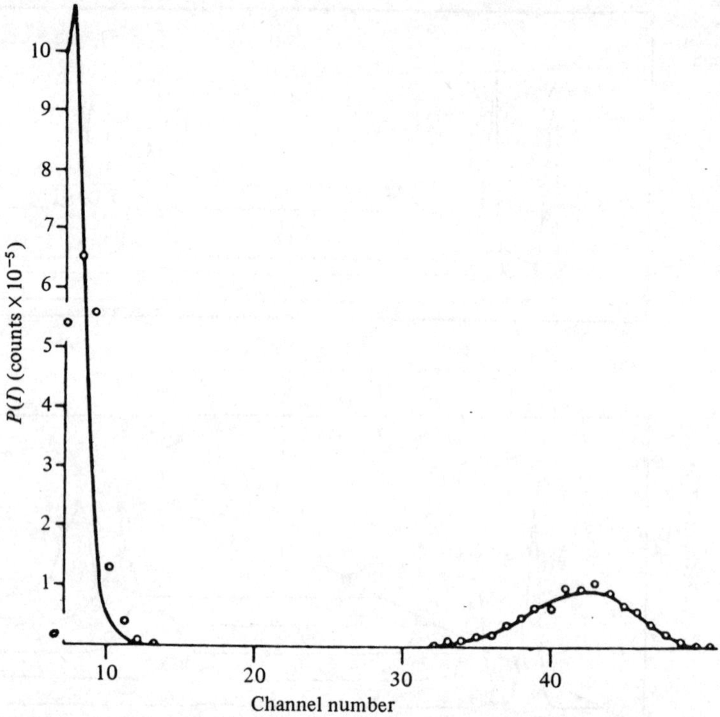

Figure 5.16. Histogram of the output voltage occurrences (open circles) for $A = 58.9$ in the McCall *et al.* experiment. Zero voltage is at about channel 6; the lower state is at about channel 8; the upper state is at about channel 42. The computed simulated histogram is shown as the solid curve with $A = 61.83$. From McCall *et al.* (1985).

complete theories, including, e.g., the effects of phase fluctuations of the driving laser (Hernández-Machado and San Miguel, 1986).

These theoretical models are not always exactly adapted to the experimental devices, and some modifications have to be performed in order to make numerical comparisons. Very often, the experimentalists have realized the required corrections, and have analyzed their experiments on the basis of numerical simulations. However, the aim of this chapter is not to present the theoretical analyses, already presented elsewhere in these volumes, and we expose here only the results concerning the experiments presented in the above section. The simulations mainly concern the double-peaked distribution, in the two experimental configurations where it may be observed: the steady state and the transient. Some theoretical results are also presented in the following paragraphs.

To simulate their experiment, McCall and colleagues used an approximate model which was not the correct one for their system (McCall *et al.*, 1985). With the help of this model, they fitted the steady-state distribution. They used two

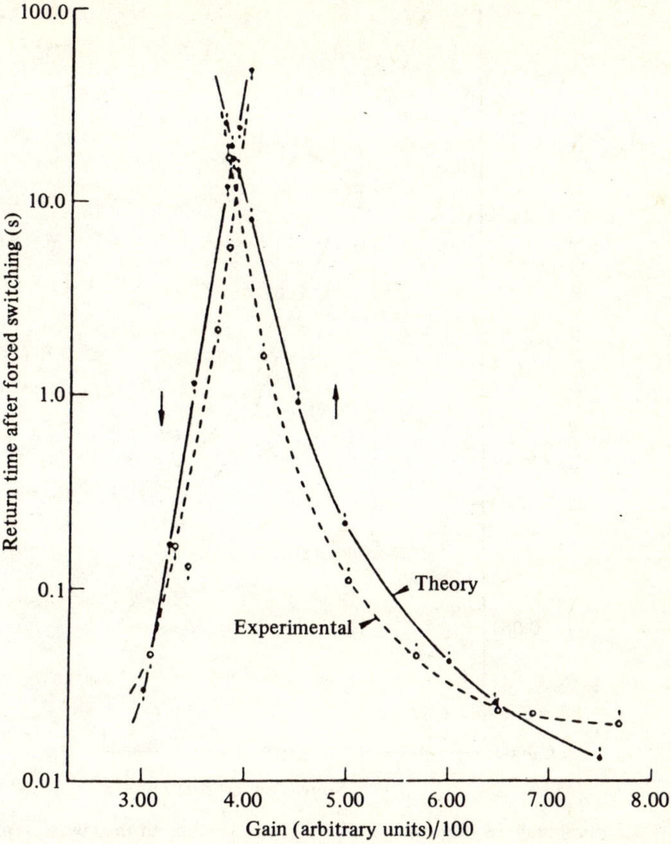

Figure 5.17. Return time after forced switching to the lower state (↓ points) and to the upper state (↑ points) as a function of feed-back gain setting (in arbitrary units) in the McCall *et al.* experiment. In the experiment, $A = 19.6$; the theoretical points are with $A = 61.83$. From McCall *et al.* (1985).

fitting parameters: the gain G of the high voltage operational amplifier, which alone determined the noise-free operating curve, and the product A of the counting rate at maximum transmission and the amplifier decay time, which characterized the amount of noise in the system. The result is presented in Figure 5.16. The return time after forced switching to the lower or the upper state was also plotted as a function of feed-back gain setting (Figure 5.17). The effect of noise on the response time versus gain curve is shown in Figure 5.18. This study shows that the use of simple models permits us to find out what happens by using an adjusted amount of noise. However, some details remain incorrect.

Concerning the TB, Lange *et al.* (1985) could not apply directly the theoretical approach of Broggi and Lugiato (1984), because the experimental and theoretical situations are quite different: Broggi and Lugiato were in the

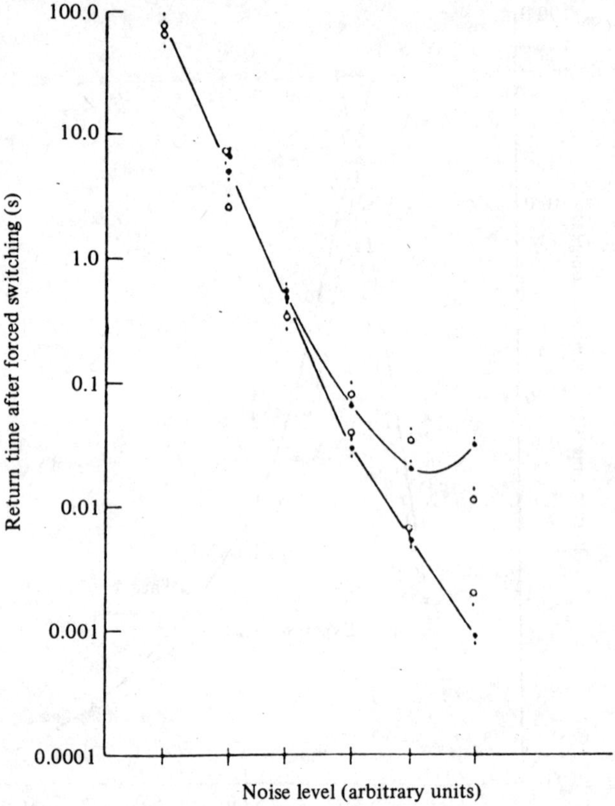

Figure 5.18. Return time after forced switching to the lower (resp. upper) bistable state as a function of the noise level in the McCall *et al.* experiment. The gain is fixed at the value 291.8 (resp. 620.0) in the same units as in Figure 5.17. The parameter A, defined in the text, characterizes the amount of noise in the system. Symbols: ○, experimental; ●, theory. From McCall *et al.* (1985).

case of the good cavity limit, and the non-linear medium consisted of two-level atoms; Lange *et al.* performed their experiment in the bad cavity limit, and the non-linear mechanism was due to Zeeman pumping of the sodium ground state. Developing a simple model, Lange *et al.* computed histograms of output intensities (Figure 5.19) and of switching delays (Figure 6.12b of Chapter 6 of this volume), which are in good qualitative agreement with the experimental results. In particular, the double-peaked distribution is obtained, as well as the asymmetric distribution of τ_D shifting to shorter times with increasing noise level. They could also obtain distributions of τ_D for very low noise levels that were not accessible in the experiment. They found nearly symmetric distributions, with a width increasing with the noise level. It seems from these analyses that the long tails observed in the experimental distribution of τ_D are related to colored noise distributions, like, e.g., uncontrolled low-frequency fluctuations of acoustic origin. The Lange group have also found similar results in the dynamic behavior of a bistable electric circuit (Mitschke *et al.*, 1985, and see next chapter).

An important difference appears when the LSA measurements are compared to the theory of Broggi and Lugiato or to the electronic circuit observations of the Lange group. The noise level required in the LSA experiment to modify the double-peak structure of the probability distribution is much larger than that considered in that theory or applied in that experiment. An explanation may be that the noise is not purely additive and has a multiplicative component. However, the response of the LSA may be interpreted only on the basis of a proper LSA theoretical treatment that is not available at the present time.

5.3 Noise and chaos

5.3.1 Introduction

Many optical systems present a period-doubling route to chaos, e.g. the hybrid optical-bistable devices with delayed feed-back (e.g. Chrostowski, Vallee and Delisle, 1983; Derstine, Gibbs, Hopf and Kaplan, 1982), the LIMs (Arecchi *et al.*, 1982; Midavaine, Dangoisse and Glorieux, 1985), the I_2 laser, etc. The first ones are similar to the hybrid optical-bistable devices presented in Section 5.2.1, with a modification suggested by Ikeda (Ikeda, Daido and Akimoto, 1980), in the aim to produce instabilities (Figure 5.20a). LIMs may be compared to the active devices presented in the previous section, but the modulation applied to the intracavity crystal produces a phenomenon of GB (Figure 5.20b) (Arecchi *et al.*, 1982; Dangoisse, Glorieux and Hennequin, 1986a). The difference between the two systems is more serious than it appears at first glance, since in the case of passive systems the delay introduces an infinite number of degrees of freedom (Y. Pomeau, 1986, private communication). On the other hand, the number of degrees of freedom of the LIM, whence the dimension of the phase space associated to the system is very low, and the reconstruction of the attractor associated to the regime is possible (Dangoisse, Glorieux and Hennequin, 1987; Puccioni *et al.*, 1985). In this last case, it has also been possible to show the simultaneous existence of several attractors producing GB (Dangoisse, Glorieux and Hennequin, 1987).

The different attractors appearing in the two cases present the same evolution, described by the period-doubling route from a periodic regime to a chaotic one. The term 'chaotic' refers to deterministic chaos, i.e. a deterministic regime with a very low correlation time, leading to a continuous spectrum (see, e.g., Berge, Pomeau and Vidal, 1984; or Schuster, 1984). The characterization of such a regime – and its differentiation from noise – may be done by the study of its onset. Many routes to chaos have been observed (Eckmann, 1981), and

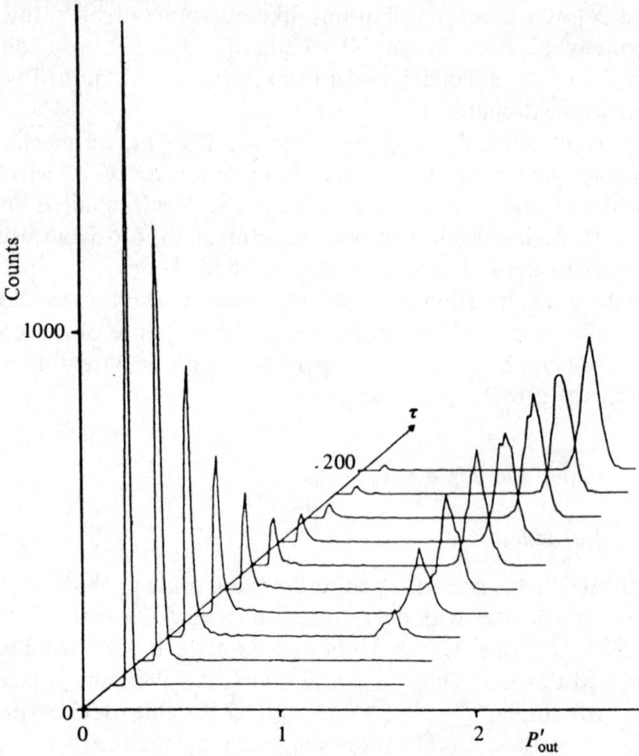

Figure 5.19. Calculated histograms of the transmitted light P'_{out} for different τ after the switch-on of the light in the Hannover experiment. Root mean square noise level $\sigma = 0.02$. From Lange *et al.* (1985).

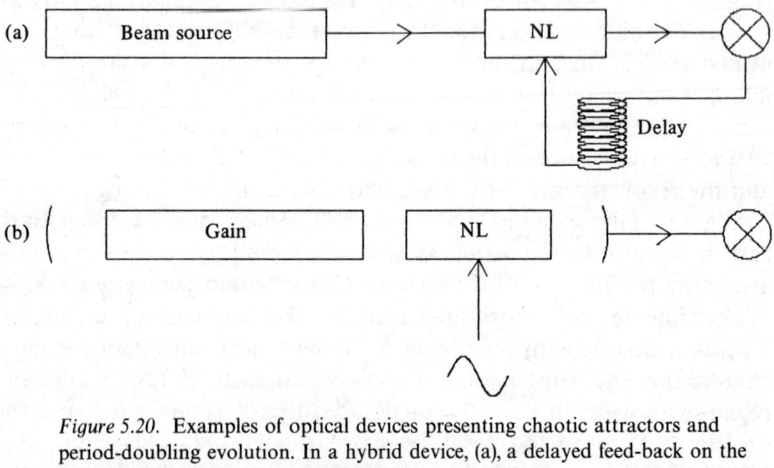

Figure 5.20. Examples of optical devices presenting chaotic attractors and period-doubling evolution. In a hybrid device, (a), a delayed feed-back on the non-linear element produces instabilities. In the LIM, (b), a modulation applied inside the cavity produces instabilities.

the one studied here is the 'Feigenbaum' route, which occurs, for instance, in the logistic map (see Figure 5.21) (Feigenbaum, 1979, 1980). The regime of the system, which is at first T-periodic, undergoes a succession of period-doubling bifurcations, and becomes $2T$-periodic, then $4T$, $8T$, and so on. Finally, the regime becomes chaotic, exhibiting the inverse cascade: the chaotic regime is similar to a noisy nT-periodic regime, with n decreasing from infinity to 8, 4, 2, and, finally, the regime loses any periodicity and becomes fully chaotic.

The influence of fluctuations may be studied in different ways. One can first study the influence of noise in the stationary regime, i.e. the noise-induced changes in the attractor. The noise may alter the route to chaos and the interaction between different coexisting attractors. Eventually, noise may also give rise to new effects in the transient, just as it induced TB in OB.

Concerning the chaos itself, important modifications of its characteristics by adding noise have never been predicted or experimentally observed, except in a recent theoretical study of R. Lefever (1987, private communication) on the Brusselator, which is a model for an oscillatory chemical reaction exhibiting the period-doubling route to chaos (Kai, 1981). In this model, it appears that the addition of noise stabilizes states that are non-existent in the regime without noise. To our knowledge, there has not been any experimental observation of this feature.

We will concentrate here on the influence of noise on the period-doubling sequence. Crutchfield and Huberman (1980) have predicted a bifurcation gap: the higher order regimes disappear in presence of noise. The experimental observation of such an effect occurred before its prediction: no more than five bifurcations have ever been experimentally observed before the chaotic regime (i.e. no regime with period higher than 32 times the basic period). However, the original work taking noise into consideration came from the theoretical side. The first experimental study of this feature was effected on a non-linear semiconductor oscillator (Perez and Jeffries, 1982). Other studies have been done on the properties of the bifurcations themselves by additive or multiplicative noise, using various models (e.g., Chrostowski, 1982; Linz and Lücke, 1986; Napiorkowski, 1985; Neumann, Koch, Schmidt and Haug, 1984). But the universality of the period-doubling scenario is such that there is no qualitative difference between the results, and it seems that a result obtained in a particular case may be expected to be valid for a very large set of systems. More recently, Linz and Lücke (1986) have been interested in the influence of noise on the first bifurcations of the period-doubling scenario, but there has been no experimental evidence of these properties.

Concerning the interaction between two attractors, Arecchi *et al.* (1984a, b) have demonstrated that noise may induce transitions from one attractor to the other, a phenomenon equivalent to steady-state bimodality in OB. These so-called noise-induced crises are at the present time under experimental investigation.

In the transient, the regime is the transient chaos: the trajectory remains on

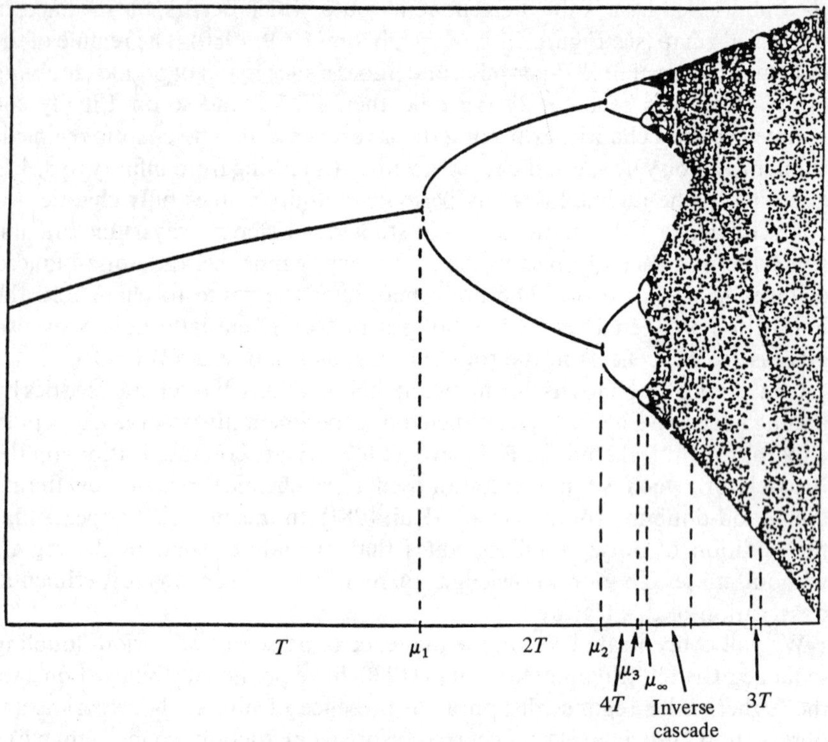

Figure 5.21. Bifurcation diagram of the logistic map. The asymptotic solutions of the logistic map $x_{n+1} = 4\mu x_n(1 - x_n)$, $x \in [0, 1]$, are represented as a function of the value of the control parameter μ for $0.5 \leqslant \mu \leqslant 1$. For $\mu < \mu_1$ ($\mu_1 = 0.75$), there is only one asymptotic solution. At $\mu = \mu_1$, a bifurcation occurs, after which the single limit point is no longer stable; rather there are two values which the system visits alternately. Thus the system returns to either one of them in a time that is twice as long as before the bifurcation: the period of the system has doubled. A second period-doubling bifurcation arises at $\mu_2 = 0.862$, after which the period is $4T$. At $\mu_3 = 0.886$, the system passes from the $4T$ regime to the $8T$ regime. There is an infinite series of values μ_n for which the regime passes from a $2^{n-1}T$-periodic regime to a $2^n T$ one. The difference $\mu_{n+1} - \mu_n$ is not constant, but is reduced each time by a factor 4.669. So there is a μ_∞, convergence point of the period-doubling cascade, where the regime is $2^n T$-periodic, where n tends to infinity. Above μ_∞, the system becomes chaotic, but the asymptotic solutions remain situated inside 2^n intervals, n decreasing from infinity to 2 when one moves away from the accumulation point. This evolution is called the inverse cascade. Inside the chaotic region there are periodic windows, where the system describes the period-doubling cascade from a period multiple of T. The $3T$ window is indicated on the figure; there also exists an infinity of other windows, narrower than that at $3T$ and invisible at this scale, which compose the 'universal sequence'.

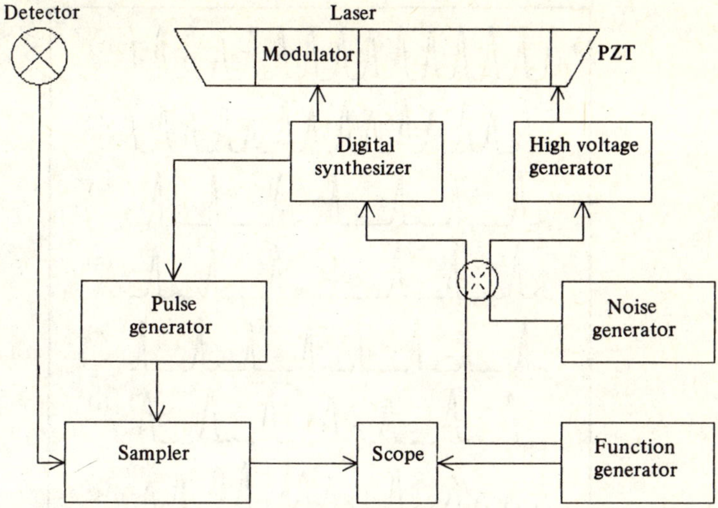

Figure 5.22. Set-up of the laser with internal modulation.

the first attractor until it reaches its border, where an unstable cycle ejects the trajectory on the second attractor. At the present time, neither theoretical nor experimental work has studied the influence of noise on this behavior.

We present here a review of the present status of the experiments realized on the influence of noise on the Feigenbaum route to chaos in the optical systems.

5.3.2 *Experimental arrangement – steady-state behavior*

The experimental device consists essentially of a sealed-off waveguide CO_2 laser, in the cavity of which a ZnSe crystal is inserted (Figure 5.22). This device has been described in detail elsewhere (Dangoisse, Glorieux and Hennequin, 1987). The ZnSe crystal has elasto-optic properties: the application of a stress changes its optical properties, particularly its refractive index and optical axes. That results firstly in a change in the optical length of the cavity, and secondly in a change of the losses in the cavity (Midavaine, Dangoisse and Glorieux, 1985). Therefore, modulating the squeeze on a crystal placed inside the laser cavity produces a modulation of the amplitude (AM) or of the frequency (FM) of the laser output. The experiments described here correspond to modulation of the losses at a frequency of 330 kHz. The ratio between the mean output power of the laser for central tuning (maximum emission) and the internal noise is in excess of 60 dB.

The Feigenbaum trees have been observed using two different control parameters: (i) the modulation amplitude of the constraint, and (ii) the laser frequency, i.e. the laser cavity detuning. The results presented here used the

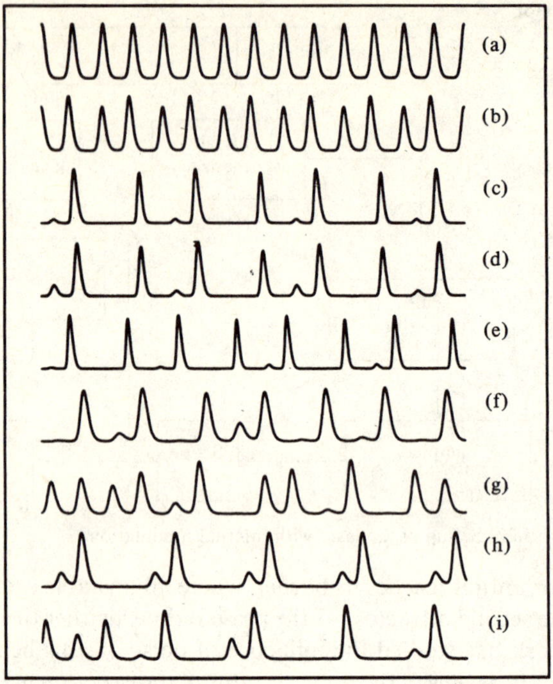

Figure 5.23. Period-doubling sequence in the LIM. On increasing the modulation amplitude the laser response is first T-periodic, (a), then successively (b) $2T$-periodic, (c) $4T$-periodic, (d) $8T$-periodic, (e) $4T$-chaotic, (f) $2T$-chaotic, (g) fully chaotic, (h) $3T$-periodic, (i) $3T$-chaotic.

modulation amplitude V as control parameter. In this case, on increasing V the laser exhibits the period-doubling route to chaos: without modulation, the emission is cw. For a weak modulation, the laser responds linearly (Figure 5.23a), i.e. the laser output intensity is a time sinusoid with the same period T of the excitation. On increasing V, the response loses its linearity but remains periodic, with period $2T$, i.e. with period twice that of the excitation one (Figure 5.23b). Successive period-doubling bifurcations arise (Figure 5.23c, d) before the laser response loses all its periodicity, and the regime becomes chaotic (Figure 5.23e). On increasing V again, the system remains chaotic (Figure 5.23f, g, i), sometimes with windows where a new periodic regime appears, of period, e.g., $3T$ (Figure 5.23h), $5T$, ...

To describe the complete evolution of the laser regime as a function of the control parameter, a stroboscopic method is used; this allows us to obtain bifurcation diagrams in real time on an oscilloscope. A sample and hold module is synchronized on the modulation signal and stores a signal proportional to the laser output intensity at some definite time of the modulation sinusoid. As it is synchronized on the modulation, if the response of the system is T periodic the sampler delivers a single value output. In the

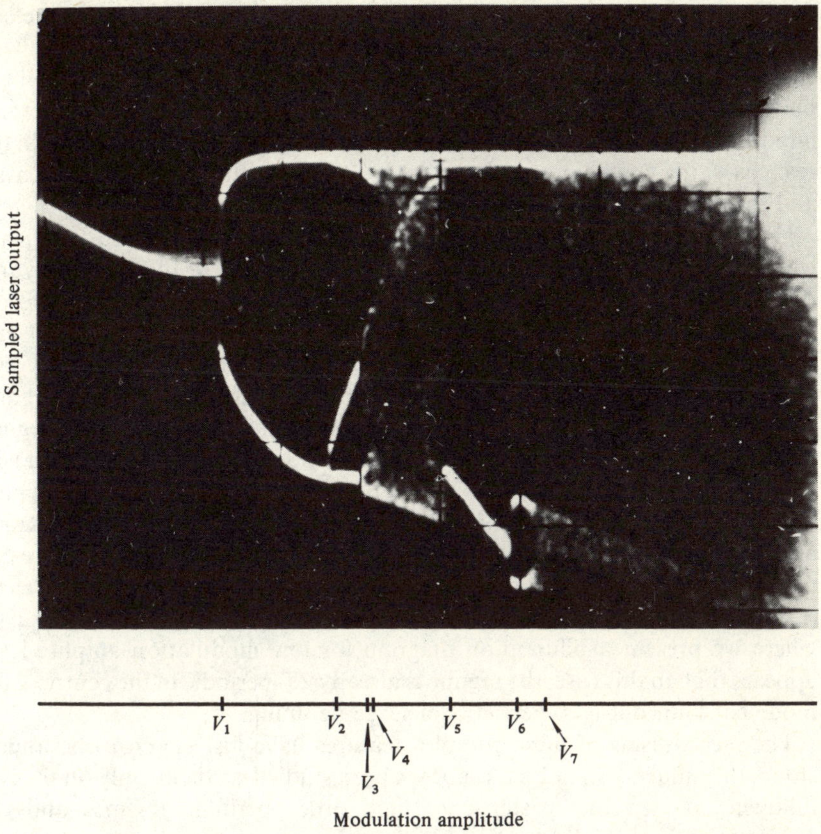

Figure 5.24. Bifurcation diagram of the LIM with the modulation amplitude of the crystal as control parameter. The regime is T-periodic below V_1, then $2T$ until V_2, $4T$ from V_2 to V_3, $8T$ between V_3 and V_4. Above V_4, it becomes chaotic until V_5 where a new periodic regime appears, with period $3T$. Then, the regime is $6T$-periodic between V_6 and V_7, and then again becomes chaotic.

case of an nT-periodic response, n different values are successively available at the sampler output. When the laser is chaotic, any periodicity in the sampler is lost, and a series of apparently random values which reflect the statistics of chaos are obtained. A display of the sampler output as a function of the modulation amplitude produces the bifurcation diagram. The example given in Figure 5.24 shows the very large analogy of the laser behavior with the results of the logistic map.

These results are similar to those obtained using hybrid optical devices (e.g. Chrostowski, Vallee and Delisle, 1983; Derstine *et al.*, 1982). A typical hybrid optical arrangement has been shown in Figure 5.20(a). The delay may be optical or electrical. Derstine *et al.* (1982) introduce an optical delay: the beam crossing an electro-optic crystal passes through a very long optical fibre, and then is detected by a photomultiplier, whose output is amplified and impressed on the crystal. On the other hand, Vallee, Delisle and Chrostowski (1984) have chosen an electrical device: the beam crossing a Bragg cell is detected by a photodiode, whose output, after being delayed by passage through several hundred metres of coaxial cable, is then returned to the Bragg cell driver. In the two cases, the control parameter is proportional to the input laser intensity and the feed-back amplification.

However, in the LIM the bifurcation diagrams also reveal some divergence with the logistic map which have not been shown up in the case of hybrid devices. It appears through the GB that different attractors coexist (Arecchi *et al.*, 1982; Arecchi, 1986; Dangoisse, Glorieux and Hennequin, 1986a), and collisions between these attractors lead to complex phenomena (Dangoisse, Glorieux and Hennequin, 1986b). These so-called 'crises' (Grebogi, Ott and Yorke, 1982, 1983) appear as, for example, the destruction of a chaotic regime replaced by a $3T$- or $5T$-periodic one, or also the sudden appearance of a new chaotic regime superimposed on the first one (Dangoisse, Glorieux and Hennequin, 1987). Here, it must be remarked that the laser cavity detuning and the mean laser losses play an important role in these different features; in particular, the modulation may become less efficient in certain cases (Dangoisse, Glorieux and Hennequin, 1987). This is seen in Figure 5.29(a) where we present a bifurcation diagram for low modulation amplitude: it appears that, in this case, the regime is already $2T$-periodic in the centre of the mode contour, but is T-periodic for large detuning.

The mechanisms of these complex features have just begun to be understood; the influence of noise has not yet been studied on them, only on the two following points: the existence of high order periodic regimes and the characteristics of the bifurcations themselves.

Two procedures have been used: the increasing of the relative level of internal noise by modifying the laser conditions, such as gain, detuning, etc., and the superimposition of external noise on the control parameter using a Gaussian noise generator as was presented in the first part of this chapter.

The corresponding experimental results are presented in the next section, as are the ones obtained by other groups and other devices.

5.3.3 *Influence of noise on chaotic attractors*

The succession of period-doubling bifurcations is theoretically infinite, although the bifurcations are closer and closer when the control parameter μ is increased. Experimentally, the maximum period observed is $16T$ in the case of a LIM, and often only $8T$ or $4T$ with other systems (e.g., Gibbs, Hopf, Kaplan and Shoemaker, 1981; Hopf *et al.*, 1982; and in non-optical systems, e.g., Lauterborn and Cramer, 1981; Linsay, 1981). The higher order chaotic regimes are not observed in the stationary regime. Crutchfield and Huberman have demonstrated that such a gap may originate from the internal noise of the system. In the case of the LIM, the relative amount of internal noise may be increased by modifying some laser parameters, in particular the cavity detuning. Thus, going from central tuning to the edge of the mode, i.e. close to threshold, we observe the progressive disparition of the higher periodic regimes, until just one bifurcation remains (Figure 5.25b). In the extreme case, the regime goes directly from a T-periodic regime to chaos (Figure 5.25c).

More complete studies have been done by Derstine *et al.* (1982) on their system (see the previous section), where the internal noise originates from the shot noise of the photomultiplier, which may be modified in amplitude using an attenuation of the light that reaches it, then by increasing the dynode's voltage to return the signal to its previous level. In this experiment, the Feigenbaum evolution is observed with a gap after three period-doublings. To determine the bifurcation points in the presence of noise, the power spectra of the waveforms were observed in real time on a spectrum analyzer. Figure 5.26 summarizes the experimental spectra as a plot of the power of the $2T$, $4T$ and $8T$ frequency peaks and the chaotic background as a function of the control parameter μ. Figure 5.27 shows the evolution of the bifurcation points versus the noise level. A new feature appears here: it is clear from Figure 5.27 that more noise is needed to eliminate a chaotic waveform than is needed to eliminate the corresponding periodic one. The bifurcation gap presented by Crutchfield and Huberman was symmetric: the periodic and chaotic waveforms were eliminated at the same noise level. Two differences must be emphasized: they concern first the nature of the noise which is multiplicative for Derstine *et al.* and additive for Crutchfield and Huberman, and secondly the nature of systems themselves. It seems that asymmetry does not come from the nature of noise: the studies of Vallee, Delisle and Chrostowski (1984) have revealed the same behavior for both additive and multiplicative noise (Figure 5.28). However, their system is described by delay-differential equations instead of the non-linear differential equations of Crutchfield and Huberman.

The effect of noise can be understood as a kind of dynamical average of the structure of attractors over a range of nearby parameters. According to this view, the averaging of a periodic orbit with an adjacent chaotic orbit tends to lower the transition to chaos. On the other hand, in the case of the transition from a periodic orbit to the next one, the averaging does not produce a shift of the bifurcation. This is illustrated in Figure 5.29 in a simple bifurcation diagram, with just one bifurcation, where the laser cavity detuning is used as the control parameter. The whole mode of the LIM is swept, and it appears very clearly that the bifurcation point remains globally unchanged. Another bifurcation is given in Figure 5.30 in the case of the Vallee, Delisle and Chrostowski (1984) experiments.

Figure 5.31 permits to analyse with more details the influence of noise on the first $T \to 2T$ bifurcation in the case of the LIM (Dangoisse, Glorieux and Hennequin, 1986a). It appears in Figure 5.31(a) that a small noise level has more influence on the $2T$ part of the bifurcation, as if the laser responded

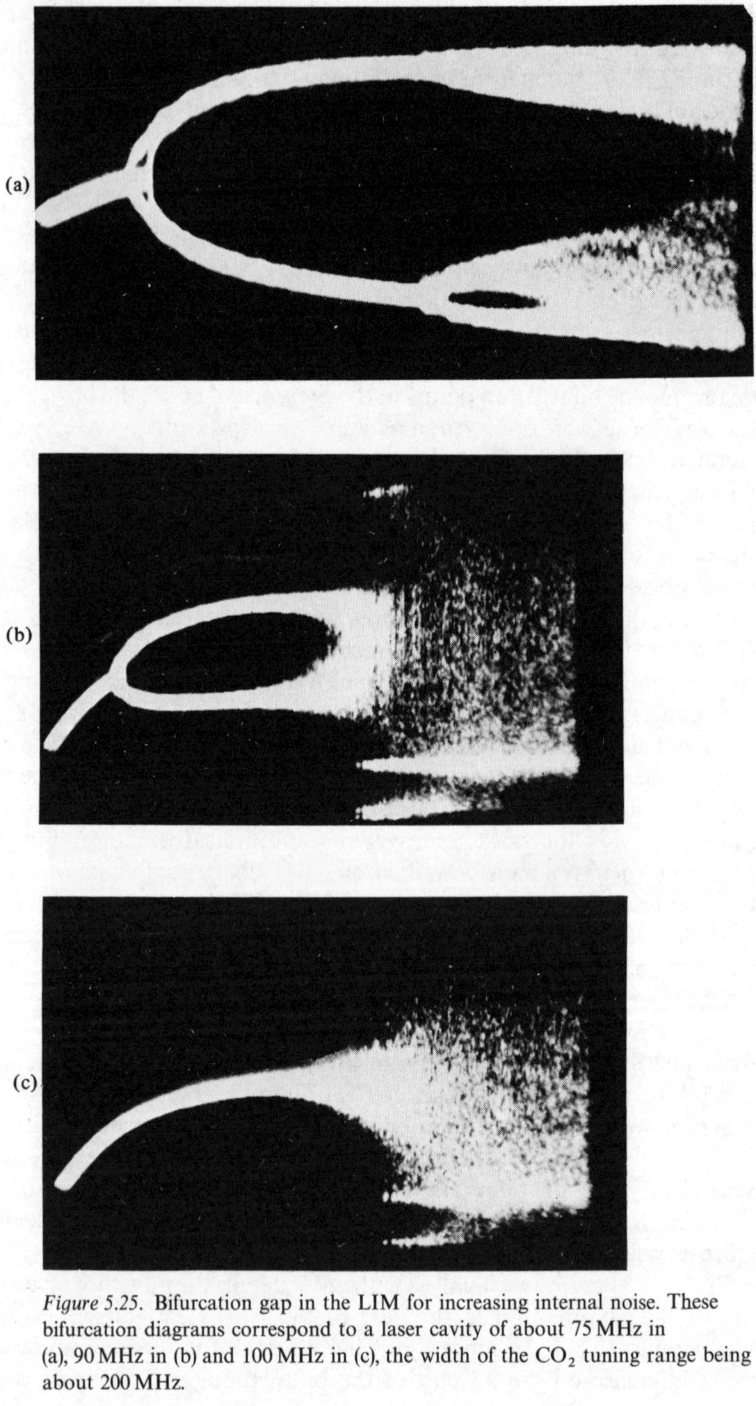

Figure 5.25. Bifurcation gap in the LIM for increasing internal noise. These bifurcation diagrams correspond to a laser cavity of about 75 MHz in (a), 90 MHz in (b) and 100 MHz in (c), the width of the CO_2 tuning range being about 200 MHz.

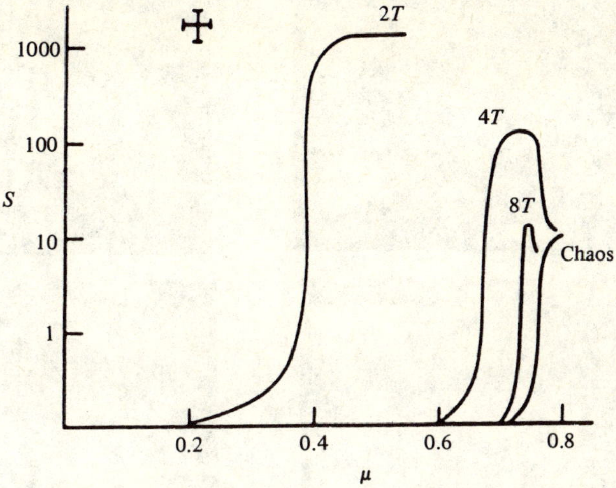

Figure 5.26. Behavior of the Derstine *et al.* hybrid device. S, the log of the power density of the powerful spectral components of period $2T$, $4T$, $8T$ and chaos, is represented as a function of the control parameter μ. This figure was obtained for a 0.3% noise level. From Derstine *et al.* (1982).

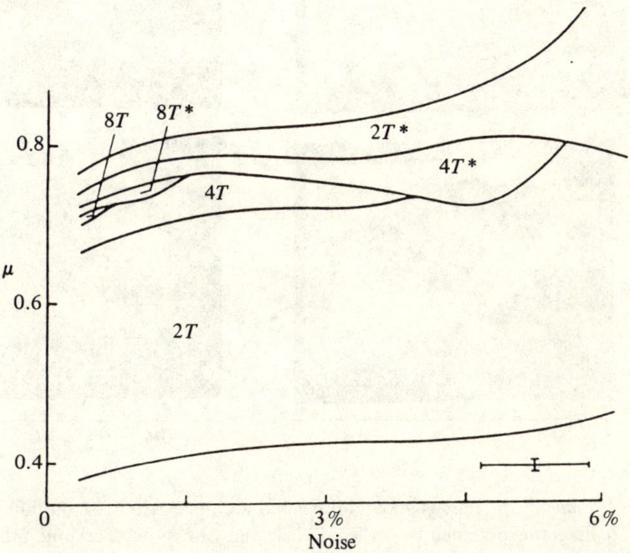

Figure 5.27. Bifurcation gap in a hybrid device for increasing internal noise. The lines are the bifurcation points between different waveforms. The numbers denote the period of the waveform. The starred numbers denote chaotic waveforms. From Derstine *et al.* (1982).

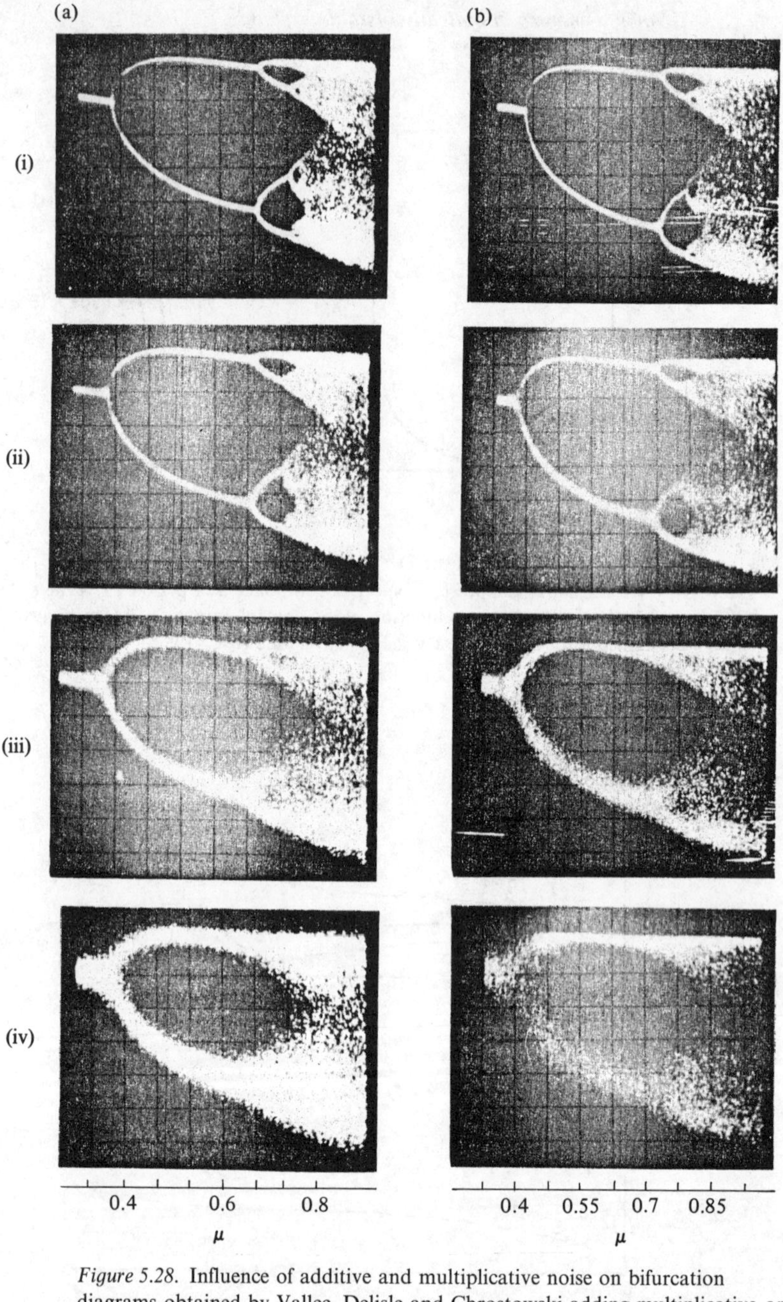

Figure 5.28. Influence of additive and multiplicative noise on bifurcation diagrams obtained by Vallee, Delisle and Chrostowski adding multiplicative or additive noise in their system. (a) shows the effect of additive noise for different amplitudes: (i) 0.0004, (ii) 0.001, (iii) 0.005, (iv) 0.008. (b) shows the effect of multiplicative noise for different amplitudes: (i) 0, (ii) 0.01, (iii) 0.04, (iv) 0.15. A residual, mainly additive, noise was always present with an amplitude of 0.0004. From Vallee, Delisle and Chrostowski (1984).

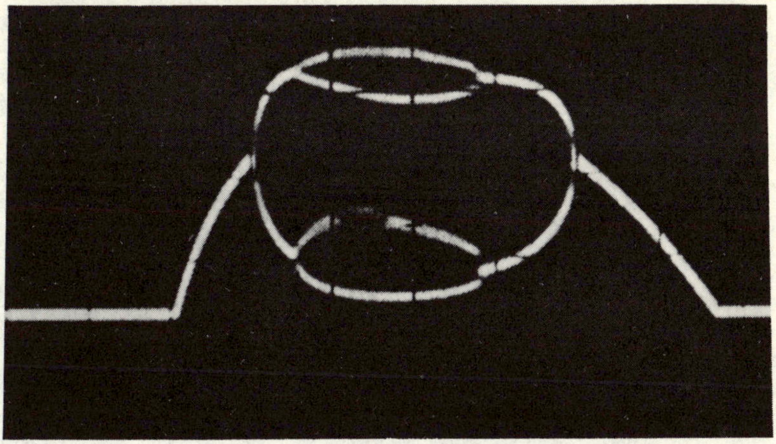

(a)

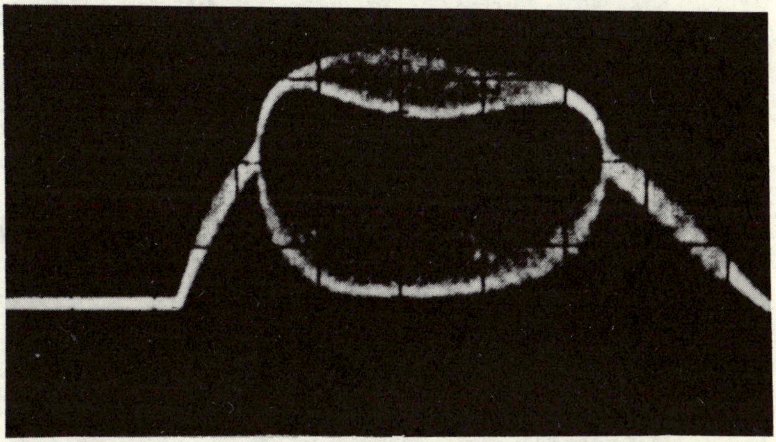

(b)

Figure 5.29. Influence of noise on a bifurcation diagram with laser cavity detuning as control parameter. (a) Without external noise. (b) With external noise. From Dangoisse, Glorieux and Hennequin (1986a).

adiabatically to the noise-induced modulation. But, with increasing noise level, this asymmetry disappears (Figure 5.31b), and another arises: in Figure 5.31(c), it becomes evident that, close to the bifurcation point, the upper branch is much less affected than the lower one.

All the results described here demonstrate that the chaotic attractors appear to be very stable, and that the noise has practically no effect on these regimes.

We present now the numerical simulations associated with these experimental studies.

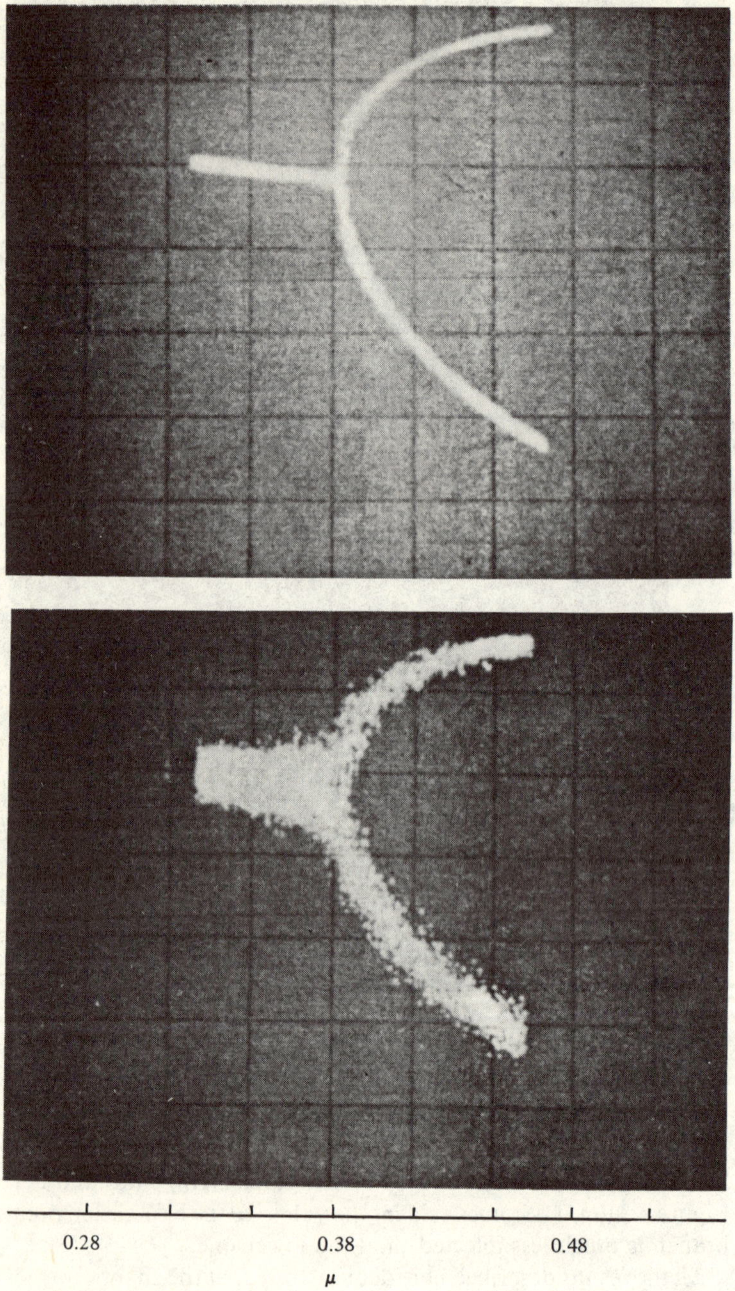

Figure 5.30. Influence of noise on the $T \rightarrow 2T$ bifurcation location in the case of the Vallee, Delisle and Chrostowski experiment. (a) shows the bifurcation without external noise, and (b) shows it with a multiplicative noise of amplitude 0.03. From Vallee, Delisle and Chrostowski (1984).

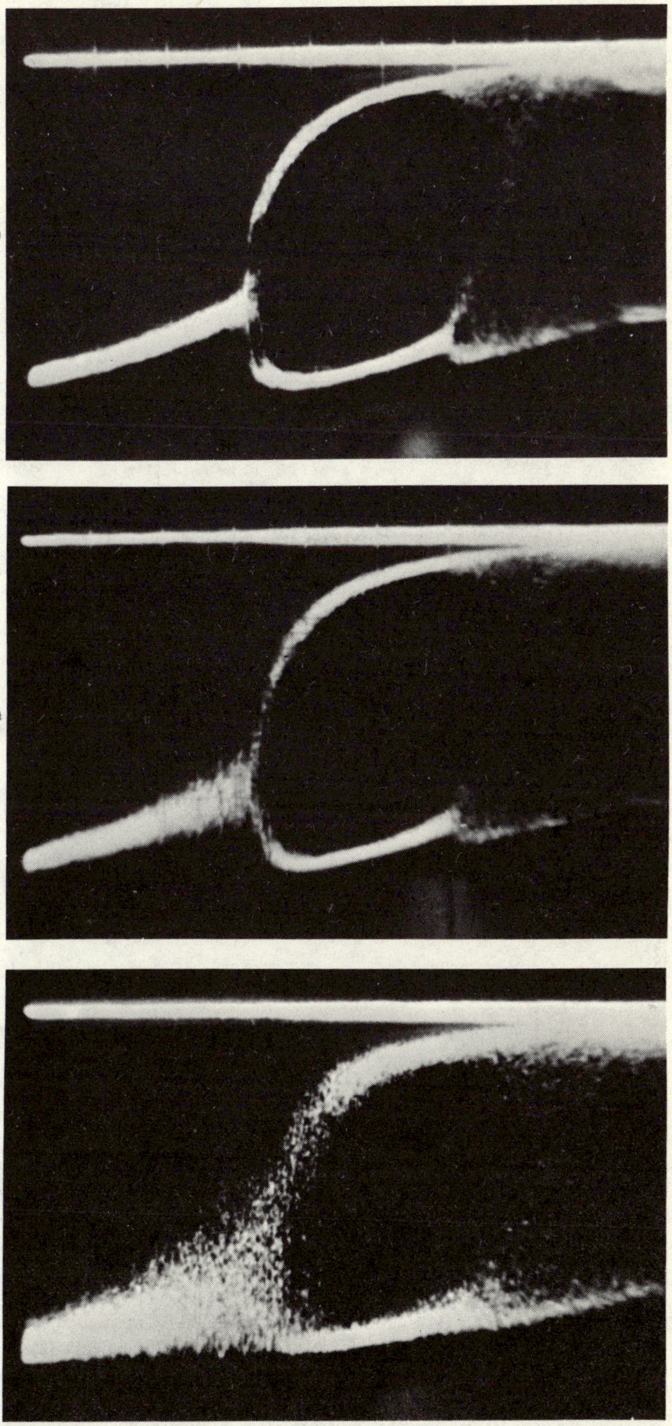

(a)

(b)

(c)

Figure 5.31. Effects of noise on a regime close to the $T \to 2T$ bifurcation. (a) Without external noise. (b) and (c) With external noise. From Dangoisse, Glorieux and Hennequin (1986a).

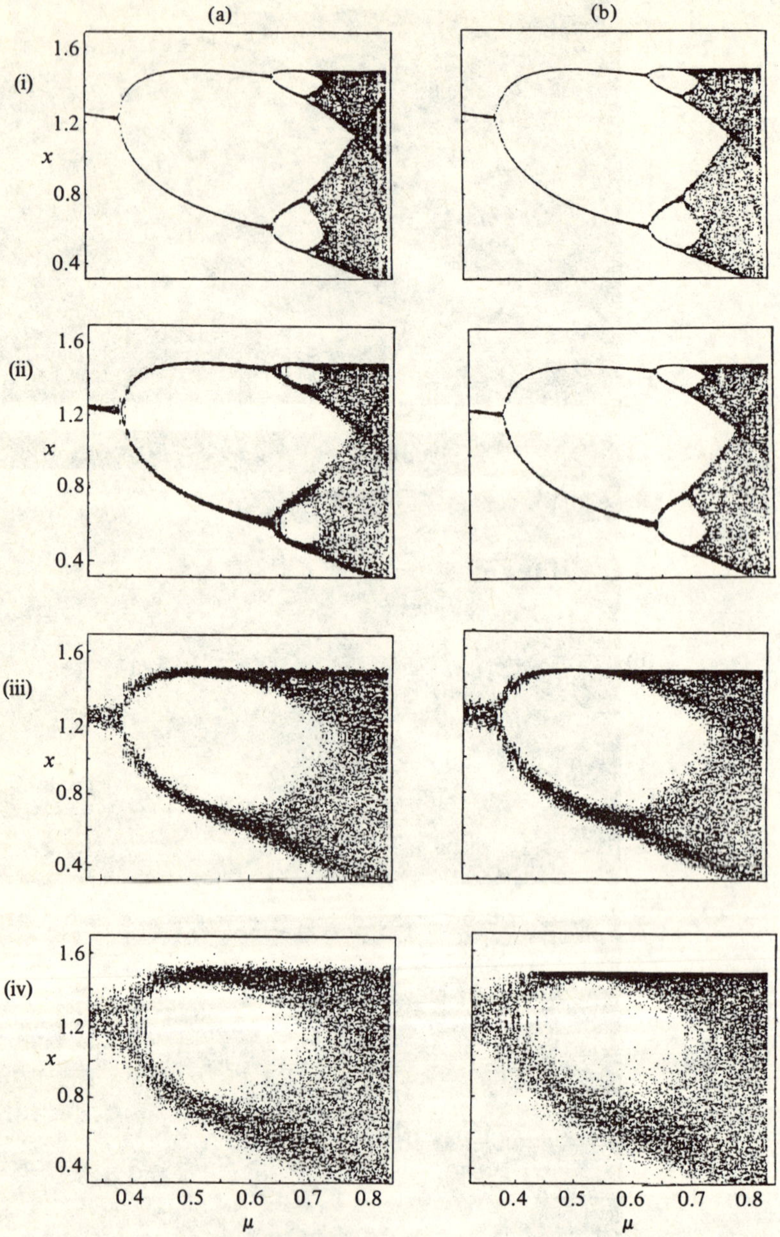

Figure 5.32. Bifurcation diagrams obtained by numerical simulations in the case of the Vallee, Delisle and Chrostowski experiments. (a) Effect of additive noise with amplitude of (i) 0.0003, (ii) 0.0011, (iii) 0.0045, (iv) 0.0083. (b) Effect of multiplicative noise with amplitude of (i) 0.0015, (ii) 0.0025, (iii) 0.02, (iv) 0.08. From Vallee, Delisle and Chrostowski (1984).

5.3.4 Numerical simulations

As we discussed in the case of OB, the purpose of this chapter is not to detail the different theories concerning laser instabilities, but to present the results of simulations made by the experimentalists, just mentioning the models used.

Since the first works of Crutchfield and Huberman (1980), the effect of both additive and multiplicative noise in both colored and white noise approximations has been extensively studied in many systems (e.g., Chrostowski, 1982; Crutchfield, Farmer and Huberman, 1982; Geisel, Heldstab and Thomas, 1984; Heldstab, Thomas, Geisel and Radons, 1982; Linz and Lücke, 1986; Napiorkowski, 1985; and see also relevant chapters of these volumes). However, the numerical simulations of models leading to semi-quantitative comparisons with the experimental results remain scarce.

In fact, Vallee, Delisle and Chrostowski (1984) are the only ones to have performed numerical simulations of their experiments. They have analyzed a noisy difference equation which describes with a crude approximation, their system. These results are presented in Figure 5.32, which have to be compared with the experimental ones of Figure 5.28. A very good qualitative agreement is obtained, particularly in the case of additive noise. The behavior of the system corresponds to the one observed experimentally: (i) disparition of higher order periodic regimes before the higher chaotic ones – this is not the case for the logistic map; (ii) very weak influence on the chaotic regime; and (iii) no shift of the periodic bifurcations.

5.4 Conclusion

The studies presented above describe the influence of external noise or high level internal noise on optically bistable systems. The importance of low-level internal noise on the dynamics of the systems may be deduced from these studies. Thus, it seems that the attractors themselves – a point in OB and a periodic or strange attractor in GB – are not very influenced by noise. In the case of LSA, a high-level noise is necessary to produce an effect on the cw regime. In GB, we have seen that the attractors appear to be very stable. On the other hand, the transient behavior is very sensitive to noise, even at very weak levels. Therefore, noise appears as a decisive element in the transient behavior of optical-bistable systems.

However, the influence of nature and properties of noise have not been extensively studied experimentally, neither in OB nor in GB, although some theoretical studies have shown the importance of, e.g., the frequency distribution or the correlation function of the applied noise. These experimental studies should represent the following step in the experimental investigations.

Concerning OB, the above experiments have always used a noise applied to the control parameter itself. In practice, fluctuations appear on the other parameters, and it would be interesting to investigate them. In fact, this is linked to the previous point, since the used parameter determines the nature – multiplicative or additive – of the noise.

Concerning GB, the measurements have been performed only on the attractors themselves. Experiments in the future will concern the interaction between the attractors in both the steady-state (noise-induced crises) and the transient regimes. For these two points, the results acquired in the OB case lead us to predict a very large influence of the applied noise.

In many cases, simple models permit the description of what happens in the systems. This is not so in the case of the LSA, and simulations have to be performed on the full system of equations. It would be interesting to locate the origin of the relatively weak response of the LSA to the noise. The LIM is described by a very simple and effective model, and simulations are being carried out at the present time.

Acknowledgements

The research in Pisa was carried out in the framework of the European Joint Optical Bistability Project, an operation launched by the Commission of the European Communities under the experimental phase of European Community Stimulation Action.

Concerning the Lille group, the financial support of D.R.E.T. and La Région Nord Pas de Calais are gratefully acknowledged. Experiments were made possible by the loan of materials from the Société Anonyme de Télécommunications.

References

Arecchi, F. T. 1986. In *Optical Chaos* (J. Chrostowski and N. B. Abraham, eds.). *Proc. SPIE* **667**, 139. Bellingham: SPIE.

Arecchi, F. T., Badii, R. and Politi, A. 1984a. *Phys. Lett.* **103A**, 3.

Arecchi, F. T., Badii, R. and Politi, A. 1984b. *Phys. Rev. A* **29**, 1006.

Arecchi, F. T., Badii, R. and Politi, A. 1985. *Phys. Rev. A* **32**, 402.

Arecchi, F. T. and De Giorgio, V. 1971. *Phys. Rev. A* **3**, 1108.

Arecchi, F. T., Meucci, R., Puccioni, G. P. and Tredicce, J. R. 1982. *Phys. Rev. Lett.* **49**, 1217.

Arimondo, E., Bootz, P., Glorieux, P. and Menchi, E. 1985. *JOSA B* **2**, 193.

Arimondo, E., Casagrande, F., Lugiato, L. A. and Glorieux, P. 1983. *Appl. Phys. B* **30**, 57.

Arimondo, E., Dangoisse, D. and Fronzoni, L. 1987. *Europhys. Lett.* **4**, 287.

Arimondo, E., Gabbanini, C., Menchi, E. and Zambon, B. 1986. In *Optical Chaos* (J. Chrostowski and N. B. Abraham, eds.). *Proc. SPIE* **667**, 234. Bellingham: SPIE.

Arimondo, E. and Menchi, E. 1985. *Appl. Phys. B* **37**, 55.

Baras, F., Nicolis, G., Malek Mansour, M. and Turner, J. W. 1983. *J. Stat. Phys.* **32**, 1.

Berge, P., Pomeau, Y. and Vidal, C. 1984. *L'ordre Dans le Chaos*. Paris: Hermann.

Bonifacio, R. and Lugiato, L. A. 1978. *Phys. Rev. Lett.* **40**, 1023.

Broggi, G. and Lugiato, L. A. 1984. *Phys. Rev. A* **29**, 2949.

Broggi, G., Lugiato, L. A. and Colombo, A. 1985. *Phys. Rev. A* **32**, 2803.

Chrostowski, J. 1982. *Phys. Rev. A* **26**, 3023.

Chrostowski, J., Vallee, R. and Delisle, C. 1983. *Can. J. Phys.* **61**, 1143.

Chung, H. K., Lee, J. B. and DeTemple, T. A. 1981. *Opt. Comm.* **39**, 105.

Crutchfield, J. P., Farmer, J. D. and Huberman, B. A. 1982. *Phys. Rep.* **92**, 45.

Crutchfield, J. P. and Huberman, B. A. 1980. *Phys. Lett.* **77A**, 407.

Dangoisse, D., Glorieux, P. and Hennequin, D. 1986a. In *Optical Chaos* (J. Chrostowski and N. B. Abraham, eds.). *Proc. SPIE* **667**, 242. Bellingham: SPIE.

Dangoisse, D., Glorieux, P. and Hennequin, D. 1986b. *Phys. Rev. Lett.* **57**, 2657.

Dangoisse, D., Glorieux, P. and Midavaine, T. 1986. In *Optical Instabilities* (R. W. Boyd, M. G. Raymer, L. M. Narducci, eds.), p. 293. Cambridge University Press.

Dangoisse, D., Glorieux, P. and Hennequin D. 1987. *Phys. Rev. A* **36**, 4775.

Derstine, M. W., Gibbs, H. M., Hopf, F. A. and Kaplan, D. L. 1982. *Phys. Rev. A* **26**, 3720.

Eckmann, J. P. 1981. *Rev. Mod. Phys.* **53**, 643.

Feigenbaum, M. J. 1979. *J. Stat. Phys.* **21**, 669.

Feigenbaum, M. J. 1980. *Los Alamos Science* **1**, 4.

Frankowitz, M., Malek Mansour, M. and Nicolis, G. 1984. *Physica* **125A**, 237.

Francowitz, M. and Nicolis, G. 1983. *J. Stat. Phys.* **32**, 595.

Geisel, T., Heldstab, J. and Thomas, H. 1984. *Z. Phys. B* **55**, 165.

Gibbs, H. M. 1985. *Optical Bistability: Controlling Light by Light.* NY: Academic Press.

Gibbs, H. M., Hopf, F. A., Kaplan, D. L. and Shoemaker, R. L. 1981. *Phys. Rev. Lett.* **46**, 474.

Gioggia, R. S. and Abraham, N. B. 1983. *Phys. Rev. Lett.* **51**, 650.

Glass, C., Guevara, M. R. and Shrier, A. 1983. *Physica* **7D**, 89.

Glorieux, P. and Dangoisse, D. 1987. In *Handbook of Molecular Lasers, Vol I* (P. K. Cheo, ed.). NY: Marcel Dekker.

Grebogi, C., Ott, E. and Yorke, J. A. 1982. *Phys. Rev. Lett.* **48**, 1507.

Grebogi, C., Ott, E. and Yorke, J. A. 1983. *Physica* **7D**, 181.

Harrison, R. G., Firth, W. J., Emshary, C. A. and Al-Saidi, I. A. 1983. *Phys. Rev. Lett.* **51**, 562.

Heldstab, J., Thomas, H., Geisel, T. and Radons, G. 1983. *Z. Phys. B* **50**, 141.

Hernández-Machado, A. and San Miguel, M. 1986. *Phys. Rev. A* **33**, 2481.

Hopf, F. A., Kaplan, D. L., Gibbs, H. M. and Shoemaker, R. L. 1982. *Phys. Rev. A* **25**, 2172.

Ikeda, K., Daido, H. and Akimoto, O. 1980. *Phys. Rev. Lett.* **45**, 709.

Kai, T. 1981. *Phys. Lett.* **86A**, 263.

Lange, W., Mitschke, F., Deserno, R. and Mlynek, J. 1985. *Phys. Rev. A* **32**, 1271.

Lauterborn, W. and Cramer, E. 1981. *Phys. Rev. Lett.* **47**, 1445.

Lefebvre, M., Bootz, P., Dangoisse, D. and Glorieux, P. 1983. In *Laser Spectroscopy VI* (H. P. Weber and W. Luthy, eds.), p. 282. Berlin: Springer-Verlag.

Libchaber, A. and Maurer, J. 1982. *Nonlinear Phenomena at Phase Transitions and Instabilities.* NATO Advanced Study Institute, p. 259. NY: Plenum Press.

Linsay, P. S. 1981. *Phys. Rev. Lett.* **47**, 1349.

Linz, S. J. and Lücke, M. 1986. *Phys. Rev. A* **33**, 2694.
Lugiato, L. A., Colombo, A., Broggi, G. and Horowicz, R. J. 1986. *Phys. Rev. A* **33**, 4469.
McCall, S. L., Gibbs, H. M., Hopf, F. A., Kaplan, D. L. and Ovadia, S. 1982. *Appl. Phys. B* **28**, 99.
McCall, S. L., Ovadia, S., Gibbs, H. M., Hopf, F. A. and Kaplan, D. L. 1985. *IEEE J. Quant. Electron.* **QE21**, 1441.
Midavaine, T., Dangoisse, D. and Glorieux, P. 1985. *Phys. Rev. Lett.* **55**, 1989.
Mitschke, F., Deserno, R., Mlynek, J. and Lange, W. 1985. *IEEE J. Quant. Electron.* **QE21**, 1435.
Moore, S. M. 1985. *Lett. Nuovo Cimento* **44**, 129.
Moore, S. M. 1986. *Phys. Rev. A* **33**, 1091.
Nakatsuka, H., Asaka, S., Itoh, H., Ikeda, K. and Matsuoka, M. 1983. *Phys. Rev. Lett.* **50**, 109.
Napiorkowski, M. 1985. *Phys. Lett.* **112A**, 357.
Neumann, R., Koch, S. W., Schmidt, H. E. and Haug, H. 1984. *Z. Phys. B* **55**, 155.
Perez, J. and Jeffries, C. 1982. *Phys. Rev. B* **26**, 3460.
Puccioni, G. P., Poggi, A., Gadomski, W., Tredicce, J. R. and Arecchi, F. T. 1985. *Phys. Rev. Lett.* **55**, 339.
Roux, J. C., Rossi, A., Bachelart, S. and Vidal, C. 1980. *Phys. Lett.* **77A**, 391.
Schuster, H. G. 1984. *Deterministic Chaos.* Berlin: Physik-Verlag.
Tachikawa, M., Tanii, K., Kajita, M. and Shimizu, T. 1986. *Appl. Phys. B* **39**, 83.
Vallee, R., Delisle, C. and Chrostowski, J. 1984. *Phys. Rev. A* **30**, 336.
Wascat, J., Dangoisse, D., Glorieux, P. and Lefebvre, M. 1983. *IEEE J. Quant. Electron.* **QE-19**, 92.

Note added in proof

Since the manuscript was prepared, some major progress has been made in the areas of LSA and LIM. In particular, period-doubling and chaos have been observed in LSA (D. Dangoisse, A. Bekkali, F. Papoff and P. Glorieux, *Europhys. Lett.*, to be published; D. Hennequin, F. DeTomasi, B. Zambon and E. Arimondo, *Phys. Rev. A* **37**, 2243, 1988). In the latter paper, an analysis of the phase space trajectories has also been presented. Modifications of the LSA chaos by internal or external added noise have been investigated (D. Hennequin, F. DeTomasi, L. Fronzoni, B. Zambon and E. Arimondo, in preparation). In the case of LIM, the interactions between coexisting attractors have been studied theoretically (H. G. Solari, E. Eschenazi, R. Gilmore and J. R. Tredicce, *Opt. Comm.* **64**, 49, 1987; I. B. Schwartz, *Phys. Lett. A* **126**, 411, 1988; I. B. Schwartz, *Phys. Rev. Lett.* **60**, 1359, 1988).

6 Use of an electronic model as a guideline in experiments on transient optical bistability

W. LANGE

6.1 Introduction

The transient behavior of nonlinear optical resonators which display bistability and its reaction to fluctuations of system parameters are of extreme importance in all possible applications of these devices. Moreover, however, studies on these processes may give insight into some more general aspects of the behavior of nonlinear dynamical systems. In fact it has been emphasized in the early stages of the development of the field that optically bistable systems, in a suitable model, may exhibit a *first-order phase transition* (Bonifacio and Lugiato, 1978). Since the threshold behavior of a laser can be described as an example of a *second-order phase transition* (Haken, 1964, 1975; see also Haken, 1977), optical bistability in (passive) nonlinear resonators complements nicely lasers as nonlinear dissipative systems in optics. Both types of optical systems are easily accessible to measurements, and the time scales coming into play are convenient for detailed experimental studies of the dynamics.

It is worth mentioning that there is a close connection between laser operation and optical bistability in nonlinear (passive) resonators (see Lugiato, 1984). The standard single-mode laser model with distributed losses (Haken, 1970; Sargent, Scully and Lamb, 1974) can be generalized to include the effect of an external field that continuously drives the atoms within the optical resonator. The standard model of absorptive bistability, the Bonifacio–Lugiato model, is a special case of this generalized model. It represents the situation of a pumping mechanism being absent. Thus it is not surprising that first-order phase transitions and optical bistability also occur in lasers with injected signal (Chow, Scully and van Stryland, 1975), which may be regarded as 'active' nonlinear resonators. Such behavior also occurs in lasers with saturable absorbers (Scott, Sargent and Cantrell, 1975; see also Chapter 5). This chapter only refers to 'passive' systems, i.e. to resonators filled by a nonlinear medium, whose only energy input is the coherent input light beam (see Figure 6.1a).

Excellent reviews of the theory of optical bistability have been given

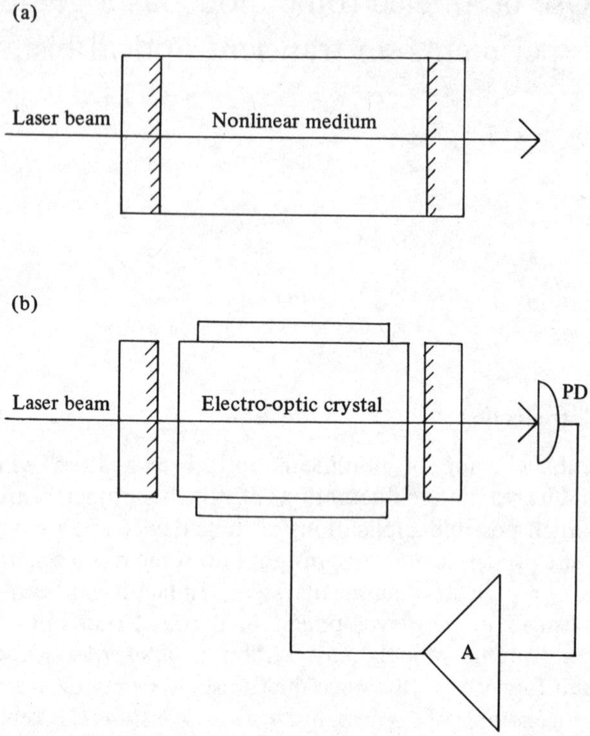

Figure 6.1. Schematic of (a) intrinsic and (b) hybrid optical bistable device. A = amplifier; PD = photodiode. The amplifier can be biased in order to determine the point of operation of the device.

(Carmichael, 1986; Lugiato, 1984) and a monograph on the field is available (Gibbs, 1985). It should not be necessary to give any details here, but it should be mentioned that the models have to be grossly simplified in order to be solvable. For example, the atoms filling the optical resonator are assumed to be two-level atoms; any substructure of the levels is neglected, the transitions are assumed to be homogeneously broadened, and diffusion of the atoms in space and velocity space and diffusion of radiation are not taken into account. In order to arrive at simple equations, variables have to be eliminated adiabatically; in the standard model ('good cavity case') the relaxation of the atomic variables is assumed to be much faster than the energy dissipation of the resonator, and consequently these are eliminated.

In principle, however, the most serious approximation is the description of the system by *ordinary* differential equations. In the standard model a uniform light field is assumed ('mean field approximation'), which can be described by its complex amplitude. In the experiments, however, which are performed by means of laser beams of fairly restricted diameter, some influence of the variation of the electric field in the direction perpendicular to the direction of the laser beam ('transverse effects') is always present, and in extreme cases the experimental results can be severely influenced by self-focusing and -defocusing (for a review see, e.g., Gibbs, 1985). Moreover the propagation of the laser beam in the direction of the axis has to be taken into account. The most obvious effect related to this topic is the formation of a standing-wave light pattern in a linear ('Fabry–Perot') resonator; it can be avoided by using ring resonators, which are assumed in most theoretical treatments, although they are by no means used in all experiments. Even in a ring resonator, however, propagation effects can play a major role. This can be concluded from the results of Ikeda and coworkers, who treat the extreme situation of a resonator with discrete mirrors and a medium whose time constant is smaller than the round-trip time of the resonator (Ikeda, 1979). In this case a simplified theoretical description yields a delay differential equation, and the instabilities typical for nonlinear delay differential equations and carefully studied in control theory can, of course, occur in nonlinear optical resonators.

In the deterministic case the consequences of the approximations and possible improvements of the theoretical description have been carefully studied by many authors (see Gibbs, 1985), even though there are many open questions, for example with respect to transverse effects. In the stochastic case, however, the situation is much worse, and a reasonably complete discussion of the possible consequences of the simplifications of the models does not seem to be feasible at present.

On the other hand, the influence of stochastic processes is by no means trivial, even in the Bonifacio–Lugiato model. Thus, a difference between experimental results and theory might be due either to inadequate solutions in treating the stochasticity or to the use of an inadequate description of the optical system. In other words, the real optical systems are probably too complicated to provide a rigid test of theoretical approaches developed to treat noise in nonlinear dynamical systems, at least in the present state of theory and experiment.

These problems inherent in 'intrinsic' optical bistability along with a lot of experimental problems can largely be avoided by use of so-called 'hybrid' devices.

A well-known example (Smith and Turner, 1977) is shown in Figure 6.1(b). In this scheme the light transmitted by the Fabry–Perot resonator is detected by a photodetector; the output signal is fed to an amplifier, whose output drives an electro-optic crystal, i.e. a device varying the optical path length between the mirrors. The operation of the apparatus is very similar to an intrinsic nonlinear optical resonator operating in the 'dispersive' regime, i.e. to a nonlinear resonator operating on the basis of the saturation of dispersion, instead of the saturation of absorption used in the standard system. The dependence of the optical path length on the intensity within the nonlinear medium is simulated in this case by external electronic means; it can be varied conveniently by choice of the amplifier gain. The important conceptual difference to Figure 6.1(a) is that the length variation depends only on the *total* optical power passing to the photodetector, i.e. the device is not sensitive to any spatial dependence of the intensity within the resonator, regardless of whether in the transverse or longitudinal direction. The temporal response can be varied over a wide range, and it is even possible to introduce a delay in the length variation by electronic (Gibbs, Hopf, Kaplan and Shoemaker, 1981a, b) or optical (Hopf, Derstine, Gibbs and Rushford, 1984) means, thus introducing the possibility of studying Ikeda-type instabilities. It is possible to give equations of motion governing the system very accurately, and these equations are ordinary differential equations in the normal case. By use of a hybrid device, the influence of shot-noise fluctuations has been studied experimentally (McCall *et al.*, 1981, 1985).

Hybrid systems seem to be very attractive at first sight, but it is hard to see the special role played by the optical components in the set-up, and there are serious doubts as to whether any information can be obtained which is not accessible by *purely* electronic means equally well, or probably even better.

From the experimental point of view, one of the big advantages of electronic systems is that an ample supply of different components is available, which are engineered extremely well; the noise figures and all types of drift problems have been optimized for years. From the theoretical point of view, the equations of motion are known very accurately; there is no need to use grossly simplified models in describing the physical system. As long as the frequencies involved in the operation of the electronic devices are restricted to less than a few hundreds of megahertz, propagation problems can safely be neglected in the normal case, i.e. the whole structures can be small with respect to the wavelength of the electromagnetic radiation field, while optical devices are typically much larger than the wavelength.

It should be possible to simulate the simplified model of absorptive or dispersive optical bistability by electronic means, i.e. to construct an electronic 'special purpose analogue computer' solving the corresponding equations of motion. This approach based on electronic simulation can be very successful in nonlinear dynamics (see Chapters 8 and 9 of this volume). It has also successfully been used with respect to stochastic problems in optics (Arecchi, Politi and Ulivi, 1982). However, though electronic circuits are, surely, physical objects, results of electronic simulation tend not to be regarded as 'experimental', thus leaving experiments desirable.

The approach to be described here is somewhat different. In parallel to experiments in an intrinsic optical device, studies in a widely used electronic standard circuit, the 'Schmitt trigger' circuit (Schmitt, 1938), have been performed. The optical system and the electronic one are two *different physical systems*; they will be shown to display very similar physical behavior, thus indicating that it is *typical* for a whole class of systems and that it is not related to special properties of optical bistability.

Due to the experimental convenience of the electronic system, it gives results much faster than the optical one. It should be emphasized that it also gives results much faster, and considerably cheaper, than a digital computer programmed to solve the stochastic problem, either on the basis of a Monte Carlo type approach or an analytic approximation. Thus it was possible to use the electronic device as a *guideline*: The electronic device gave results originally not predicted by theory, and it was possible to observe a behavior which has not yet been observed in the optical system, since the experimental problems are much harder.

In this chapter the discussion will be restricted to experiments on 'noise-induced transient bistability' (Broggi and Lugiato, 1984a, b). After a short description of the phenomenon and of the experimental set-ups in Section 6.2, results will be presented, in Section 6.3, which have been obtained by application of white noise. In Section 6.4 a few results obtained with colored noise will be presented.

The purpose of the chapter is twofold: first, experimental results on the influence of noise on the transient behavior of bistable devices are to be given; second, it is hoped to encourage an even more widespread use of electronic circuits in the discussion of nonlinear dynamics.

6.2 Transient noise-induced bimodality

6.2.1 The phenomenon of transient noise-induced bimodality

In the Bonifacio–Lugiato model the equation of motion is given by

$$\mathrm{d}x/\mathrm{d}t = y - x - 2cx/(1 + x^2). \qquad (6.2.1)$$

Here y represents the suitably normalized field strength of the incident light field; the state variable x represents the normalized field strength within the resonator; the time t is normalized; and c is the 'cooperativity parameter'. x is proportional to the field strength of the transmitted field, of course. Bistability can occur for $c \geqslant 4$. The behavior of the system can conveniently be interpreted by defining a potential U by means of the equation

$$\mathrm{d}x/\mathrm{d}t = -\,\mathrm{d}U/\mathrm{d}x. \qquad (6.2.2)$$

The typical S-shaped state curve corresponding to the steady state solution of (6.2.1) is shown in Figure 6.2(a). The value of y_{cr} will be called 'threshold' in the following, and the paper refers to a choice of y in the vicinity of y_{cr}. At values of y slightly below y_{cr}, the potential has the shape shown in Figure 6.2(b), while the shape slightly beyond y_{cr} is given by Figure 6.2(c). Most of this chapter refers to the situation of Figure 6.2(c).

If y is suddenly switched from zero to a value slightly beyond y_{cr} at $t = 0$, then the initial value of x is $x(0) = 0$, but the system moves fast into the vicinity of the turning point x_t. Here the motion becomes very slow deterministically, if y is only slightly larger than y_{cr}, with the slope becoming very small in the

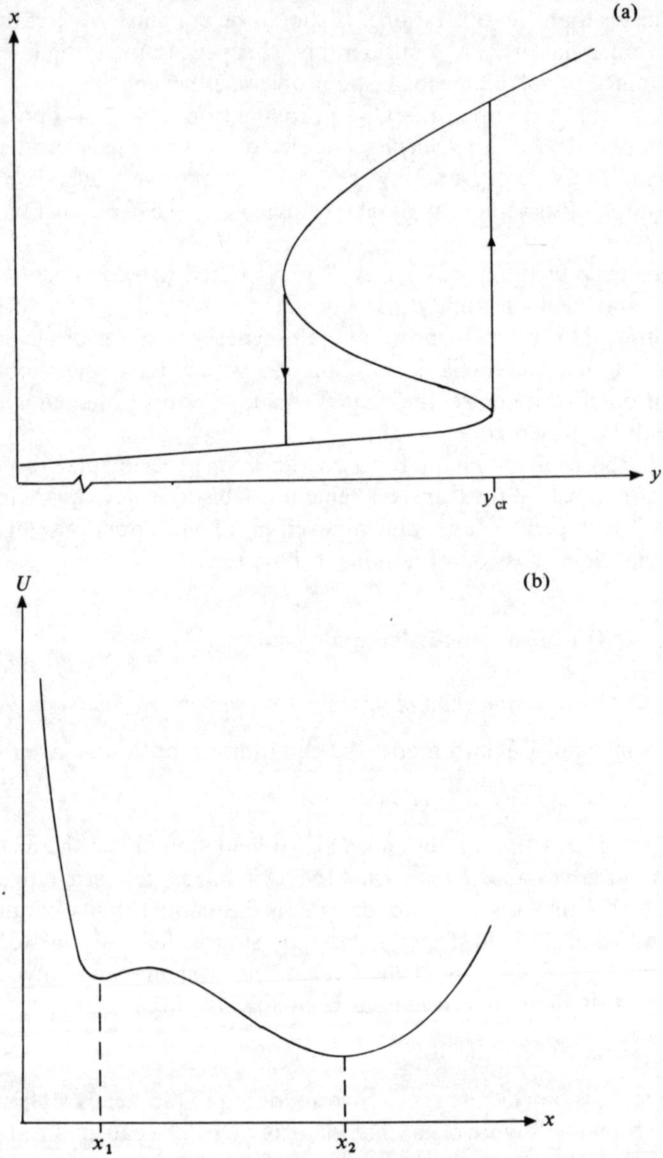

Figure 6.2. (a) State curve of absorptive bistability ($C = 8$). (b) and (c) Potential for a value of y slightly below or slightly above threshold y_{cr}, respectively.

turning point. As soon as the region of the turning point has been passed, the motion becomes fast again and the system moves quickly into the minimum at x_2. Thus, in an experiment, a fast initial increase in the transmitted intensity is observed when the input is switched on. It is followed by a long 'lethargic stage', during which nothing seems to happen in the system, until it finally

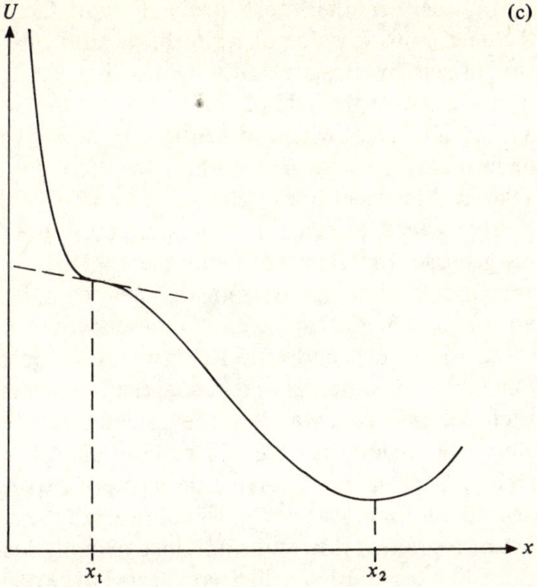

switches to the 'high transmission state' belonging to x_2 after a 'delay time' τ_D. This *critical slowing down* behavior was first predicted by Bonifacio and Lugiato (1976, 1978) and has been the object of intense theoretical study (Benza and Lugiato, 1979; Bonifacio and Meystre, 1978; Grynberg and Cribier, 1983; Mandel and Erneux, 1982). Its experimental observation has been attained in a variety of systems: hybrid (Garmire, Marburger, Allen and Winful, 1979); nonlinear microwave resonator (Barbarino *et al.*, 1982); nonlinear optical resonator containing sodium atomic beams (Grant and Kimble, 1983), or filled by sodium vapor (Mitschke, Deserno, Mlynek and Lange, 1983), or rubidium (Cribier, Giacobino and Grynberg, 1983).

A stochastic description has been given by Broggi and Lugiato (1984a, b) and in more detail by Broggi, Lugiato and Colombo (1985) and by Lugiato, Colombo, Broggi and Horowicz (1986). In this treatment the starting point is a very narrow probability distribution $p(x, t)$ of x at $t = 0$ which drifts fairly fast to x_t. Evidently this distribution has to develop into a single-peaked distribution centred at $x = x_2$ for $t \to \infty$. The analysis reveals that this evolution occurs by a steady decrease of the peak at $x = x_t$, while, simultaneously, a peak at $x = x_2$ develops, provided that there are fluctuations of y. Thus, in the presence of fluctuations, there are *two* peaks in p for a sizeable time interval. The double-peaked character of p is the signature of *bimodality* or *bistability*, which is *noise-induced* and *transient* in this case.

As a consequence of the double-peaked character of p, the variance of x is very large for intermediate times. Moreover, there is a strong variation of the delay time τ_D, and the analysis reveals that the probability distribution of τ_D is very asymmetric and neither the most probable nor the mean value corresponds to the deterministic value of τ_D in the general case.

It has been emphasized by Broggi and Lugiato that the phenomenon discussed here is not specific for the field of optics, but that the same behavior is to be expected in all cases where potentials of the type shown in Figure 6.2(c) come into play. In fact, these authors refer to a paper of Nicolis and coworkers (Baras, Nicolis, Malek Mansour and Turner, 1983) that discusses an explosive chemical reaction and predicts the same type of phenomenon. This generality is the reason why the choice of the electronic system is fairly unrestricted and why an optical system completely different from the Bonifacio–Lugiato model can be used in the experiments. Perhaps it should be mentioned that the phenomenon is even more general than the interpretation by means of the potential might suggest. In our optical system, conditions can be chosen in such a way that the potential *cannot* be defined, but nevertheless the phenomenon is observed without alterations.

From the papers of Broggi and Lugiato, the type of experiment to be performed is quite evident. First, it is necessary to measure the state variable for a fixed time T. A histogram of the results obtained in many successive runs represents the probability distribution, which is expected to be double-peaked. Second, the delay time τ_D has to be determined; the corresponding histogram represents the distribution of delay times.

6.2.2 The optical experiment

Experimentally a very convenient bistable optical device is a Fabry–Perot resonator filled with sodium atoms; as a light source a dye laser tuned to one of the D-lines of sodium can be used. This was the system used in the first experiment, by Gibbs and coworkers, on optical bistability to become widely known (Gibbs, McCall and Venkatesan, 1976). Unfortunately the analysis is tremendously complicated by the presence of Doppler broadening and by the fact that the sodium levels involved have a hyperfine structure. (For a full discussion see Gibbs, 1985.)

In principle these complications could largely be avoided by using a buffer gas pressure sufficient to introduce collisional broadening larger than the Doppler effect and the hyperfine splitting of the sodium ground state. Under these conditions, however, a dye laser would not supply sufficient power in order to saturate the sodium transition. Fortunately, Zeeman pumping within the sodium ground state can be used as a very efficient nonlinear mechanism (Arecchi, Giusfredi, Petriella and Salieri, 1982; Mlynek, Mitschke, Deserno and Lange, 1982; see also Mlynek, Mitschke, Deserno and Lange, 1984). In this case the time constant of the medium is determined by the relaxation of the *orientation* of the sodium ground state. The corresponding time-constant is very large, since the orientation is very insensitive to collisions with rare gas atoms. In this situation it would be very impractical, or even impossible in practice, to work in the 'good cavity limit', but it is natural to work in the 'bad cavity limit'. In this case the time-constant of the medium, τ_m, has to be much larger than the time-constant of the cavity, τ_c. This requirement is excellently fulfilled in the experiments described here ($\tau_m \simeq 10\,\mu s$, $\tau_c \simeq 10\,ns$). Under these conditions the electric field can be eliminated adiabatically and the state variable turns out to be the *ground state orientation.*

In order to be specific, the quantity

$$z = \rho_{11} - \rho_{22} \tag{6.2.3}$$

is defined; here ρ represents the density matrix, and 1 and 2 designate the Zeeman sublevels $m = -1/2$ and $m = 1/2$ of the sodium ground state, respectively. z can be called the 'z-component of the orientation'; it is proportional to the z-component of the expectation value of the magnetization of the simple; the direction of the laser beam has been chosen as the z-axis. The equation of motion is

$$\mathrm{d}z/\mathrm{d}t = -(1+P)z + P \tag{6.2.4}$$

(Lange, Mitschke, Deserno and Mlynek, 1985; for a full discussion see Mitschke, Deserno, Lange and Mlynek, 1986). The time is measured in units of τ_m, and P is normalized to $1/\tau_m$. P is the *pump rate* introduced by the light field within the resonator, which is assumed to be left-hand circularly polarized; P is proportional to the *intensity* of the light field, of course. (The Bloch equations describing the interaction between the atoms and the radiation field reduce to a rate equation due to the strong collisional broadening; see Mitschke *et al.*, 1986.) P can be written in the form

$$P = P_{in} A(z). \tag{6.2.5}$$

P_{in} represents the pump rate without resonator; it is proportional to the intensity of the input beam. $A(z)$ is the Airy function; it depends on z, since the optical properties of the resonator depend strongly on dispersion and absorption of the sodium vapor, which in turn are determined by z. In the *purely absorptive* case, which occurs if the laser is exactly tuned to the atomic resonance, $A(z)$ is given by

$$A(z) = (1-R)[1 - R + AR(1-z)]^{-2} \tag{6.2.6}$$

in the cavity resonance. Here R is the reflection of the front mirror, and it is assumed that the back mirror has a reflection of nearly unity; A is the absorption of the sample, and $A \ll 1$ is assumed. Equation (6.2.6) can easily be generalized to less restrictive conditions, and a corresponding formula can be given for the *dispersive case* or for the *mixed case* incorporating contributions from absorption *and* dispersion.

In the experiment (Lange *et al.*, 1985), whose arrangement is shown schematically in Figure 6.3, broadband amplitude noise in the input is introduced by modulating the laser beam by means of an electro-optic

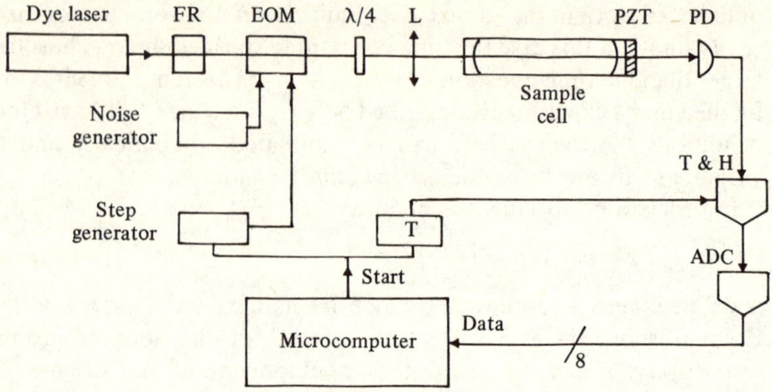

Figure 6.3. Schematic of the optical experiment. FR = Faraday rotator; EOM = electro-optic modulator; L = mode matching lens; PZT = piezoelectric translator; PD = photodiode; T & H = track-and-hold circuit; ADC = analog-to-digital converter. The noise generator is based on a hard-wired pseudo-random number generator; the frequency spectrum of the noise is nearly Lorentzian with a band width of 1 MHz, if not otherwise stated; it is reduced to 700 kHz by the amplifier driving the EOM (not shown). (From Lange *et al.*, 1985.)

modulator fed by a broadband noise generator. The band width of the noise modulation of the laser beam is 700 kHz, which has to be put into relation to $\tau_m \simeq 10\,\mu s$. By adding a suitable filter between the noise generator and the modulator the properties of the noise modulation can be modified over a wide range. The electro-optic modulator is also used in order to produce a step input of light. The intensity transmitted by the nonlinear resonator is detected by means of a photodiode and is sampled by means of a track and hold circuit; the microcomputer is used for data collection in producing the histogram of transmitted intensities. The set-up can be modified for the measurement of the distribution of delay times in an obvious way.

It should be clear that in the experiment the state variable is not directly determined, but the measurements refer to a quantity related to it in a well-defined way. Moreover it should be evident that noise is *not purely* additive in the experiment, since P in (6.2.4) depends on the state variable z via (6.2.5) and (6.2.6).

While a potential leading to (6.2.4) can easily be given, the situation becomes more complicated if a transverse magnetic field is applied. In this case the x- and y-components of the orientation come into play and the dynamics of the system can become very complicated (see Mitschke *et al.*, 1986). There are, however, still parameter ranges in which the transient bimodality is only slightly modified, while a potential does not exist.

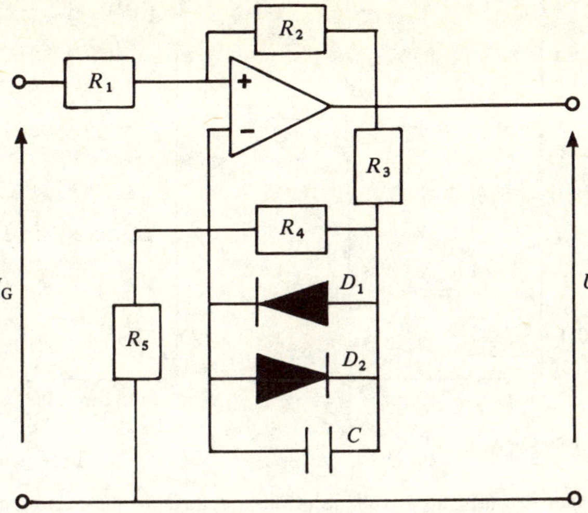

Figure 6.4. Schematic of the electronic bistable device. $R_1 = 3.28\,\mathrm{k\Omega}$, $R_2 = 97.9\,\mathrm{k\Omega}$, $R_3 = 11.9\,\mathrm{k\Omega}$, $R_4 = 680\,\mathrm{k\Omega}$, $R_5 = 3.16\,\mathrm{k\Omega}$, $C = 57.5\,\mathrm{pF}$, $D_1 = D_2 = \mathrm{BA}\ 182$. The diode capacities are negligible in comparison to C. The 'cooperativity parameter' c and the time τ_0 defined in the text are calculated to be $c = 27270$ and $\tau_0 = 147\,\mathrm{ms}$, respectively.

6.2.3 The electronic system

It is quite obvious that an electronic device that is very familiar to experimentalists, the *Schmitt trigger circuit* (Schmitt, 1938) has much in common with an optical bistable system. It is bistable, has a threshold behavior and displays hysteresis; it is also well known that it displays the phenomenon called *critical slowing down* in the field of optical bistability; in fact, the phenomenon has been studied in some detail in the field of engineering (Mejerowitsch and Selitschenko, 1960), but the methods of either deterministic nonlinear dynamics or statistical physics do not seem to have been applied to the interpretation of the behavior of the device.

Schmitt trigger circuits are available as integrated circuits; most modern textbooks on electronics explain the principle of operation and also show the realization of the circuit by means of an operational amplifier. The circuit used in the experiment to be described here (Mitschke, Deserno, Mlynek and Lange, 1985) is shown in Figure 6.4. In order to have a well-defined equation of motion, a slight modification with respect to the standard design is used. While normally the nonlinearity is determined by saturation of the operational amplifier, and the temporal response by its internal time constants, the two antiparallel diodes in the feedback loop are introduced in the present case. They saturate with a time-constant which can be adjusted by a proper choice of the parallel capacitor C. The equation of motion can be determined by

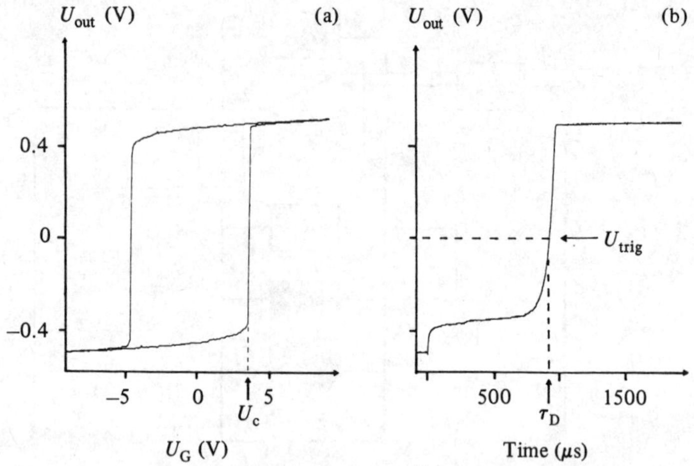

Figure 6.5. (a) Output voltage vs. input voltage for the electronic device shown in Figure 6.4, with no noise signal added. (b) Transient response of the electronic device. At time $t = 0$ the input voltage was switched from − 10 V to $U_c + 0.1$ V; U_c was 3.56 V. (From Mitschke *et al.*, 1985.)

straightforward methods to be (Lange, Mitschke, Deserno and Mlynek, 1986a)

$$dx/dt = y + cx - \sinh(x) \tag{6.2.7}$$

with the 'cooperativity parameter'

$$c = \{R_1/(R_2R_5 - R_1R_3) - 1/R_4\}U_T/(2I_0)$$

and the 'control parameter'

$$y = U_GR_2/(2I_0(R_2R_5 - R_1R_3)).$$

The state variable x is the voltage across the diodes measured in units of U_T. U_G is the voltage supplied to the input, and the quantities U_T and I_0 stem from the description of the characteristics of the diode by the equation

$$I(U) = I_0(\exp(U/U_T) - 1);$$

as an experimental complication U_T depends on temperature. The time is to be measured in units of $\tau_0 = CU_T/(2I_0)$ in (6.2.7).

When the input voltage U_G is scanned, then the output voltage U_{out}, which is clearly connected to x, displays the symmetric hysteresis loop shown in Figure 6.5(a). Switching occurs at $U_G = \pm U_c$. When the input voltage is suddenly switched from large negative values to a value slightly above U_c, then the critical slowing down behavior shown in Figure 6.5(b) is observed.

In the experiment the noise signal is superimposed onto the step function of the input voltage, i.e. y contains the noise in an *additive* way in (6.2.7). Thus (6.2.7) can easily be converted into a Fokker–Planck equation, if *white noise* is assumed (see, e.g., Risken, 1984). In the experiment the noise spectrum was restricted to 1 MHz; for reasons given in Section 6.4 this is believed to be an excellent approximation of white noise in the present case. The time-dependent Fokker–Planck equation was solved by a standard computer code; at very low noise levels, however, numerical problems arose, and generally the application of a Monte Carlo type approach was more satisfactory.

6.3 Experimental results obtained with white noise

6.3.1 Bimodality

Histograms of the output signals for different times after switching on the input are shown in Figure 6.6 for the optical system. As discussed in Section 6.2 the histogram is directly related to the probability distribution of the state variable $p(x,t)$ for different fixed times $t = T_n$, i.e. we perform a *stroboscopic* observation of the evolution of the system. It is clearly seen that the phenomenon of bimodality exists.

By varying the level of the external noise the histograms are changed. At high noise levels the time interval for which bimodality persists shrinks, since the peak representing the system in the final state reaches its maximum value after a short time. The histograms obtained in the electronic system look very similar and show a similar dependence on the noise level.

In the optical system we cannot prove that the bimodality is noise-induced, since it is observed even if the external noise modulation is switched off. The reason is due to the presence of 'intrinsic' noise, i.e. of noise which cannot be controlled in our experiment. In the electronic system, however, this intrinsic noise level is very low. In fact, this feature is one of the big advantages of the electronic system.

If the equivalent to Figure 6.6 is displayed for the electronic system *without* external noise, results as displayed in Figure 6.7(a) are obtained. Obviously there is *no* bimodality. In Figure 6.7(b) the behavior is shown for intermediate times with higher temporal resolution. Apparently, at first the peak in the probability distribution broadens by *diffusion*, while no *drift* of its centre can be observed (curves 1 to 3 in Figure 6.7b). This corresponds to the temporal evolution of the system in the *turning point* of the potential. The rapid motion of the centre of the peak displayed by curves 4 and 5 is an indication of the fast *drift* on the steep slope of the potential. During this rapid motion, there is a considerable influence of *diffusion*, but obviously the drift dominates. The drift ends when the deep minimum of the potential is reached, and since the slope is large the peak narrows very quickly.

Figures 6.8(a) and (b) correspond to Figures 6.7(a) and (b), respectively, but a very small noise signal is added. It can be seen now that the influence of the diffusion is much stronger. There is a very pronounced *asymmetry* of the peaks, especially in curve 6 of Figure 6.8(b), but there are *not* two maxima, i.e. there is

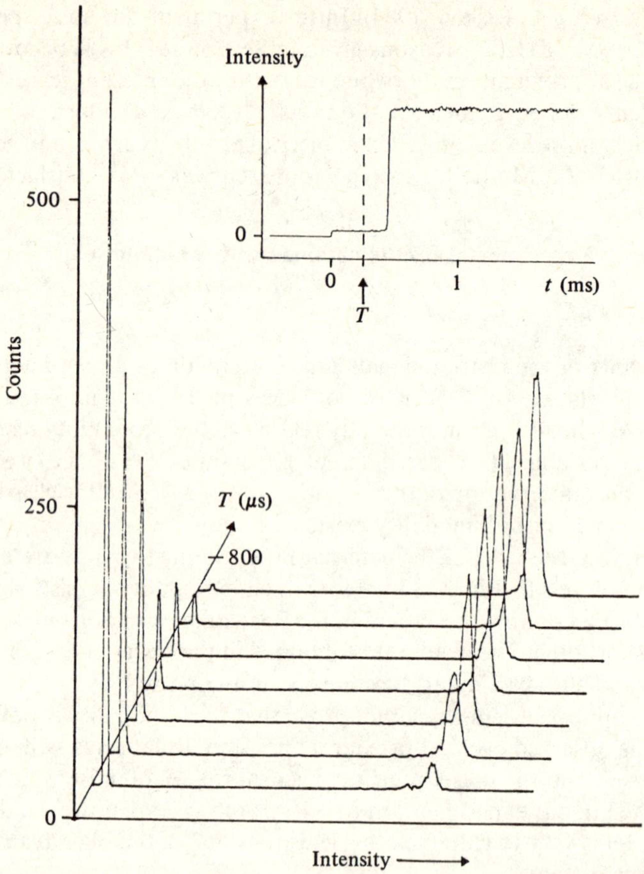

Figure 6.6. Measured histograms of output intensities for different times T after switch on of the light. Experimental parameters: average input power $P_{in} = 12\,\text{mW}$; rms noise level $= 0.2\,P_{in}$; number of intensity values on the abscissa = 256. The inset shows the temporal behavior of the output intensity for a single shot. (From Lange *et al.*, 1985.)

no bimodality, even though there is a non-negligible noise level. (It should be noted that curve 1 in Figures 6.7(a) and 6.8(a) displays a broader peak than curve 2, since the centre does not yet correspond completely to the position of the turning point of the potential, but the system is still on the lower end of the steep slope connecting the starting and the turning points. The initial rapid motion is not resolved here.)

In Figures 6.9(a) and (b) the noise level is further increased, but it is still fairly low. It can clearly be seen how the long tails already present in Figure 6.8(b) develop a second maximum, i.e. now the noise level is sufficient to induce bimodality.

It should be emphasized that Figures 6.7 to 6.9 show that a *minimum noise level is required in order to obtain bimodality.* This has been expected by

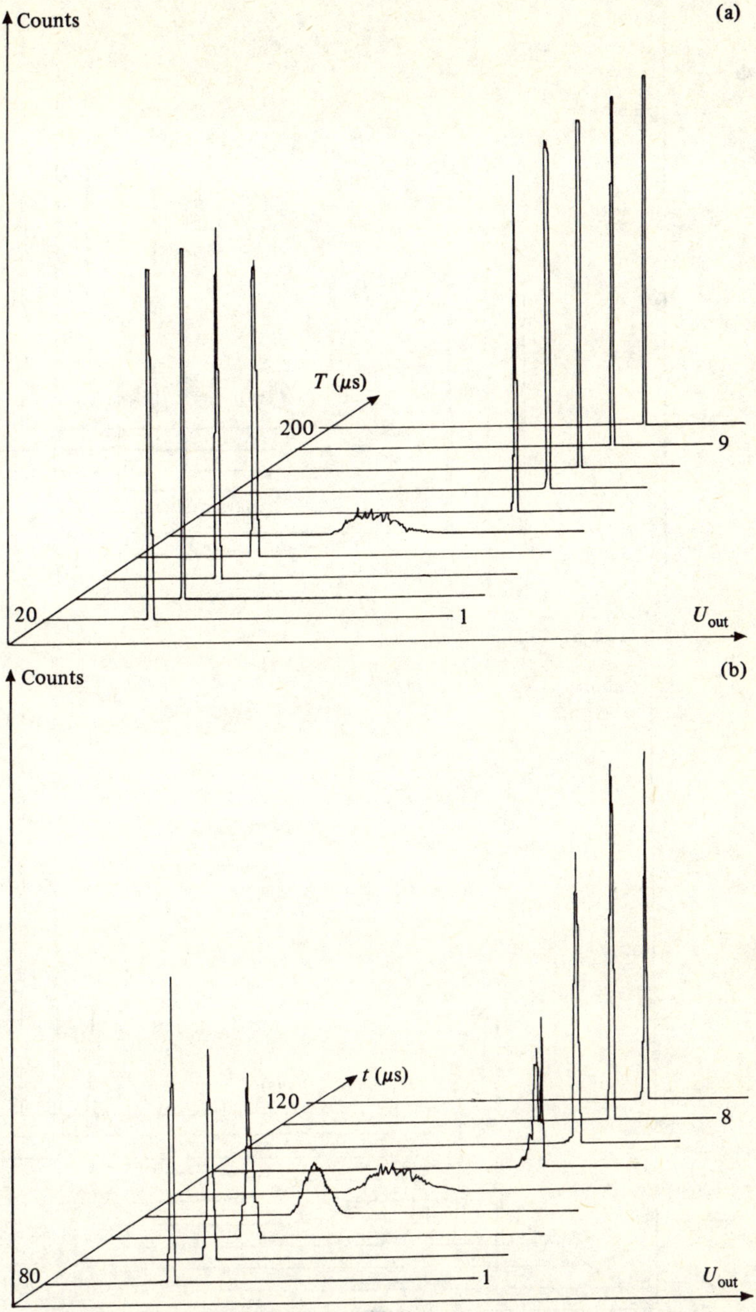

Figure 6.7. (a) Measured histograms of output voltage of the electronic device for different times T after switching the input voltage U_G to a value slightly above threshold. No external noise was added. Each histogram is averaged over 5000 single shots. (b) Histograms for medium times in enhanced temporal resolution. (After Oertel, E. 1986. *Staatsexamensarbeit*. Unpublished.)

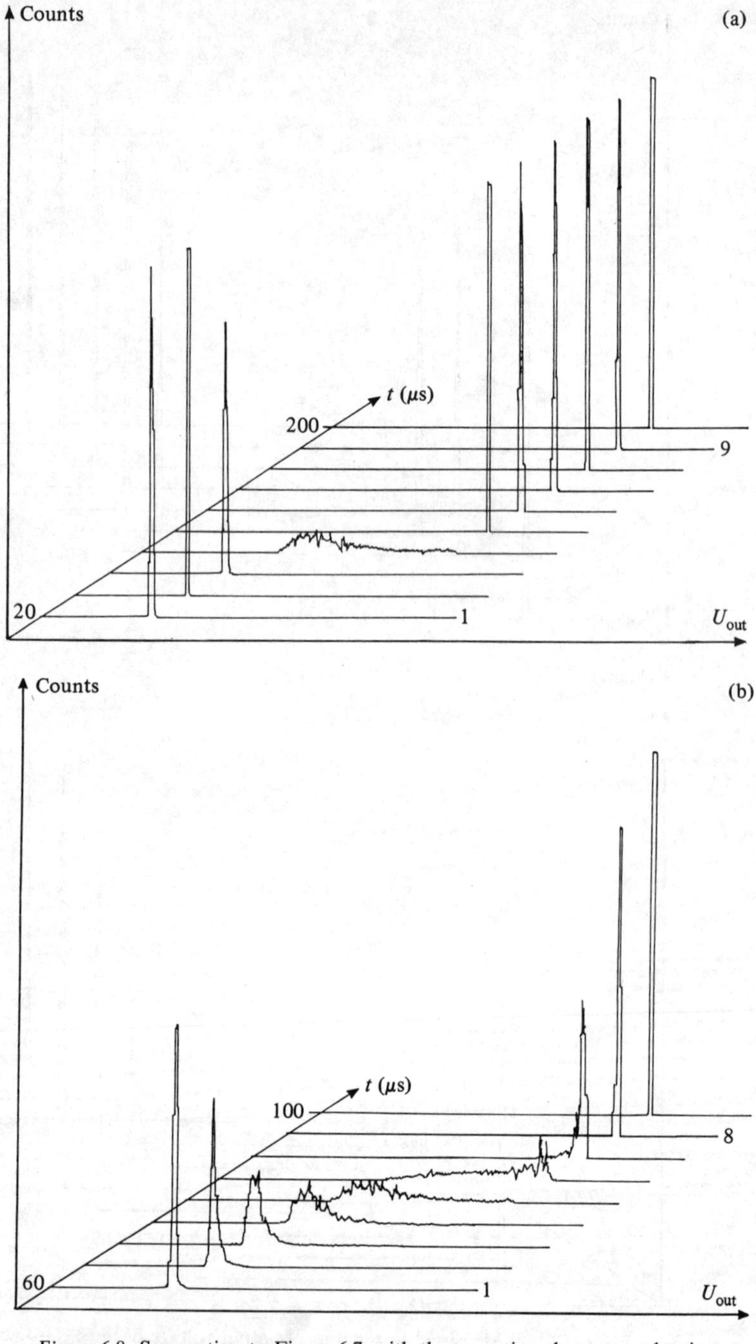

Figure 6.8. See caption to Figure 6.7, with the exception that external noise was added from the source described in the caption to Figure 6.3 (noise band width 1 MHz). The rms noise level relative to the average input voltage was $\sigma = 0.0026$.

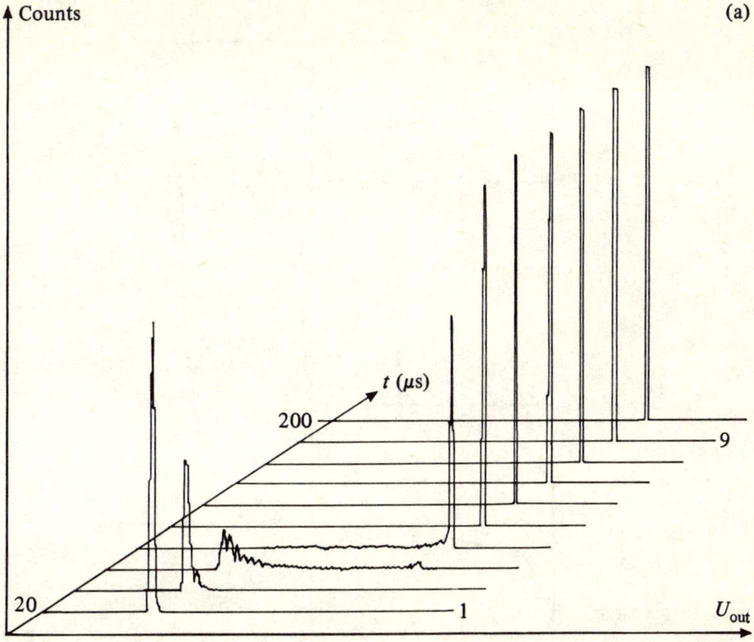

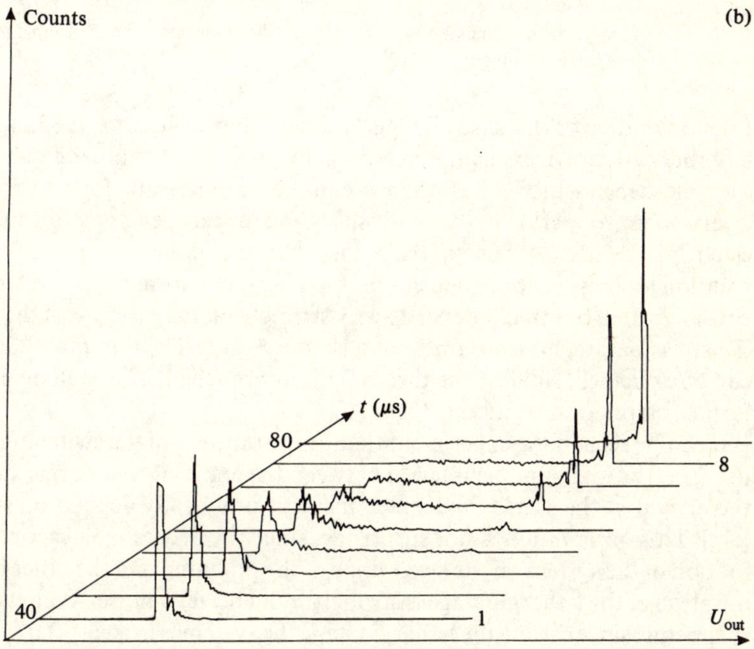

Figure 6.9. See caption to Figure 6.8, with the exception that the relative rms noise level was increased to $\sigma = 0.017$.

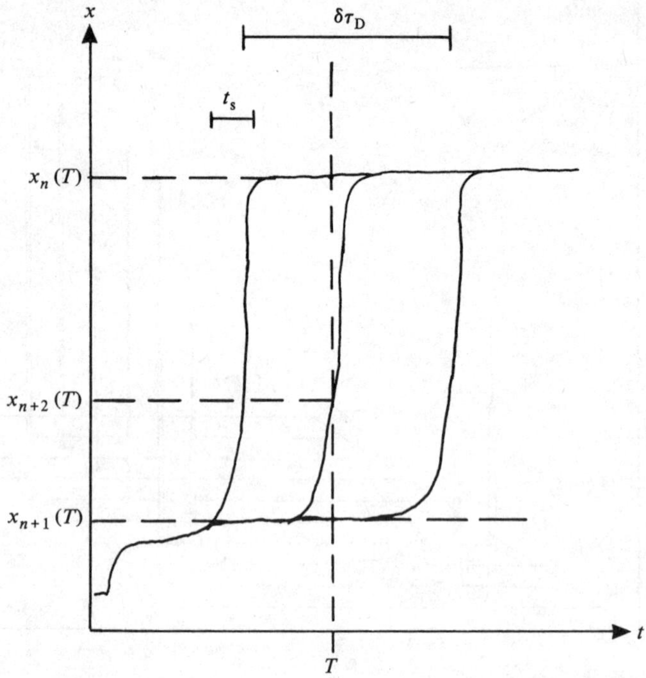

Figure 6.10. Schematic plot of temporal dependence of state variable in three successive switching events.

Lugiato and coworkers (see Broggi, Lugiato and Colombo, 1985), though it has not been shown explicitly, since it seems to be much more difficult to solve the time-dependent Fokker–Planck equation numerically for very low noise levels than to perform the corresponding experiments by means of an electronic device. It has to be noted that the minimum noise level just mentioned does not only depend on the physical nature of the system under consideration, but that it depends very strongly on the distance of the value of the stress parameter from threshold, i.e. on $y - y_{cr}$. The minimum noise level can be extremely small, if the threshold is approached. This statement will be further clarified in Section 6.3.2.

It can be seen in the experiment that the duration t_s of the switching process itself, i.e. the length of the interval between the end of the 'lethargic stage' and the arrival in the stable final state, does not practically depend on the noise level. This observation is not surprising, since the process corresponds to the motion of the system on the steep slope of the potential. On the other hand the duration of the lethargic stage is strongly influenced by noise. Again this is not surprising, since during the lethargic stage the system is in a very flat part of the potential. The 'jitter', $\delta\tau_D$, in the duration of the lethargic stage, τ_D, is negligible far beyond threshold: τ_D is vanishing. On the other hand the fluctuations

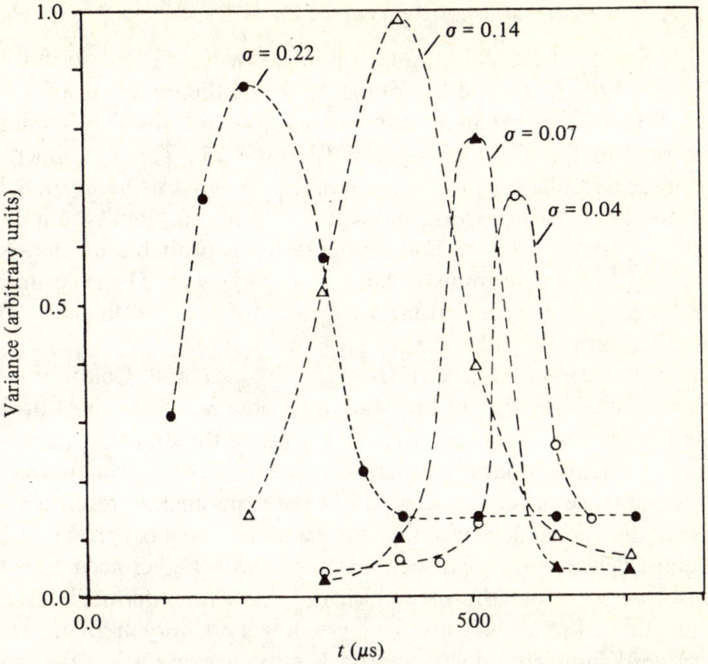

Figure 6.11. Variance of output intensities as a function of time, derived from measured distributions of the type shown in Figure 6.6. The parameter σ indicates the rms value of input noise relative to average intensity. (From Lange *et al.*, 1987.)

of τ_D become large, if the threshold is approached. Deterministically τ_D should *diverge*, while the expectation value is *finite* in the presence of arbitrarily small amounts of noise (see Mitschke *et al.*, 1985).

The situation of the experiment is further illustrated by Figure 6.10. With $t_s \to 0$ the result of a measurement will nearly always be $x \simeq x_t$ or $x \simeq x_2$, i.e. there is a very small probability of finding the system in the switching process. With the inclusion of noise there is a finite time interval in which the result may be *either* $x \simeq x_t$ *or* $x \simeq x_2$, i.e. bimodality will be observed as soon as there is some jitter of τ_D. With larger values of t_s, however, a fairly broad distribution of experimental values $x(T)$ will be obtained, even if the fluctuations of τ_D are very small, but obviously in the latter case bimodality is not observed. It is an open question whether transient bimodality can be observed at all if t_s is large.

It is quite evident that the variance of the output signals is large if the probability distribution is double-peaked. Nevertheless it is illuminating to plot the variance in dependence on time (Figure 6.11). The figure shows that the largest variances do *not* occur for the largest, but for intermediate, noise levels. This had been stressed by Lugiato and coworkers (Broggi, Lugiato and Colombo, 1985).

6.3.2 Distribution of delay times

A histogram of the delay times obtained in the optical set-up is shown in Figure 6.12(a). It should be compared to the theoretical result obtained in a Monte Carlo type calculation shown in Figure 6.12(b). Very similar results are obtained in the electronic system (Figure 6.13). The experimental and the theoretical results have in common that there is a *distribution* of delay times of considerable width which indicates a '*jitter*' in the switching, and this distribution is *asymmetric* and neither the most probable nor the average delay is related to the deterministic value in a simple way. The experimental results further confirm the prediction (Broggi and Lugiato, 1984a, b) that the width of the distribution shrinks at very large noise levels.

It has been pointed out (Broggi, Lugiato and Colombo, 1985) that, approximately, a straight line should be obtained if the most probable value of τ_D is plotted versus $\ln(1/q)$, with q being the diffusion parameter in the Fokker–Planck equation, which is proportional to our quantity σ^2 characterizing the noise power. In Figure 6.14 the experimental result is shown in the case of the optical device; in the figure not only the most probable, but also the average delays, which can be determined with higher accuracy, are plotted. It can be seen that the prediction is correct for intermediate values of σ^2, while there are deviations for very low and very high noise levels: for very weak noise the delay approaches its deterministic value, as expected; for very strong noise the delay approaches the value also obtained very far beyond threshold. Obviously this limiting value is related to the time t_s needed for the switching process itself, which is not strongly influenced by noise as explained in the preceding section.

In Figure 6.15 the dependence of the average switching delay on the stress parameter is displayed for two different noise levels in the case of the optical system. It can be seen that the difference between the results increases drastically when threshold is approached. It should be emphasized that the difference between the deterministic value of τ_D and its mean value in the presence of noise becomes infinitely large at threshold, simply because arbitrarily small amounts of noise are sufficient to remove the divergence at threshold.

In the optical system it is very difficult to perform measurements in the immediate vicinity of threshold, since the problem of a small drift of experimental parameters becomes severe. The experimental situation is much more favorable in the case of the electronic device, and it has been feasible to perform reliable measurements, even far *below* threshold. An example is shown in Figure 6.16.

In the bistable region the experiment obviously measures the first passage time, and it can be shown that the straight lines occurring in the semi-logarithmic plot of Figure 6.16 are a nice demonstration of Kramers' predictions on the *escape rate* (Kramers, 1940; see also Risken, 1984). It should be possible to perform a rather detailed experimental test of theoretical

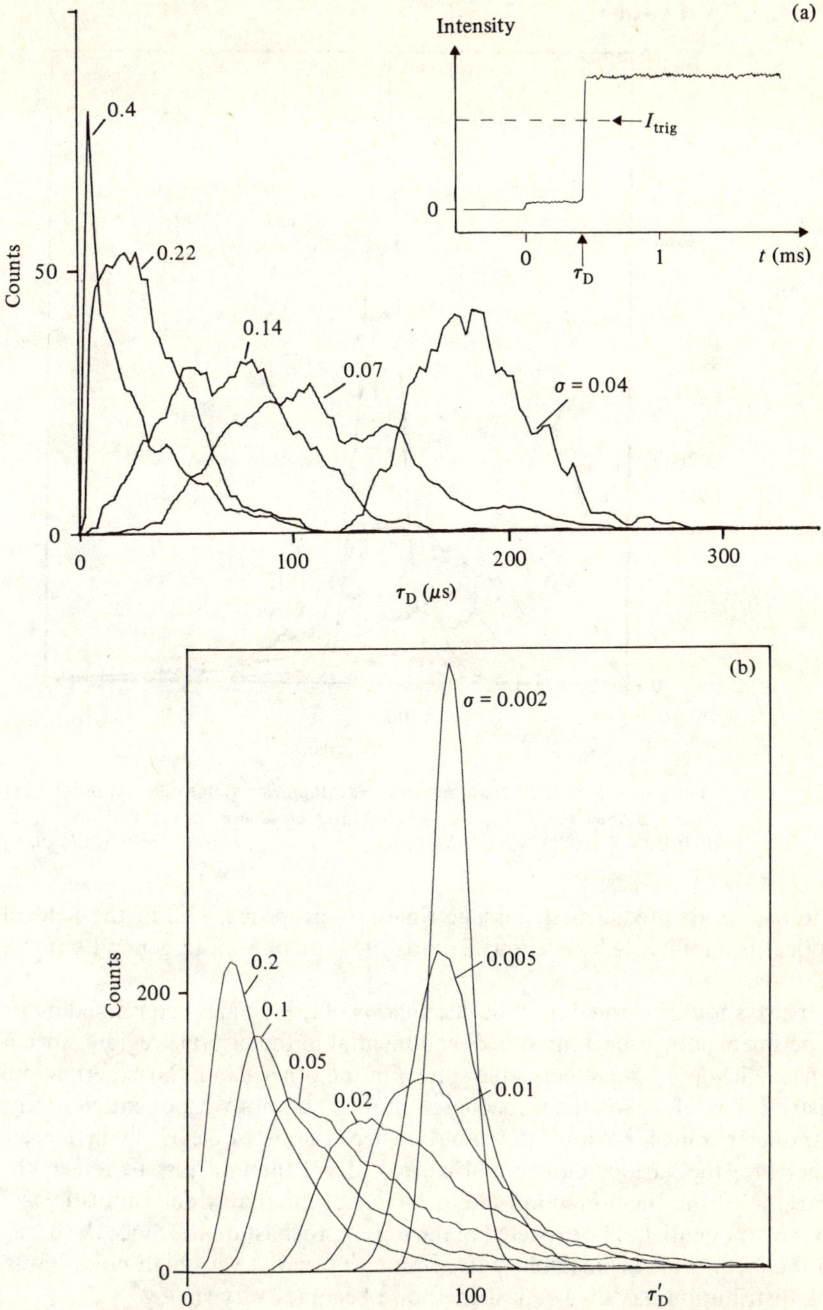

Figure 6.12. Measured histograms of switching delay in the optical device in the presence of broadband noise. σ characterizes the rms noise level relative to the average intensity. (a) Experimental result (from Lange, Deserno, Mitschke and Mlynek, 1986b). (b) Calculated result; τ'_D is the switching delay relative to the time constant τ_m of the medium. (From Lange *et al.*, 1985.)

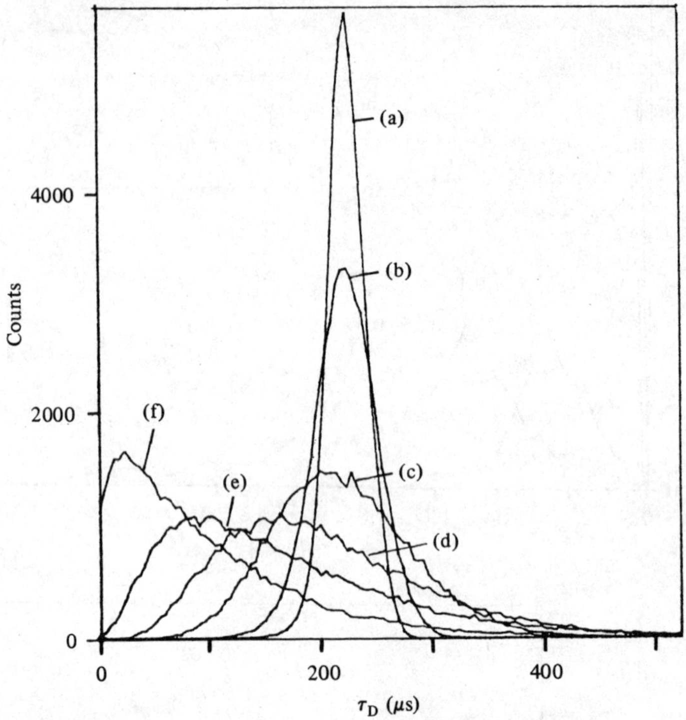

Figure 6.13. Measured histograms of switching delay τ_D for different noise levels. Parameters are: $U_G = U_c + 0.34$ V; rms noise levels are (a) 0.016 V, (b) 0.032 V, (c) 0.08 V, (d) 0.16 V, (e) 0.32 V, (f) 0.8 V. (From Mitschke *et al.*, 1985.)

approaches to predict first passage times, as proposed, e.g., in the field of optical bistability (Schenzle and Brand, 1979) or in a more general context (Shenoy, 1984).

It also should be noted that the *distribution* of delay times determined in the experiment should be a (modified) exponential in the bistable region, since it reflects the decay of a metastable state. On the other hand, the experimental distribution of delay times becomes more and more exponential in the parameter range beyond threshold, when the noise level is increased. Therefore the various curves in Figure 6.12 are thought just to reflect the transition from the δ-function-like behavior of the purely deterministic case to the exponential to be expected in the purely stochastic case. Since the decay of the metastable or unstable states becomes rapid at very high noise levels, the distribution has to shrink if the noise becomes very strong.

In the optical system, with the noise not being purely additive, the position of the threshold might be changed by noise (see, e.g., Horsthemke and Lefever, 1984), but the analysis reveals that this effect is not detectable in the experiment, since the exact position of threshold has only academic meaning

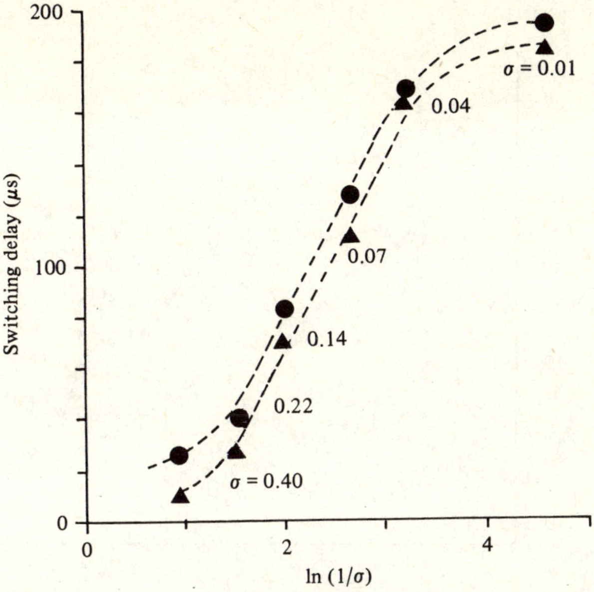

Figure 6.14. Average and most probable delay (points and triangles, respectively) in dependence on ln $(1/\sigma)$ in the optical device. (From Lange *et al.*, 1987.)

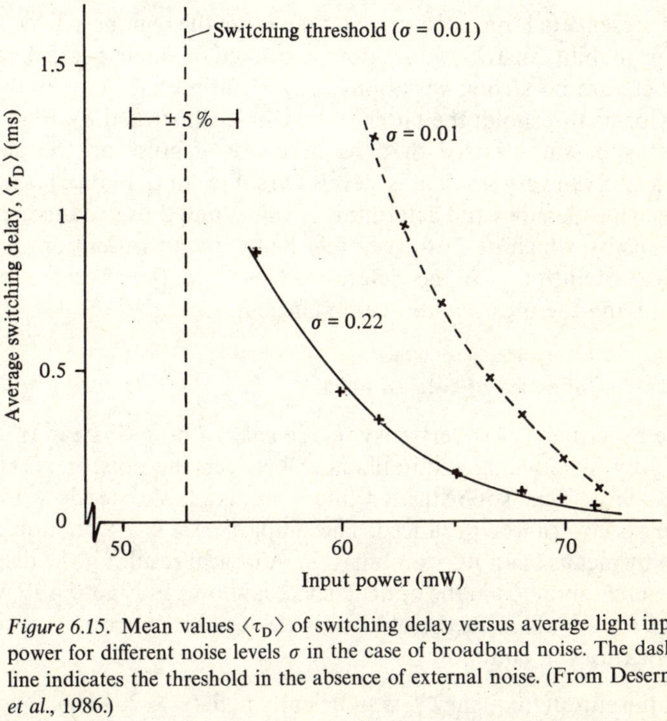

Figure 6.15. Mean values $\langle\tau_D\rangle$ of switching delay versus average light input power for different noise levels σ in the case of broadband noise. The dashed line indicates the threshold in the absence of external noise. (From Deserno *et al.*, 1986.)

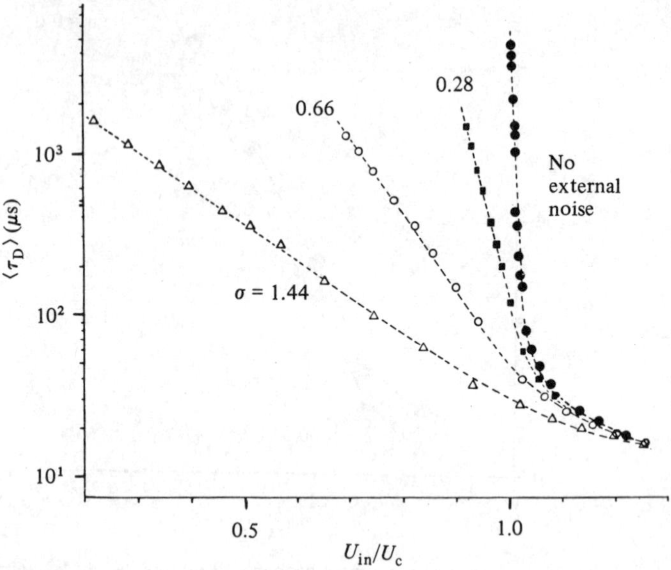

Figure 6.16. Mean values $\langle\tau_D\rangle$ of switching delay versus normalized average input voltage for different levels of broadband noise for the electronic device. (From Deserno *et al.*, 1986.)

in the presence of large noise levels. It separates the regions in which the steady state probability distribution is double-peaked or single-peaked, respectively, but there are no strong variations of expectation values or of the temporal behavior at threshold; the latter statement is illustrated by Figure 6.16.

It may be summarized that the influence of noise on the delay time is dramatic. Even very small noise levels are sufficient to induce large differences between mean values and deterministic values and thus produce a problem for experiments, which in most cases rely on the tacit assumption that the 'true' physical quantities can be determined even in the presence of noise by determining the mean values with sufficient accuracy.

6.4 Influence of colored noise

In the experiment it is very easy to use *colored* noise instead of white noise simply by introducing a suitable filter between the noise generator and the device. Since very sophisticated filters are available, nearly arbitrary noise power spectra can be produced. The simplest case is a reduction of the band width by means of an RC-combination. A typical result for the distribution of delay times obtained in the optical device is shown in Figure 6.17; very similar results are obtained in the electronic device. All experimental results have some features in common:

(1) If the cut-off frequency ν_c is sufficiently high ($\nu_c \gg 20\,\text{kHz}$), then the results

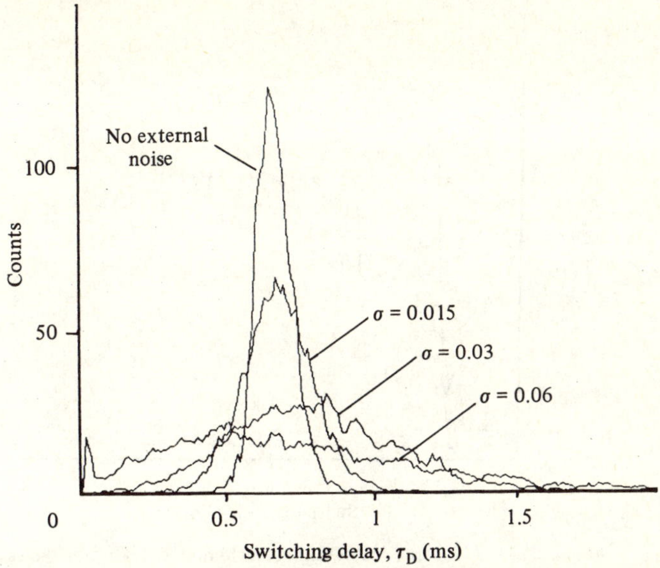

Figure 6.17. Measured histograms of switching delay τ_D for different relative rms noise levels σ for the case of colored noise in the optical system. Cut-off frequency of first-order low-pass filter = 250 Hz. (From Deserno *et al.*, 1986.)

do not differ from the results without the filter. This fact is regarded as a justification to treat the band width limited noise used in the experiments of Section 6.3 ($\nu_c = 700$ kHz and 1 MHz, respectively) as 'white noise'.

(2) If the cut-off frequency is low ($\nu_c \ll 20$ kHz), then the distribution of delay times develops a pronounced tail to long times, and as a consequence the average delay may be *lengthened*, while it is always shortened in the case of white noise.

In a similar way it is found by means of a high-pass filter that noise has no measurable effect at all if the cut-off frequency is sufficiently high, while the average delay is shortened more efficiently than by white noise if the cut-off frequency of the high-pass filter is in the order of 10 kHz.

Though the superposition principle clearly does not hold in nonlinear systems, the general behavior of the devices can be traced back to the influence of the *spectral components of noise*. This can be concluded from experiments in which the noise input is replaced by a *sinusoidal signal* superimposed on the step input. In this case the only stochastic element is the phase of the sinusoidal signal with respect to the step in the input signal. Distributions of delay times obtained with sine-modulation are displayed in Figure 6.18. From the measurements it can be concluded that:

(1) high frequency components ($\nu \gtrsim 300$ kHz) do not have any influence at all;

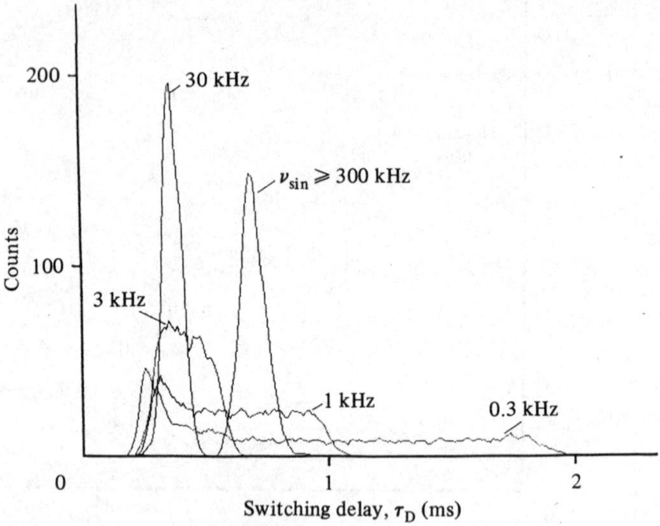

Figure 6.18. Measured histograms of switching delay τ_D for the case of sinusoidal perturbation by different frequency components in the optical system. The rms level of the sine wave relative to the average intensity was $\sigma = 0.07$ for all curves. (From Deserno *et al.*, 1986.)

(2) in an intermediate frequency range (30 to 100 kHz) the distribution function clearly shifts to shorter times;
(3) for low frequencies ($\nu \lesssim 10\,$kHz) the distributions broaden and develop long tails to long switching times.

From the results displayed in Figure 6.18 the mean switching delay $\langle \tau_D \rangle$ can be calculated; the result is shown in Figure 6.19 in dependence on the modulation frequency. It can be seen that $\langle \tau_D \rangle$ passes a pronounced minimum in the range of some 10 kHz and approaches a constant value for high frequencies.

Experiments have been performed in the optical and in the electronic devices. Here again the inherent stability and the low intrinsic noise level in the electronic device make it far superior in obtaining quantitative results. For example, the same type of measurement as in Figure 6.19 can be performed with different values of the mean input in the vicinity of threshold. The result is shown in Figure 6.20. It is clearly seen here that the mean input voltage mainly determines the asymptotic value of $\langle \tau_D \rangle$ for high modulation frequencies, while it does not influence the position of the minimum.

The results obtained with sine-modulation are in excellent agreement with numerical calculations; it is nearly quantitative in the case of the electronic device.

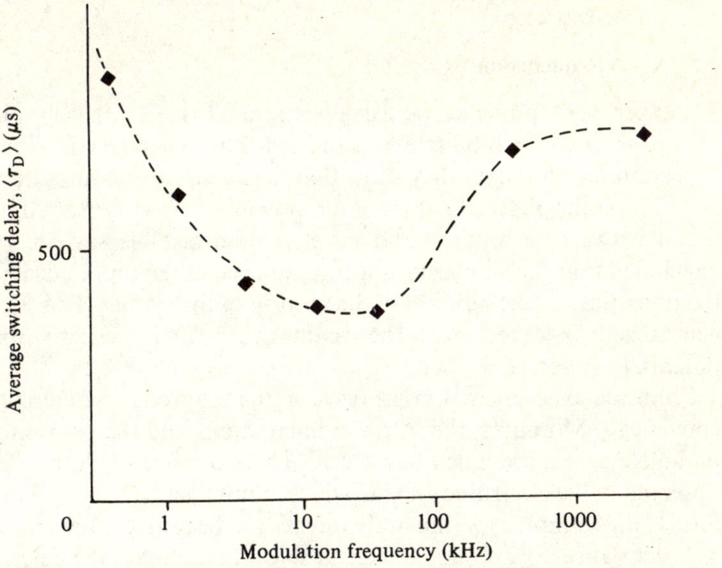

Figure 6.19. Average switching delay $\langle \tau_D \rangle$ versus modulation frequency as calculated from the data of Figure 6.18. For high modulation frequencies the curve approaches the value of $\langle \tau_D \rangle$ without modulation. (From Deserno *et al.*, 1986.)

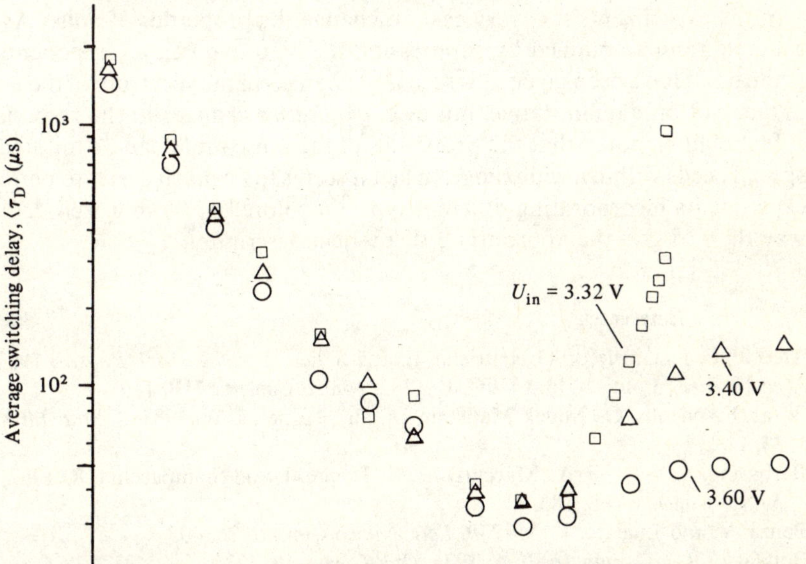

Figure 6.20. Average switching delay $\langle \tau_D \rangle$ versus modulation frequency for different average voltages below (3.32 V) and above (others) threshold in the electronic system. (From Deserno *et al.*, 1986.)

6.5 Conclusion

The experiments presented here demonstrate very clearly that the phenomenon of noise-induced bistability exists and that it is in agreement with theoretical expectations. Moreover they show that the *qualitative* features do not depend strongly on the physical nature of the system, i.e. no important differences are found between an optical and an electronic bistable system. It might be concluded that the similarity is a consequence of the topological structure of the potential surface describing the motion of the system. This idea has to be generalized, however, since the qualitative behavior is also observed if a potential cannot be defined.

From the experimental point of view, the electronic system is much more convenient. Moreover, the intrinsic fluctuations and the problems of drift of parameters, etc., are much less severe. Therefore some features can easily be observed in the electronic system, which cannot yet be observed in the optical one. As an example experimental evidence has been given that bimodality does *not* occur at *very weak* noise levels. A second example is the determination of the average switching delay in the immediate vicinity of switching threshold; these observations, which are only possible in the electronic system up to now, illustrate the close relation between the so-called 'critical slowing down' and first passage time problems.

In the experiments it is very easy to change the properties of noise. As an example results obtained by suppressing the high frequency components of white noise have been given; it is seen that changes in the spectrum of the noise induce not only quantitative, but even *qualitative* changes in the response.

It should be noted that the properties of the noise input can be tailored to special needs within a wide range; in fact it seems to be much easier to perform experiments incorporating different types of colored noise in a well-defined way than to give the appropriate theoretical description.

References

Arecchi, F. T., Giusfredi, G., Petriella, E. and Salieri, P. 1982. *Appl. Phys. B* **29**, 79.
Arecchi, F. T., Politi, A. and Ulivi, L. 1982. *Nuovo Cimento* **71B**, 119.
Baras, F., Nicolis, G., Malek Mansour, M. and Turner, J. W. 1983. *J. Stat. Phys.* **32**, 1.
Barbarino, S., Gozzini, A., Maccarrone, F., Longo, I. and Stampacchia, R. 1982. *Nuovo Cimento* **71B**, 183.
Benza, V. and Lugiato, L. A. 1979. *Lett. Nuovo Cimento* **26**, 405.
Bonifacio, R. and Lugiato, L. A. 1976. *Opt. Comm.* **19**, 172.
Bonifacio, R. and Lugiato, L. A. 1978. *Phys. Rev. A* **18**, 1129.
Bonifacio, R. and Meystre, P. 1978. *Opt. Comm.* **29**, 131.
Broggi, G. and Lugiato, L. A. 1984a. *Phys. Rev. A* **29**, 2949.
Broggi, G. and Lugiato, L. A. 1984b. *Phil. Trans. R. Soc. Lond. A* **313**, 425.
Broggi, G., Lugiato, L. A. and Colombo, A. 1985. *Phys. Rev. A* **32**, 2803.

Carmichael, H. J. 1986. In *Optical Instabilities* (R. W. Boyd, M. G. Raymer and L. M. Narducci, eds.), p. 111. Cambridge University Press.

Chow, W. W., Scully, M. O. and van Stryland, E. W. 1975. *Opt. Comm.* **15**, 6.

Cribier, S., Giacobino, E. and Grynberg, G. 1983. *Opt. Commun.* **47**, 170.

Deserno, R., Kumme, R., Mitschke, F., Lange, W. and Mlynek, J. 1986. *IOCC-1986 International Optical Computing Conference* (J. Shamir, A. A. Friesem and E. Marom, eds.), *Proc. SPIE 700*, p. 83. Bellingham: SPIE.

Garmire, E., Marburger, J. H., Allen, S. D. and Winful, H. G. 1979. *Appl. Phys. Lett.* **34**, 374.

Gibbs, H. M. 1985. *Optical Bistability: Controlling Light With Light.* Orlando: Academic Press.

Gibbs, H. M., Hopf, F. A., Kaplan, D. L. and Shoemaker, R. L. 1981a. *Phys. Rev. Lett.* **46**, 474.

Gibbs, H. M., Hopf, F. A., Kaplan, D. L. and Shoemaker, R. L. 1981b. *J. Opt. Soc. Am.* **71**, 367.

Gibbs, H. M., McCall, S. L. and Venkatesan, T. N. C. 1976. *Phys. Rev. Lett.* **36**, 1135.

Grant, D. E. and Kimble, H. J. 1983. *Opt. Comm.* **44**, 415.

Grynberg, G. and Cribier, S. 1983. *J. de Phys. Lett.* **44**, 4449.

Haken, H. 1964. *Z. Phys.* **181**, 96.

Haken, H. 1970. *Handbuch der Physik*, vol. XXV/2C. Berlin: Springer.

Haken, H. 1975. *Rev. Mod. Phys.* **47**, 47.

Haken, H. 1977. *Synergetics.* Berlin: Springer.

Hopf, F. A., Derstine, M. W., Gibbs, H. M. and Rushford, M. C. 1984. In *Optical Bistability 2* (C. M. Bowden, H. M. Gibbs and S. L. McCall, eds.), p. 67. New York: Plenum.

Horsthemke, W. and Lefever, R. 1984. *Noise-Induced Transitions.* Berlin: Springer.

Ikeda, K. 1979. *Opt. Comm.* **30**, 257.

Kramers, H. A. 1940. *Physica* **7**, 284.

Lange, W., Mitschke, F., Deserno, R. and Mlynek, J. 1985. *Phys. Rev. A* **32**, 1271.

Lange, W., Mitschke, F., Deserno, R. and Mlynek, J. 1986a. In *Optical Instabilities* (R. W. Boyd, M. G. Raymer and L. M. Narducci, eds.), p. 364. Cambridge University Press.

Lange, W., Deserno, R., Mitschke, F. and Mlynek, J. 1986b. In *Optical Bistability III* (H. M. Gibbs, P. Mandel, N. Peyghambarian and S. D. Smith, eds.), p. 213. Berlin: Springer.

Lange, W., Deserno, R., Kumme, R., Mitschke, F. and Mlynek, J. 1987. In *From Optical Bistability Towards Optical Computing – The EJOB Report* (P. Mandel, S. D. Smith and B. S. Wherrett, eds.), pp. 230–7. Amsterdam: North Holland.

Lugiato, L. A. 1984. *Progress in Optics*, vol. XXI, p. 69. Amsterdam: North-Holland.

Lugiato, L. A., Colombo, A., Broggi, G. and Horowicz, R. J. 1986. *Frontiers in Quantum Optics*, p. 231. Bristol: Adam Hilger.

McCall, S. L., Gibbs, H. M., Hopf, F. A., Kaplan, D. L. and Ovadia, S. 1982. *Appl. Phys. Lett. B* **28**, 99.

McCall, S. L., Ovadia, S., Gibbs, H. M., Hopf, F. A. and Kaplan, D. L. 1985. *IEEE J. Quant. Electron.* **QE-21**, 1441.

Mandel, P. and Erneux, T. 1982. *Opt. Comm.* **42**, 362.
Mejerowitsch, L. A. and Selitschenko, L. G. 1960. *Impulstechnik.* Stuttgart: Berliner Union.
Mitschke, F., Deserno, R., Lange, W. and Mlynek, J. 1986. *Phys. Rev. A* **33**, 3219.
Mitschke, F., Deserno, R., Mlynek, J. and Lange, W. 1983. *Opt. Comm.* **46**, 135.
Mitschke, F., Deserno, R., Mlynek, J. and Lange, W. 1985. *IEEE J. Quant. Electron.* **QE-21**, 1435.
Mlynek, J., Mitschke, F., Deserno, R. and Lange, W. 1982. *Appl. Phys. B* **28**, 135.
Mlynek, J., Mitschke, F., Deserno, R. and Lange, W. 1984. *Phys. Rev. A* **29**, 1297.
Risken, H. 1984. *The Fokker–Planck Equation.* Berlin: Springer.
Sargent III, M., Scully, M. O. and Lamb, Jr., W. E. 1974. *Laser Physics.* Reading, MA: Addison-Wesley.
Schenzle, A. and Brand, H. 1979. *Opt. Comm.* **31**, 401.
Schmitt, O. H. 1938. *J. Sci. Instrum.*, p. 24.
Scott, J. F., Sargent III, M. and Cantrell, C. D. 1975. *Opt. Comm.* **15**, 13.
Shenoy, S. R. 1984. *Phys. Rev. A* **30**, 2849.
Smith, P. W. and Turner, E. H. 1977. *Appl. Phys. Lett.* **30**, 280.

7 Computer experiments in non-linear stochastic physics

RICCARDO MANNELLA

7.1 Introduction

This chapter is devoted to developing digital techniques to integrate multi-dimensional stochastic differential equations. A variety of different spectral densities of the stochastic forcing can be achieved with the algorithms presented.

Very often (for example, when several dimensions are involved, or an external deterministic forcing is present, or fluctuation–dissipation relations do not hold for the physical system modelled by the set of stochastic differential equations) a digital simulation is the only viable way of extracting the detailed information one is looking for. Also, digital techniques can simulate the theoretical model in a sort of ideal environment, where everything is under control and no non-idealities are present. It is actually the author's personal belief that digital simulation in stochastic physics should be thought of as a theoretical tool, like, say, Padé approximations or steepest descents, and it should be considered as the natural complement of modelling.

The chapter is organized as follows: in Section 7.2 a general algorithm for integrating sets of autonomous differential equations in the presence of just one external stochastic forcing is derived. In Section 7.3 the algorithm is generalized to include non-autonomous differential equations and multi-dimensional external forcings. Sections 7.4 and 7.5 contain applications to, respectively, overdamped and underdamped non-linear oscillators: what is presented in these two sections should be regarded as examples of digital simulation capabilities, and papers cited there should be consulted for more details and references.

In this chapter 'recipes' for dealing with colored (rate of fluctuation comparable to the time scales of the systems) or white noises are given in terms of different algorithms. An appendix (Section 7.7) is devoted to the case where one insists in wanting to take the limit of white noise in the algorithm for colored noise. In the same appendix the problem of non-linear stochastic forcing is briefly addressed, too.

7.2 Basic numerical algorithm

Several algorithms to integrate autonomous or non-autonomous stochastic differential equations have been proposed: predictor–corrector schemes (Borkovec, Straub and Berne, 1986; Fronzoni *et al.*, 1986; Schneider and Stoll 1978; Schöbinger, Koch and Abraham, 1986; Straub and Berne, 1985; Straub, Borkovec and Berne, 1985, 1986; Straub, Hsu and Berne, 1985), Runge–Kutta approaches (Drummond, Duane and Horgan, 1984, 1987; Drummond, Hoch and Horgan, 1986; Drummond and Horgan, 1986; Greensite and Helfand, 1981; Klauder and Petersen, 1985), one-step collocations (Munakata, 1986; Rao, Borwankar and Ramkrishna, 1974; Sancho, San Miguel, Katz and Gunton, 1982; Smythe, Moss, McClintock and Clarkson, 1983). In this section we shall develop a one-step collocation to solve systems of autonomous stochastic differential equations: the method stems directly from Rao, Borwankar and Ramkrishna (1974), and is a generalization to multidimensional systems.

Before deriving the algorithm, we would like to discuss briefly why we opted for a one-step collocation scheme. As far as predictor–corrector methods are concerned, the basic assumption is that the 'noise' (the stochastic term) is thought of as a constant during each integration time step, to be updated via a suitable rule before the following step.

Given the simple stochastic differential equation,

$$\begin{aligned} &\dot{x} = f(x) + \xi(t) \\ &\xi(t) \text{ white, Gaussian} \end{aligned} \tag{7.2.1}$$

and a simple explicit predictor, the value of $x_n \equiv x(nh)$ (h is the integration time step) is computed as

$$x_n = \sum_{i=1}^{N} a_i x_{n-i} + \sum_{i=1}^{N'} b_i \dot{x}_{n-i}, \tag{7.2.2}$$

where a_n and b_n are determined by making (7.2.2) 'right' for an x-dependence on time of the form $x(t) \sim t^m$ (with $m = N + N' - 1$) (Lapidus and Seinfeld, 1971): it is clear that high-order predictors cannot be adopted for (7.2.1), because they would lead to a continuous and differentiable $x(t)$, whereas $\dot{x}(t)$ in (7.1.1) can only be differentiated in a distributional sense because of the random process present on the right hand side.

As far as Runge–Kutta algorithms are concerned, we have more than one objection. First, when dealing with numerical integration, it is impossible not to introduce a coarse-graining of the 'time', through the 'integration time step' (h). Now in the presence of a stochastic process, we would like to think of those 'instants' spaced by h in time as instants at which a 'snap-shot' photograph of the process is obtained. It is not quite clear to us why a time scale shorter than h should or could be justified.

The second objection is similar to the one we made regarding predictor–

corrector schemes: with a Runge–Kutta the value of $x_n \equiv x(nh)$ is computed via a suitable linear combination of the form

$$x_n = x_{n-1} + \sum_{i=1}^{N} w_i \dot{x}_i \tag{7.2.3}$$

where the $\dot{x}_i$ are computed at t_i with $(n-1)h \leqslant t_1 \leqslant \cdots \leqslant t_N \leqslant nh$. However, the expression for $x(t)$ in the interval $[t_i, t_{i+1}]$ is supposed to be analytical (Lapidus and Seinfeld, 1971).

The last objection is that in the time coarse-graining process it is implicit that we are computing what the average effect of the random forcing would be over the small time h; whereas in a Runge–Kutta scheme the w_i, t_i in (7.2.3) are chosen to yield the 'correct' value of the x-moments (up to, say, $O(h^2)$), the averages being computed over the whole phase space of the 'noise' (or, via the assumption of random processes, time averages computed over times much longer than h): we believe that, where a Runge–Kutta will yield a good description of quantities like probability distributions, problems may arise where quantities like dynamics at very short times must be dealt with.

Let us now turn to our one-step collocation algorithm. First we shall consider just one source of external 'noise', driving a system of autonomous differential equations

$$\begin{aligned} \dot{x}_i &= f_i(x_i(t)) + g_i(x_i(t))\xi(t) \\ \langle \xi(t)\xi(0) \rangle &= \delta(t), \end{aligned} \tag{7.2.4}$$

where ξ is Gaussian, the f_i are the deterministic forces, and the g_i are the couplings between the noise and the 'state variables' x_i. We shall interpret (7.2.4) in the Stratonovich sense (see Section 7.3) throughout this chapter, so that the rules of standard differential and integral calculus do apply. An alternative way of looking at this problem would be to shift back to white noise in the result after evaluating an ordinary integral (Arnold, 1974).

In order to integrate (7.2.4) we coarse-grain the time by introducing an 'integration time' h. A summation over repeated indices is implicit from now on. We can write (formally)

$$\begin{aligned} x_i(h) - x_i(0) &= \int_0^h \dot{x}_i(s)\,\mathrm{d}s \\ &= \int_0^h f_i(\mathbf{x}(s))\,\mathrm{d}s + \int_0^h g_i(\mathbf{x}(s))\xi(s)\,\mathrm{d}s. \end{aligned} \tag{7.2.5}$$

This equation is a formal expression, because the right hand side still depends on $x(s)$. Let us define $\delta x_i(t) \equiv x_i(t) - x_i(0)$ and Taylor expand f_i and g_i, in $x(t)$,

$$\begin{aligned} f_i(x_i(t)) &= f_i(x_i(0)) + \partial_\kappa f_i(x_i(0))\delta x_\kappa(t) \\ &\quad + \frac{1}{2}[\partial_J \partial_\kappa f_i(x_i(0))]\delta x_\kappa(t)\delta x_J(t) + \cdots \end{aligned} \tag{7.2.6a}$$

$$g_i(x_i(t)) = g_i(x_i(0)) + \partial_\kappa g_i(x_i(0))\delta x_\kappa(t)$$
$$+\frac{1}{2}[\partial_J\partial_\kappa g_i(x_i(0))]\delta x_J(t)\delta x_\kappa(t)$$
$$+\frac{1}{3!}[\partial_n\partial_J\partial_\kappa g_i(x_i(0))]\delta x_J(t)\delta x_\kappa(t)\delta x_n(t)+\cdots. \qquad (7.2.6b)$$

Let us define $f_i^0 \equiv f_i(x_i(0))$, $g_{i,J\kappa}^0 \equiv \partial_J\partial_\kappa g_i(x_i(0))$, etc. There is no assumption on the smallness of our $\delta x_i(t)$ in (7.2.6a, b). The only hypothesis we are now making is that h is small, so that an expansion in leading orders of h is possible.

We can write

$$\delta x_i(h) = \int_0^h \{f_i^0 + f_{i,\kappa}^0\delta x_\kappa(s) + \tfrac{1}{2}f_{i,\kappa J}^0\delta x_J(s)\delta x_\kappa(x)\,ds + \cdots\}\,ds$$
$$+\int_0^h \xi(s)\{g_i^0 + g_{i,\kappa}^0\delta x_\kappa(s) + \tfrac{1}{2}g_{i,\kappa J}^0\delta x_\kappa(s)\delta x_J(s)$$
$$+\frac{1}{3!}g_{i,\kappa Jn}^0\delta x_\kappa(s)\delta x_J(s)\delta x_n(s) + \cdots\}\,ds. \qquad (7.2.7)$$

The expansions in (7.2.6a, b) and (7.2.7) will be halted when the required precision in leading orders of h is achieved. In this spirit, we shall group together terms of the same order in h; the lower contributions to $\delta x_i(h)$ will be substituted back into (7.2.7) to obtain a higher order approximation of $\delta x_i(h)$, and so on, until the required order in h is reached.

At the lowest possible order $(O(h^{1/2}))$* we have from (7.2.7)

$$\delta x_i^0(h) = g_i^0 Z_1(h) + o(h) \qquad (7.2.8)$$

$$Z_1(h) \equiv \int_0^h \xi(s)\,ds, \qquad (7.2.9)$$

where $Z_1(h)$ is a Gaussian, Wiener process with zero average and standard deviation h.

Substituting (7.2.8) and (7.2.9) back into (7.2.7) $(O(h))$, we obtain

$$\delta x_i'(h) = \int_0^h f_i^0\,dt + \int_0^h g_{i,\kappa}^0\delta x_\kappa^0(s)\xi(s)\,ds + o(h^{3/2})$$
$$= f_i^0 h + g_\kappa^0 g_{i,\kappa}^0\int_0^h Z_1(s)\xi(s)\,ds. \qquad (7.2.10)$$

We are going to use the Stratonovich calculus. This means that

$$\int_0^h Z_1(s)\xi(s)\,ds = Z_1(h)Z_1(h) - \int_0^h \xi(s)Z_1(s)\,ds, \qquad (7.2.11)$$

*$O(h^\alpha) \equiv$ order α is significant; $o(h^\alpha) \equiv$ order α is not significant.

and

$$\int_0^h Z_1(s)\xi(s)\,ds = \tfrac{1}{2}[Z_1(h)]^2. \tag{7.2.12}$$

Collecting all terms (at $O(h)$)

$$\delta x_i'(h) = hf_i^0 + \tfrac{1}{2}g_\kappa^0 g_{i,\kappa}^0 [Z_1(h)]^2 + o(h^{3/2}). \tag{7.2.13}$$

Using (7.2.8) and (7.2.13) in (7.2.7) (at $O(h^{3/2})$) we have

$$\delta x_i''(h) = \int_0^h f_{i,\kappa}^0 \delta x_\kappa^0(s)\,ds + \int_0^h g_{i,\kappa}^0 \delta x_\kappa'(s)\xi(s)\,ds$$
$$+ \frac{1}{2}\int_0^h g_{i,J\kappa}^0 \delta x_J^0(s)\delta x_\kappa^0(s)\xi(s)\,ds + o(h^2). \tag{7.2.14}$$

The separate terms yield

$$\int_0^h f_{i,\kappa}^0 \delta x_\kappa^0(s)\,ds = f_{i,\kappa}^0 g_\kappa^0 \int_0^h Z_1(s)\,ds$$
$$= f_{i,\kappa}^0 g_\kappa^0 Z_2(h), \tag{7.2.15}$$

$$Z_2(h) \equiv \int_0^h Z_1(s)\,ds, \tag{7.2.16}$$

where $Z_2(h)$, a linear combination of Gaussian, Wiener processes, is again a Gaussian, Wiener process with the following properties (Rao, Borwankar and Ramakrishna, 1974):

$$\langle Z_2(h)\rangle = 0; \langle (Z_2(h))^2\rangle = \tfrac{1}{3}h^3; \langle Z_1(h)Z_2(h)\rangle = \tfrac{1}{2}h^2. \tag{7.2.17}$$

A suitable expression for $Z_2(h)$ is

$$Z_2(h) = h\left\{\frac{Z_1(h)}{2} + \frac{Y_1(h)}{2\sqrt{3}}\right\}, \tag{7.2.18}$$

where $Y_1(h)$ is another Gaussian, Wiener process with zero average, standard deviation h and statistically independent of $Z_1(h)$ (see also Sancho *et al.*, 1982). The other terms in (7.2.14) yield

$$\int_0^h g_{i,\kappa}^0 \delta x_\kappa'(s)\xi(s)\,ds = g_{i,\kappa}^0 \left\{ f_\kappa^0 \int_0^h s\xi(s)\,ds \right.$$
$$\left. + \tfrac{1}{2}g_J^0 g_{\kappa,J}^0 \int_0^h (Z_1(s))^2 \xi(s)\,ds \right\}$$
$$= g_{i,\kappa}^0 f_\kappa^0 [hZ_1(h) - Z_2(h)]$$
$$+ \tfrac{1}{6}g_J^0 g_{\kappa,J}^0 g_{i,\kappa}^0 (Z_1(h))^3, \tag{7.2.19}$$

$$\frac{1}{2}\int_0^h g^0_{i,\kappa J}\delta x^0_J(s)\delta x^0_\kappa(s)\xi(s)\,\mathrm{d}s = \tfrac{1}{2}g^0_{i,\kappa J}g^0_J g^0_\kappa \int_0^h (Z_1(s))^2\xi(s)\mathrm{d}s$$
$$= \tfrac{1}{6}g^0_J g^0_\kappa g^0_{i,J\kappa}(Z_1(h))^3, \qquad (7.2.20)$$

where in both (7.2.19) and (7.2.20) standard partial integration (i.e. the Stratonovich calculus) has been widely used.
Collecting together (7.2.15), (7.2.19) and (7.2.20) ($\mathrm{O}(h^{3/2})$)

$$\delta x''_i(h) = Z_2(h)\{g^0_\kappa f^0_{i,\kappa} - f^0_\kappa g^0_{i,\kappa}\} + hZ_1(h) f^0_\kappa g^0_{i,\kappa}$$
$$+ \tfrac{1}{6}(Z_1(h))^3\{g^0_J g^0_{\kappa,J} g^0_{\kappa,i} + g^0_J g^0_\kappa g^0_{i,J\kappa}\} + \mathrm{o}(h^2). \qquad (7.2.21)$$

Our goal is an algorithm $\mathrm{O}(h^2)$ or $\mathrm{o}(h^{5/2})$. This requires that we evaluate the next contribution to (7.2.7). The algebra is a bit more involved.

$$\delta x'''_i(h) = \int_0^h f^0_{i,\kappa}\,\delta x'_\kappa(s)\,\mathrm{d}s + \frac{1}{2}\int_0^h f^0_{i,\kappa J}\delta x^0_\kappa(s)\delta x^0_J(s)\,\mathrm{d}s$$
$$+ \int_0^h g^0_{i,\kappa}\delta x''_\kappa(s)\xi(s)\,\mathrm{d}s + \frac{1}{2}\int_0^h g^0_{i,\kappa J}\delta x^0_J(s)\delta x'_\kappa(s)\,\mathrm{d}s$$
$$+ \frac{1}{3!}\int_0^h g^0_{i,J\kappa n}\delta x^0_J(s)\delta x^0_\kappa(s)\delta x^0_n(s)\xi(s)\,\mathrm{d}s + \mathrm{o}(h^{3/2}).$$
$$(7.2.22)$$

First of all, we shall compute some quantities needed in evaluating the right hand side of (7.2.22). Define

$$Z_3(h) \equiv \int_0^h Z_2(s)\xi(s)\,\mathrm{d}s \qquad (7.2.23\mathrm{a})$$

which is a random *non-Gaussian* variable. Its moments can be computed as follows:

$$\langle Z_3(h)\rangle = 0, \quad \langle [Z_3(h)]^2\rangle = \frac{h^4}{12}$$
$$\langle Z_1(h)Z_3(h)\rangle = 0, \quad \langle Z_2(h)Z_3(h)\rangle = 0$$
$$\langle [Z_1(h)]^2 Z_3(h)\rangle = \tfrac{1}{3}h^3, \quad \langle Z_1(h)Z_2(h)Z_3(h)\rangle = \frac{h^4}{12}$$
$$\langle [Z_2(h)]^2 Z_3(h)\rangle = \frac{h^5}{15}.$$

The best approximation of $Z_3(h)$ that we have found is

$$Z_3(h) = \frac{h}{6}\{[Z_1(h)]^2 - h\}. \qquad (7.2.23\mathrm{b})$$

Using (7.2.23b) for $Z_3(h)$ there will be an error at $\mathrm{O}(h^5)$ on the moment $\langle [Z_2(h)]^2 Z_3(h)\rangle$, an error which amounts to $h^5/60$.

Other useful quantities are:

$$\int_0^h (Z_1(s))^2\,ds = Z_1(h)Z_2(h) - \int_0^h Z_2(s)\xi(s)\,ds$$
$$= Z_1(h)Z_2(h) - Z_3(h)$$
$$\int_0^h (Z_1(s))^3\,\xi(s)\,ds = \tfrac{1}{4}(Z_1(h))^4$$
$$\int_0^h Z_1(s)s\xi(s)\,ds = \frac{h}{2}(Z_1(h))^2 - \tfrac{1}{2}[Z_1(h)Z_2(h) - Z_3(h)], \qquad (7.2.24)$$

where the Stratonovich calculus has been extensively exploited to partially integrate these expressions.

Using (7.2.23a) and (7.2.24) in (7.2.22) we obtain

$$\int_0^h f^0_{i,\kappa}\,\delta x'_\kappa(s)\,ds = f^0_{i,\kappa}\left\{ f^0_\kappa \int_0^h s\,ds + \tfrac{1}{2} g^0_J g^0_{\kappa,J} \int_0^h (Z_1(s))^2\,ds \right\}$$
$$= f^0_{i,\kappa}\left\{ f^0_\kappa \frac{h^2}{2} + \tfrac{1}{2} g^0_J g^0_{\kappa,J}[Z_1(h)Z_2(h) - Z_3(h)] \right\}, \qquad (7.2.25)$$

$$\frac{1}{2}\int_0^h f^0_{i,\kappa J}\,\delta x^0_\kappa(s)\,\delta x^0_J(s)\,ds = \tfrac{1}{2} g^0_\kappa g^0_J f^0_{i,J\kappa} \int_0^h (Z_1(s))^2\,ds$$
$$= \tfrac{1}{2} g^0_\kappa g^0_J f^0_{i,J\kappa}[Z_1(h)Z_2(h) - Z_3(h)], \qquad (7.2.26)$$

$$\int_0^h g^0_{i,\kappa}\,\delta x''_\kappa(s)\xi(s)\,ds = g^0_{i,\kappa}\{g^0_J f^0_{\kappa,J} - f^0_J g^0_{\kappa,J}\} \int_0^h Z_2(s)\xi(s)\,ds$$
$$+ g^0_{i,\kappa} g^0_{\kappa,J} f^0_J \int_0^h sZ_1(s)\xi(s)\,ds$$
$$+ \tfrac{1}{6} g^0_{i,\kappa}\{g^0_J g^0_{n,J} g^0_{n,\kappa} + g^0_J g^0_n g^0_{\kappa,Jn}\}$$
$$\times \int_0^h Z_1(s))^3\,\xi(s)\,ds$$
$$= g^0_{i,\kappa}\{g^0_J f^0_{\kappa,J} - f^0_J g^0_{\kappa,J}\} Z_3(h) + g^0_{i,\kappa} g^0_{\kappa,J} f^0_J$$
$$\times \left[\frac{h}{2}(Z_1(h))^2 - \tfrac{1}{2} Z_1(h)Z_2(h) + \tfrac{1}{2} Z_3(h) \right]$$
$$+ \frac{1}{4!} g^0_{i,\kappa}\{g^0_J g^0_{n,J} g^0_{n,\kappa} + g^0_J g^0_n g^0_{\kappa,Jn}\}(Z_1(h))^4, \qquad (7.2.27)$$

$$\frac{1}{2}\int_0^h g^0_{i,J\kappa}\,\delta x^0_J(s)\,\delta x'_\kappa(s)\xi(s)\,ds$$
$$= \tfrac{1}{2} g^0_{i,J\kappa} g^0_J f^0_\kappa \int_0^h sZ_1(s)\xi(s)\,ds$$

$$+\tfrac{1}{4}g^0_{i,J\kappa}g^0_n g^0_J g^0_{\kappa,n}\int_0^h (Z_1(s))^2\xi(s)\,ds$$

$$=\tfrac{1}{2}g^0_{i,J\kappa}g^0_J f^0_\kappa\left[\frac{h}{2}(Z_1(h))^2-\tfrac{1}{2}(Z_1(h)Z_2(h)-Z_3(h))\right]$$

$$+\tfrac{1}{12}g^0_{i,J\kappa}g^0_{\kappa,n}g^0_n g^0_J(Z_1(h))^3, \qquad (7.2.28)$$

$$\frac{1}{3!}\int_0^h g^0_{i,Jn\kappa}\delta x^0_J(s)\delta x^0_\kappa(s)\delta x^0_n(s)\xi(s)\,ds$$

$$=\frac{1}{3!}g^0_{i,Jn\kappa}g^0_J g^0_n g^0_\kappa\int_0^h (Z_1(s))^3\xi(s)ds$$

$$+\frac{1}{4!}g^0_J g^0_n g^0_\kappa g^0_{i,Jn\kappa}(Z_1(h))^4, \qquad (7.2.29)$$

and finally at $O(h^2)$

$$\delta x'''_i=(7.2.25)+(7.2.26)+(7.2.27)+(7.2.28)$$
$$+(7.2.29)+o(h^{5/2}). \qquad (7.2.30)$$

The final algorithm reads

$$x_i(h)=x_i(0)+\delta x^0_i(h)+\delta x'_i(h)+\delta x''_i(h)$$
$$+\delta x'''_i(h)+o(h^{5/2}), \qquad (7.2.31)$$

and it will be generally used in this form.

7.3 Comments and generalizations

The algorithm devised in Section 7.2 can be cast in a very simple form in the presence of purely additive stochastic forcing (i.e. $g_{i,J}=0\ \forall_J$). From (7.2.31) we find

$$x_i(h)=x_i(0)+g^0_i Z_1(h)+hf^0_i+f^0_{i,J}g^0_J Z_2(h)+\frac{h^2}{2}f^0_\kappa f_{i,\kappa}$$
$$+\tfrac{1}{2}g^0_\kappa g^0_J f^0_{i,\kappa J}[Z_1(h)Z_2(h)-Z_3(h)]+o(h^{5/2}). \qquad (7.3.1)$$

It must be stressed that the discrete version of (7.2.4) at $O(h)$ (from 7.2.13) is

$$x_i(h)=x_i(0)+g^0_i Z_i(h)+hf^0_i+\tfrac{1}{2}g^0_\kappa g^0_{i,\kappa}[Z_1(h)]^2. \qquad (7.3.2a)$$

In the literature, instead of (7.3.2a), one often finds the following equation:

$$x_i(h)=x_i(0)+g^0_i Z_1(h)+hf^0_i, \qquad (7.3.2b)$$

which would be correct only for additive stochastic forcing.

It is straightforward to extend the algorithm of the previous section to include non-autonomous f_is and g_is. The contributions from the time derivatives at $O(h^{3/2})$ and $O(h^2)$ are

$$O(h^{3/2}) \quad \dot{g}_i^0(hZ_1(h) - Z_2(h)) \tag{7.3.2c}$$

$$O(h^2) \quad \frac{h^2}{2}\dot{f}_i^0 + \tfrac{1}{2}g_\kappa^0 \dot{g}_{i,\kappa}^0[h(Z_1(h))^2 - (Z_1(h)Z_2(h) - Z_3(h))]$$
$$+ \tfrac{1}{2}\dot{g}_\kappa^0 g_{i,\kappa}^0[h(Z_1(h))^2 - (Z_1(h)Z_2(h) + Z_3(h))]. \tag{7.3.3}$$

To generalize (7.2.31) to a multi-dimensional stochastic process is not trivial. We have found an algorithm $O(h)$, but we were unable to obtain a higher order one-step collocation. Let us consider

$$\dot{x}_i = f_i(\mathbf{x}(t)) + g_i^J(\mathbf{x}(t))\xi^J(t),$$
$$\langle \xi^i \rangle = 0, \quad \langle \xi^i(t)\xi^J(0) \rangle = \delta_{iJ}\delta(t), \tag{7.3.4}$$
$$\xi^i \text{ Gaussian}, \quad J = 1, \ldots, N, \quad i = 1, \ldots, M.$$

We expand f_i and g_i as in (7.2.6a, b) and substitute back into (7.3.4). At $O(h^{1/2})$ (7.3.4) yields

$$\delta x_i(h) = g_i^{0J} Z_1^J(h) + o(h), \tag{7.3.5}$$

where, (7.2.9), $Z_1^J(h) \equiv \int_0^h \xi^J(s)\,ds$ is a Wiener process with zero average and standard deviation h.

At $O(h)$ we find

$$\delta x_i'(h) = hf_i^0 + \tfrac{1}{2}g_\kappa^{0J} g_{i,\kappa}^{0m} \int_0^h Z_1^J(s)\xi^m(s)ds + o(h^{3/2}). \tag{7.3.6}$$

Let us define

$$Z^{Jm}(h) \equiv \int_0^h Z_1^J(s)\xi^m(s)ds. \tag{7.3.7}$$

$Z^{Jm}(h)$ is not Gaussian: clearly for $J = m$ we have

$$Z^{JJ}(h) = \tfrac{1}{2}[Z_1^J(h)]^2. \tag{7.3.8}$$

The case $J \neq m$ is more complicated. The idea we shall follow is to compute the moments of $Z^{Jm}(h)$ up to, say, h^3, and with this information to find a random variable which has the same (statistical) properties of $Z^{Jm}(h)$. From (7.3.7) it follows that, when $J \neq m$,

$$\langle Z^{Jm}(h) \rangle = 0 \langle [Z^{Jm}(h)]^2 \rangle = \frac{h^2}{2}$$

$$\langle Z^{Jm}(h) Z_1^i(h) Z_1^\kappa(h) \rangle = \delta_{iJ}\delta_{m\kappa}\frac{h^2}{2},$$

$$\langle Z^{Jm}(h)Z_2^i(h)Z_1^\kappa(h)\rangle = \delta_{iJ}\delta_{m\kappa}\frac{h^3}{3} + \delta_{mi}\delta_{J\kappa}\frac{h^3}{6}, \tag{7.3.9}$$

$$\langle Z^{Jm}(h)Z_2^i(h)Z_2^\kappa(h)\rangle = \delta_{Ji}\delta_{m\kappa}\mathrm{o}(h^4),$$

$$\langle [Z^{Jm}(h)]^4\rangle = \mathrm{o}(h^4).$$

We have another condition to satisfy, namely – via partial integration from (7.2.7)–

$$Z^{Jm}(h) + Z^{mJ}(h) = Z_1^J(h)Z_1^m(h). \tag{7.3.10}$$

A possible random variable which satisfies both (7.3.9) and (7.3.10) might be

$$Z^{Jm}(h) = \tfrac{1}{2}Z_1^J(h)Z_1^m(h) + \frac{1}{h}\{Z_2^J(h)Z_1^m(h) - Z_1^J(h)Z_2^m(h)\} + \eta_{Jm} + \mathrm{o}(h^2), \tag{7.3.11}$$

where η_{Jm} is a Wiener process with zero average, standard deviation $h^2/12$, statistically independent of $Z_1^J(h), Z_1^m(h), Z_2^J(h), Z_2^m(h)$ and with the property

$$\eta_{Jm} = -\eta_{mJ}. \tag{7.3.12}$$

$Z_2^J(h)$ and $Z_2^m(h)$ in (7.3.9) are, by analogy with (7.2.16), time integrals of, respectively, $Z_1^J(h)$ and $Z_1^m(h)$.

Note that the derivation of (7.3.11) is very much empirical, and probably other expressions for $Z^{Jm}(h)$ could be found. Unfortunately the identification of stochastic terms which appear at the next order is almost impossible (moreover, it cannot be systematically pursued) and we are forced to limit ourselves to

$$x_i(h) = x_i(0) + g_i^{0J}Z_1^J(h) + hf_i^0 + \tfrac{1}{2}Z^{m\kappa}(h)g_J^{0m}g_{i,J}^{0\kappa} + \mathrm{o}(h^{3/2}). \tag{7.3.13}$$

Let us now briefly address the Ito–Stratonovich problem. The original equation (7.2.4) was to be interpreted in Stratonovich sense. This forced us to write (7.2.11) in the form

$$\int_0^h Z_1(s)\xi(s)\mathrm{d}s = Z_1(h) - \int_0^h \xi(s)Z_1(s)\mathrm{d}s \tag{7.3.14a}$$

$$\int_0^h Z_1(s)\xi(s)\mathrm{d}s = \tfrac{1}{2}[Z_1(h)]^2. \tag{7.3.14b}$$

However, in general (Arnold, 1974)

$$\int_0^h Z_1(s)\xi(s)\mathrm{d}s = \tfrac{1}{2}[Z_1(h)]^2 - \frac{ph}{2}, \tag{7.3.15}$$

where $p = 0 (= 1)$ results in the Stratonovich (Ito) form. It is very important to understand that the choice between Ito and Stratonovich forms is externally imposed on a numerical algorithm alongside the actual form of the stochastic differential equations. Indeed (7.2.31) can be adapted so as to mimic an evolution of (7.2.4) according to the Ito rule: the trick being to substitute for (7.2.4) an equivalent system of stochastic differential equations whose evolution according to the Stratonovich rule yields the Ito evolution of (7.2.4); and then apply (7.2.31) to the new set of equations (see also Arnold, 1974; Klauder and Petersen, 1985). It is straightforward to prove that, given

$$\begin{aligned} &\dot{x}_i = f_i + g_i^J \xi^J(t) \\ &\langle \xi^J(t) \rangle = 0 \quad \langle \xi^J(t) \xi^i(0) \rangle = \delta_{iJ} \delta(t) \\ &\xi^J(t) \text{ Gaussian,} \end{aligned} \tag{7.3.16}$$

and defining p via (7.3.15), the 'Stratonovich' evolution of

$$\dot{x}_i = \left(f_i - \frac{p}{2} g_\kappa^J g_{i,\kappa}^J \right) + g_i^J \xi^J(t) \tag{7.3.17}$$

is statistically equivalent to the 'p' evolution of (7.3.16).

The Gaussian random numbers entering the various algorithms can be generated in different ways. If speed is the only concern, the fastest algorithm is probably to generate 6 to 12 uniform deviates on $(-0.5, 0.5)$ whose average is returned as the Gaussian deviate. This algorithm is simple enough to be written in machine code and is quite fast. However, if the tails of the Gaussian deviates are important one must look for other algorithms.

Three different algorithms have been tried: the Box Müller formula; a transformed Box Müller (see, for instance, Knuth, 1981); and the Ziggurat method (Marsaglia and Tsang, 1984). The fastest is the last one, but unfortunately it requires a very peculiar uniform deviate, and on some machines the necessary code can only be written in machine language. FORTRAN listings of the Box Müller formula and transformed Box Müller are shown below for reference:

(7.3.18a) Box Müller:

```
      SUBROUTINE GAUS1 (Y1, Y2)
      DATA TWOPI/6.283185307/
C ROUTINE TO GENERATE X1, X2
      AUX = SQRT( - 2*ALOG(X1))
      Y2 = TWOPI*X2
      Y1 = AUX*SIN(Y2)
      Y2 = AUX*COS(Y2)
      RETURN
      END
```

(7.3.18b) Transformed Box Müller:

```
      SUBROUTINE GAUS2(Y1, Y2)
    1 CONTINUE
C ROUTINE TO GENERATE X1, X2
      Y1 = 2*X1 - 1
      Y2 = 2*X2 - 1
      AUX = Y1*Y1 + Y2*Y2
      IF(AUX.GT.1.0) GOTO 1
      AUX = SQRT(-2*ALOG(AUX)/AUX)
      Y1 = Y1*AUX
      Y2 = Y2*AUX
      RETURN
      END
```

In these listings, X1 and X2 are uniform deviates on (0, 1), Y1 and Y2 are the required Gaussian deviates, with zero average and standard deviation one. Even if calls to (slow) trigonometric functions are present, (7.3.18a) was found to be about 1.4 times faster than (7.3.18b) on a VAX 11/780 (the code for generating X1 and X2 was written in machine language). In all the following numeric experiments the routine GAUS1 was generally used.

7.4 Some applications to overdamped systems

In this section attention will be focussed on applying the algorithms devised in the previous sections to some overdamped systems of interest. The general form of the differential equation to be solved is

$$\dot{x} = f(x,t) + \text{noise}, \tag{7.4.1a}$$

which becomes

$$\dot{x} = f(x,t) + (2D)^{1/2}\xi(t) \tag{7.4.1b}$$

if $\xi(t)$ is white (and it will also be a Gaussian process with zero average and standard deviation one). If the noisy driving is to be colored the equation which will yield (7.4.1b) in the correct white noise limit can be cast in the form

$$\dot{x} = f(x,t) + y$$

$$\dot{y} = -\frac{1}{\tau_n}y + \frac{(2D)^{1/2}}{\tau_n}\xi(t), \tag{7.4.1c}$$

where τ_n is the correlation time of the colored noise and $\xi(t)$ is again a Gaussian, white noise (zero average, standard deviation one). This will ensure that

$$\langle y(t)y(s)\rangle = \frac{D}{\tau_n}e^{-|t-s|/\tau_n}$$

with $y(t)$ Gaussian, and that, in the limit $\tau_n \to 0$,

$$\langle y(t)y(s)\rangle = 2D\delta(t-s).$$

Using (7.3.1) and (7.3.2), (7.3.3), (7.4.1b) and (7.4.1c) yield

$$\begin{aligned} x_{t+h} &= x_t + (2D)^{1/2}Z_1(h) + hf(x_t,t) + \partial_x f(x_t,t)(2D)^{1/2}Z_2(h) \\ &\quad + \frac{h^2}{2}\{f(x_t,t)\partial_x f(x_t,t) + \dot{f}(x_t,t)\} \\ &\quad + 2D[Z_1(h)Z_2(h) - Z_3(h)]\,(\text{white noise}), \end{aligned} \tag{7.4.2a}$$

$$\begin{aligned} x_{t+h} &= x_t + h[f(x_t,t) + y_t] + \frac{(2D)^{1/2}}{\tau_n}Z_2(h) \\ &\quad + \frac{h^2}{2}\left[f(x_t,t)\partial_x f(x_t,t) - \frac{y}{\tau_n} + \dot{f}(x_t,t)\right] \\ y_{t+h} &= y_t + \frac{(2D)^{1/2}}{\tau_n}Z_1(h) - \frac{h}{\tau_n}y_t \\ &\quad - \frac{(2D)^{1/2}}{\tau_n^2}Z_2(h) + \frac{1}{2}\left(\frac{h}{\tau_n}\right)^2 y_t \;\; (\text{colored noise}). \end{aligned} \tag{7.4.2b}$$

The first computer simulation presented refers to the phenomenon of frequency locking in ring-laser gyroscopes. Many of the following ideas are borrowed from Vogel *et al.* (1987a, b). These two papers should be consulted for a thorough list of references. If ϕ is the phase difference between the two counterpropagating waves in a gyroscope and α is the Sagnac frequency (see, for instance, Aronowitz, 1977), the following equations of motion describe the beat note $\dot{\phi}$ between the two waves and the noise, respectively:

$$\begin{aligned} \dot{\phi} &= \alpha + \beta\sin\phi + \xi(t) \\ \langle \xi(t)\xi(s)\rangle &= \frac{D}{\tau_n}e^{-|t-s|/\tau_n}, \end{aligned} \tag{7.4.3}$$

where the term $\beta\sin\phi$, coupling the two counterpropagating waves, is due to imperfections in the cavity. Phenomenologically, this term causes (at low rotation rates, i.e. small α) $\dot{\phi}$ to disappear. The noise term is introduced to restore a finite value for $\langle\dot{\phi}\rangle$.

In the paper by Vogel *et al.* (1987b) a matrix continued fraction (MCF) technique (Risken, 1984) was applied to (7.4.3) (or, better, to its equivalent Fokker–Planck equation) to compute the value of $\langle\dot{\phi}\rangle$ in presence of noise. This theoretical analysis was tested for accuracy via a joint use of analogue and digital simulations. The full account of the digital simulation is herein presented. The numerical algorithm is given in (7.4.2a, b). The integration time step was 10^{-2} and the particle was followed for 2×10^5 time steps. The final

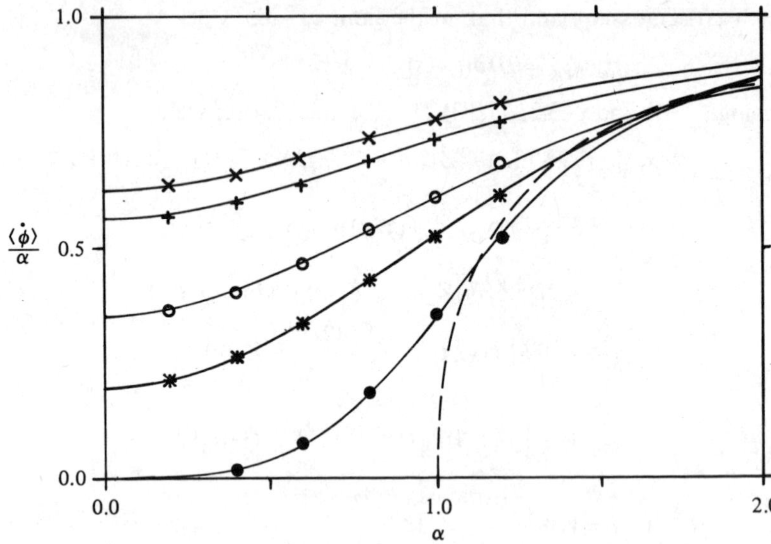

Figure 7.1. The average of the beat note $\dot{\phi}$ for the ring-laser gyroscope equations (7.4.3), normalized to the Sagnac frequency α, vs. α for different correlation times of the noise. $\beta = 1$, and the noise intensity $D = 1.0$. The symbols are digital simulations; the full lines are the theoretical predictions at the corresponding correlation times. The dashed line is the deterministic value of $\langle\dot{\phi}\rangle/\alpha$. Symbols: $\times$, $\tau_n = 0.0$; $+$, 0.2; $\circ$, 1.0; $*$, 2.0; $\bullet$, 10.0.

result for $\langle\dot{\phi}\rangle$ was obtained by averaging over five such runs. Diffusion far from the origin along the wash-board potential was prevented by reinjection on the opposite side whenever the particle left a chosen range (-2π to 2π). During a run the number of such reinjections was typically 10^2 to 10^4. Each computed value of $\langle\dot{\phi}\rangle$ required about 7×10^2 s of CPU time on a CDC7600, with GAUS1 as noise generator. The results are presented in Figure 7.1, compared with the theoretical predictions of Vogel *et al.* (1987b). Note that $\langle\dot{\phi}\rangle$ is normalized to α, the value which $\langle\dot{\phi}\rangle$ would assume if β were zero. The agreement between the MCF and digital simulation is impressive.

Before presenting the next computer experiment, it should be emphasized that (7.4.3) are of significance in many other physical systems; for example, in radio physics (see, for example, Stratonovich, 1967), Josephson junctions (see, for example, Büttiker, Harris and Landauer, 1983), charge density waves (see, for example, Grüner and Zettl, 1985). It is clear that in these systems too an alternative approach to obtain information about quantities of physical interest could well be straightforward numerical integration of the Langevin equation describing the model, in this example (7.4.3) (note that even in the MCF approach a very heavy use of the computer is necessary to invert the final matrices).

The second example considered in this section is a model of a swept parameter bifurcating system. Among other applications, it is worth noting the importance of such a model in optical bistability (see, for example, Mandel and Erneux, 1985) and in laser transitions (see, for example, Glorieaux and Dangoisse, 1985). The specific model studied is described by the equations

$$\dot{x} = -x\left[1 - \frac{A(t)}{1+x^2}\right] + y \quad \dot{y} = -\frac{1}{\tau_n}y + \frac{(2D)^{1/2}}{\tau_n}\xi(t), \tag{7.4.4a}$$

where $A(t)$ is the bifurcation parameter and obeys the equation

$$A(t) = A_0 + vt; \tag{7.4.4b}$$

$\xi(t)$ is a white, Gaussian process.

Equations (7.4.4a, b) have been already studied by Broggi, Colombo, Lugiato and Mandel (1986) as a model to represent the stochastic dynamics of a tuned, single mode, homogeneously broadened ring laser in the good cavity limit. The tool Broggi *et al.* (1986) used to study the behavior of x (the 'electrical field') was a Crank–Nicholson discretization method applied to the Fokker–Planck equation associated with (7.4.4a, b), however, taking the limit of $\tau_n \to 0$, ($y \to$ white noise). Analogue experiments with both white and colored additive forcings have been performed (Mannella, McClintock and Moss, 1987a, b). For more details, these papers and Broggi *et al.* (1986) should be consulted. Essentially (7.4.4a) describes a system which undergoes a bifurcation at $A = 1$ (for v in (7.4.4b) infinitesimally small): for $A < 1$ the stable equilibrium position is $x = 0$, whereas for $A > 1, x = 0$ becomes unstable and $x = \pm(A-1)^{1/2}$ are the stable solutions. For $v > 0$ the system exhibits a postponement in the value at which it bifurcates, depending on the intensity of the additive stochastic forcing and on the sweep rate v itself. The dependence on A_0 is washed out by the stochastic term, at the least in the limit of white noise: however, in Broggi *et al.* (1986) it has been argued that when the noise correlation time becomes comparable to the sweep rate such a dependence should be recovered.

The results of Mannella, McClintock and Moss (1987b) seem not to confirm this last prediction: that is, small or no dependence on A_0 was found even for noise correlation times comparable to the sweeping velocity, the whole scenario in presence of colored fluctuations closely resembling the picture found in presence of white fluctuations. In Mannella, McClintock and Moss (1987b) it was argued that, among other possible reasons, the small but always present internal noise (modelled as white, Mannella *et al.*, 1986) might destroy any dependence on A_0. In order to test this point, a numerical simulation seemed quite appropriate. The most sensitive indicator of the bifurcation is probably the distribution of the bifurcation points A_b (Broggi *et al.*, 1986) (or bifurcation times, $t_b \equiv A_b - A_0/v$), defining A_b as the value of $A(t)$ when $x^2(t)$ first crosses a preset threshold x_{th}^2. (As remarked in Mannella, McClintock and Moss, 1987b, this is not quite the same definition as made by Broggi *et al.* (1986) because, since their approach was to integrate the Fokker–Planck equation, *not* the Langevin equations (7.4.4a, b), they did not have available the

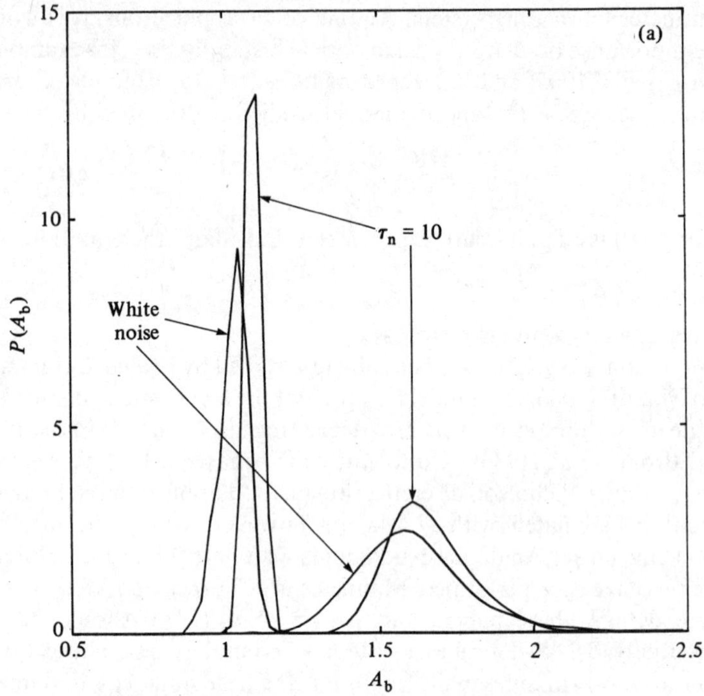

Figure 7.2. (a) Probability distributions of the bifurcation point for a model (7.4.4a, b) of optical bistability. The narrow, tall distributions are for $v = 10^{-3}$, the wide, low distributions are for $v = 10^{-1}$, and two different correlation times of the noise; $D = 10^{-3}$. (b) Probability distributions of the bifurcation point. $D = 10^{-3}$, $v = 10^{-1}$, $\tau_n = 10$. The full line is for $A_0 = 0.5$, and the dashed line is for $A_0 = -0.5$. Note there is little or no influence of different initial points (A_0) of the swept parameter.

stochastic trajectory of $x(t)$.) The computer experiments were performed with the algorithm of (7.4.2a, b): the time step was 10^{-2} and the stochastic trajectories were followed until the value $x_{th}^2 \equiv 0.1$ was reached. An average was performed over 500 to 1500 such runs. GAUS1 was used to generate the Gaussian deviates.

The starting point of the stochastic trajectory was always chosen to be the equilibrium one. The presence of the random term, of course, immediately 'kicks' x off its preset initial position, in a random way which is determined by the statistics of the stochastic forcing. Figure 7.2(a) shows the probability distribution of the bifurcation points for $D = 1 \times 10^{-3}$, $A_0 = 0.5$ and for two different v (10^{-1} and 10^{-3}) and two different correlation times of the noise (white and $\tau_n = 10$). The results for white noise are in very good agreement with both Broggi *et al.* (1986) and Mannella, McClintock and Moss (1987a), whereas the results for colored noise are in very good agreement with Mannella, McClintock and Moss (1987b). Figure 7.2(b) shows the equilibrium

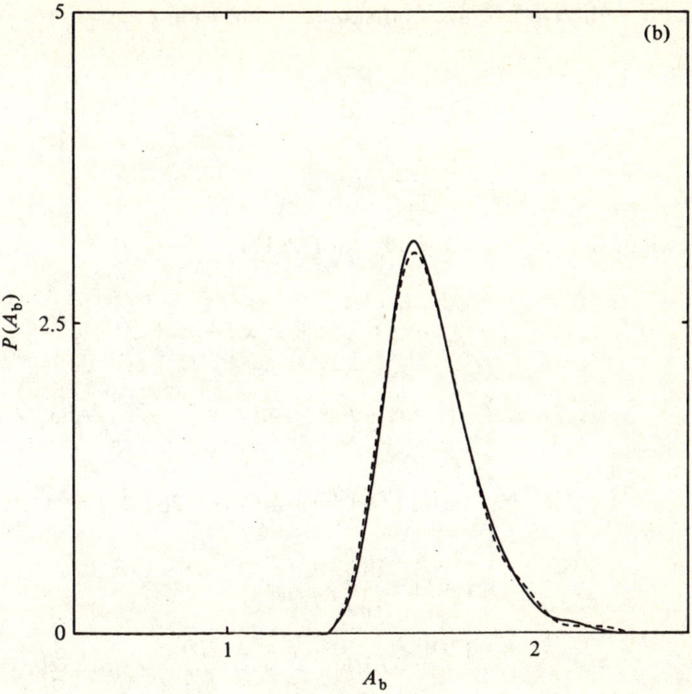

distributions obtained with $D = 10^{-3}$, $v = 10^{-1}$ and two different A_0 ($A_0 = -0.5, A_0 = 0.5$). It is clear that the two distributions are virtually the same, proving that there is little or no dependence on A_0, even in a situation where the rate of fluctuations is comparable with the sweeping time, and for which the starting point was chosen to be an equilibrium configuration. This suggests that the discrepancy found in Mannella, McClintock and Moss (1987b) between experiment and what was supposed in Broggi *et al.* (1986) cannot be accounted for in terms of internal noise.

7.5 Some applications to underdamped systems

In this section algorithm (7.2.31) is applied to some higher dimensionality systems. The general stochastic differential equation considered is (underdamped oscillator)

$$\dot{x} = v$$

$$\dot{v} = -\gamma v + F(x) + g(x)\xi(t)(2D)^{1/2}, \tag{7.5.1a}$$

where $\xi(t)$ is a Gaussian process with zero average and standard deviation one. If $\xi(t)$ is to be colored, (7.5.1a) will become (where $\xi(t)$ is still a Gaussian process with zero average and standard deviation one)

$$\dot{x} = v$$

$$\dot{v} = -\gamma v + F(x) + yg(x)$$

$$\dot{y} = -\frac{1}{\tau_n} y + \frac{(2D)^{1/2}}{\tau_n} \xi(t). \tag{7.5.1b}$$

The algorithm of (7.2.31) yields, for (7.5.1b)

$$x_{t+h} = x_t + hv_t + \frac{h^2}{2}\{-\gamma v_t + F(x_t) + y_t g(x_t)\}$$

$$\begin{aligned} v_{t+h} = {} & v_t + h\{-\gamma v_t + F(x_t) + g(x_t)y_t\} + \frac{(2D)^{1/2}}{\tau_n} g(x_t) Z_2(h) \\ & + \frac{h^2}{2}\{v_t(\partial_x F(x_t) + y_t \partial_x g(x_t)) - \gamma[-\gamma v_t + F(x_t) \\ & + y_t g(x_t)] - \frac{1}{\tau_n} g(x_t) y_t\} \end{aligned} \tag{7.5.2a}$$

$$y_{t+h} = y_t + \frac{(2D)^{1/2}}{\tau_n} Z_1(h) - \frac{h}{\tau_n} y_t - \frac{(2D)^{1/2}}{\tau_n^2} Z_2(h) + \frac{1}{2}\left(\frac{h}{\tau_n}\right)^2 y_t^2$$

and for (7.5.1a)

$$x_{t+h} = x_t + hv_t + Z_2(h) + \frac{h^2}{2}[-\gamma v_t + F(x_t)]$$

$$\begin{aligned} v_{t+h} = {} & v_t + (2D)^{1/2} g(x_t) Z_1(h) + h[-\gamma v_t + F(x_t)] \\ & + Dg(x_t)\partial_x g(x_t)[Z_1(h)]^2 \\ & - \gamma(2D)^{1/2} g(x_t) Z_2(h) + (2D)^{1/2} v_t g(x_t)[hZ_1(h) - Z_2(h)] \\ & + \frac{h^2}{2}\{-\gamma[-\gamma v_t + F(x_t)] + v_t \partial_x F(x_t)\}. \end{aligned} \tag{7.5.2b}$$

The first experiment presented is the study of the stochastic phase space portrait of a non-linear oscillator: $F(x)$ is in the form

$$F(x) = x - x^3 \tag{7.5.3a}$$

and

$$g(x) = 1 \tag{7.5.3b}$$

The theory, an analogue simulation and a preliminary digital simulation were presented elsewhere (Fronzoni *et al.*, 1986; Moss, Hänggi, Mannella and McClintock, 1986): these papers should be consulted for more details and references. Before discussing the present digital simulation, it is worth noticing that for the case $g(x)=1$ (7.5.2b) at order h is the (stochastic) analogue of the celebrated Verlet (or 'leap-frog') algorithm (Verlet, 1967), used also, for example, in Fronzoni *et al.* (1986), Moss *et al.* (1986) and Schöbinger, Koch and Abraham (1986).

At order h

$$x_{t+h}=x_t+hv_t$$
$$v_{t+h}=v_t+(2D)^{1/2}Z_1(h)-h\gamma v_t+hF(x_t) \tag{7.5.3c}$$

and, still at order h,

$$v_{t+h}=\frac{x_{t+h}-x_t}{h}\quad v_t=\frac{x_t-x_{t-h}}{h}\quad \gamma h v_t=\gamma\frac{x_{t+h}-x_{t-h}}{2} \tag{7.5.4}$$

and using (7.5.4) in (7.5.3c)

$$x_{t+h}=\frac{2}{1+(\gamma h/2)}x_t-\frac{1-(\gamma h/2)}{1+(\gamma h/2)}x_{t-h}$$
$$+\frac{h}{1+(\gamma h/2)}\{hF(x_t)+(2D)^{1/2}Z_1(h)\}. \tag{7.5.5}$$

It was found that the equilibrium distribution can be cast in the form (N is a normalization constant)

$$P_{\mathrm{eq}}(x,v,y)=N\exp(-v^2/2\sigma_v)\exp(-\{[(-x^2/2)+(x^4/4)]/\sigma_x\})$$
$$\times\exp(-(y^2/2\sigma_y)), \tag{7.5.6a}$$

where

$$\sigma_v=\frac{D}{\gamma}\frac{1}{1+\gamma\tau_{\mathrm{n}}+\tau_{\mathrm{n}}^2(3\langle x^2\rangle-1)} \tag{7.5.6b}$$

$$\sigma_x=\frac{D}{\gamma}\frac{1}{1+\tau_{\mathrm{n}}^2(3\langle x^2\rangle-1)/(1+\gamma\tau_{\mathrm{n}})} \tag{7.5.6c}$$

$$\sigma_y=\frac{D}{\tau_{\mathrm{n}}}. \tag{7.5.6d}$$

This result was obtained via a decoupling ansatz (Hänggi, Mroczkowski, Moss and McClintock, 1985). It was also shown that one-dimensional distributions are in quite good agreement with (7.5.6d) at surprisingly large τ_{n} and D (where the decoupling ansatz is bound to lose applicability). Unfortunately, the joint distributions (either y vs. x or v vs. x) are not well reproduced by (7.5.6d) as D and τ_{n} are increased. The so-called 'best Fokker–Planck' equation in leading order of τ_{n} (San Miguel and Sancho, 1980) is affected by the usual problems (negative diffusion coefficients for $\tau_{\mathrm{n}}\neq 0$ in some regions of the variables space). The full Fokker–Planck equation associated with (7.5.1b) and (7.5.3c) is

$$\dot{P}(x, v, y) = \left[-\partial_x v - \partial_v(x - x^3 - \gamma v + y) + \frac{1}{\tau_n}\partial_y\left(\frac{D}{\tau_n}\partial_y + y\right)\right] P(x, v, y), \tag{7.5.7}$$

which has the symmetry $(v, x) \to (-v, -x)$ and $(y, x) \to (-y, -x)$. This is what is found in the experiments (Figure 7.3) where the observed 'skewing' of the distributions y vs. x and v vs. x have the same symmetry (while the simpler symmetries $(x) \to (-x), (v) \to (-v)$ are violated). Equation (7.5.6a), symmetric under $(x, v) \to (-x, -v)$, is also even in each variable, failing to reproduce the mentioned skewing: note that, because of the symmetry $(x, v) \to (-x, -v)$, $\langle xv \rangle \equiv 0$. The disagreement is particularly acute as γ is increased. In the numerical simulation, the time step was $h = 10^{-2}$, the Ziggurat method (see previous sections) was used to generate the stochastic forcing and trajectories were followed for 15×10^6 time steps (about 8×10^2 s of CPU time on a VAX 11/780).

It has been argued in analogy with the one-dimensional case (J. M. Sancho and P. Grigolini, private communications) that a diffusive coefficient free from the problems of the 'best Fokker–Planck' approximation, but still depending on the state variables, could be the answer to such a discrepancy.

The problem is to solve (7.5.7) exactly, because a detailed balance is not present. Focussing only on x and v, the usual technique is to project out the noise, averaging over all its possible realizations. In Fronzoni *et al.* (1986) the evolution of $P_{red}(x, v)$ was cast in the form

$$\dot{P}_{red}(x, v) = [-\partial_x v - \partial_v(x - x^3 - \gamma v) + \tilde{D}\partial^2_{v^2} + \tau_n \tilde{D}\partial_x\partial_v] P_{red}(x, v), \tag{7.5.8a}$$

where

$$\tilde{D} = \frac{D}{1 + \gamma\tau_n + (3\langle x^2 \rangle - 1)\tau_n^2}, \tag{7.5.8b}$$

which yields (7.5.6a) as the equilibrium solution. If the decoupling ansatz (Hänggi *et al.*, 1985) is rejected, it can be proved (P. Grigolini, private communication; and also by generalising Fox, 1986, to multi-dimensional systems) that $\tilde{D}$ should be replaced by

$$\tilde{D}' = \frac{D}{1 + \gamma\tau_n + (3x^2 - 1)\tau_n^2}. \tag{7.5.8c}$$

This could provide the (x, v) 'mixing' which might account for the observed skewing, while the diffusion coefficient would still be positive over the whole

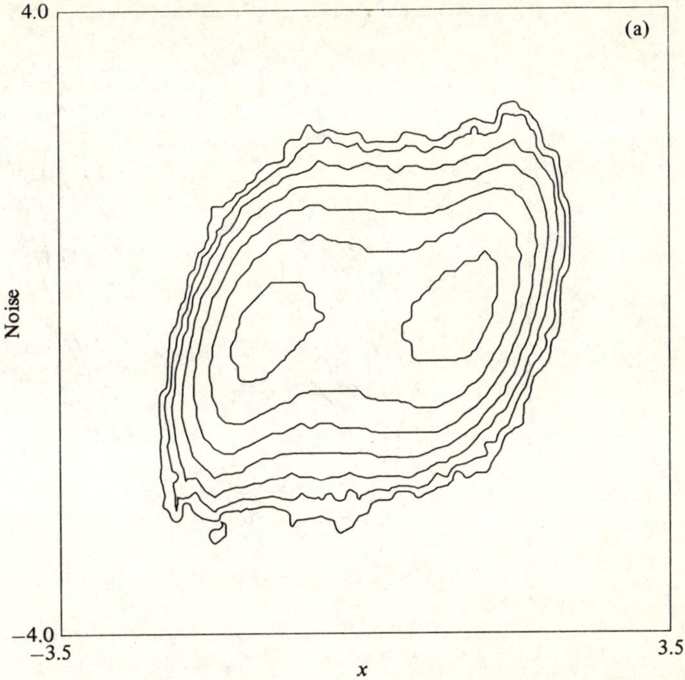

Figure 7.3. Experimental stochastic phase space portraits of the system described by (7.5.1a, b) and (7.5.3c) for $\gamma = 1.0$ and different values of D and τ_n. The natural logarithm of the equilibrium distribution is shown as a function of noise, velocity and x ('position', see 7.5.7). The contours are taken at 1/14, 3/14,..., 13/14 of the maximum. (a)–(c) Phase space portraits as functions of noise and x. (a) $D = 1.0, \tau_n = 1.0$; (b) $D = 1.0, \tau_n = 5.0$; (c) $D = 0.4, \tau_n = 5.0$. (d)–(f) Phase space portraits as functions of velocity and x. (d) $D = 1.0, \tau_n = 1.0$; (e) $D = 1.0, \tau_n = 5.0$; (f) $D = 0.4, \tau_n = 5.0$.

parameter space for a range of τ_n values. That a decoupling ansatz is not totally appropriate in situations when a suitable equivalent linear approximation (see Chapter 5, Volume 1) for the process cannot be found has been already shown (see, for example, Casademunt *et al.*, 1987; Faetti, Fronzoni, Grigolini and Mannella, 1988; Sancho, Mannella, McClintock and Moss, 1985). That in one-dimensional systems the approach of Fox (1986), which obviates some of the problems arising within a 'best Fokker–Planck' framework, can be better than the supposedly exact 'best Fokker–Planck' itself (even if it is 'wrong' when leading orders in τ_n are compared) has been proved in a variety of systems and situations (Faetti *et al.*, 1988; Jung and Hänggi, 1987; Lugiato, Mannella, McClintock and Grigolini, 1988). Clearly, it would be most interesting to solve (7.5.8a) with $\tilde{D}'$ provided by (7.5.8c). Unfortunately a detailed balance does not hold for (7.5.8a) either and one must resort to numerical techniques to compute the equilibrium distribution. Equation (7.5.8a) has been digitized on

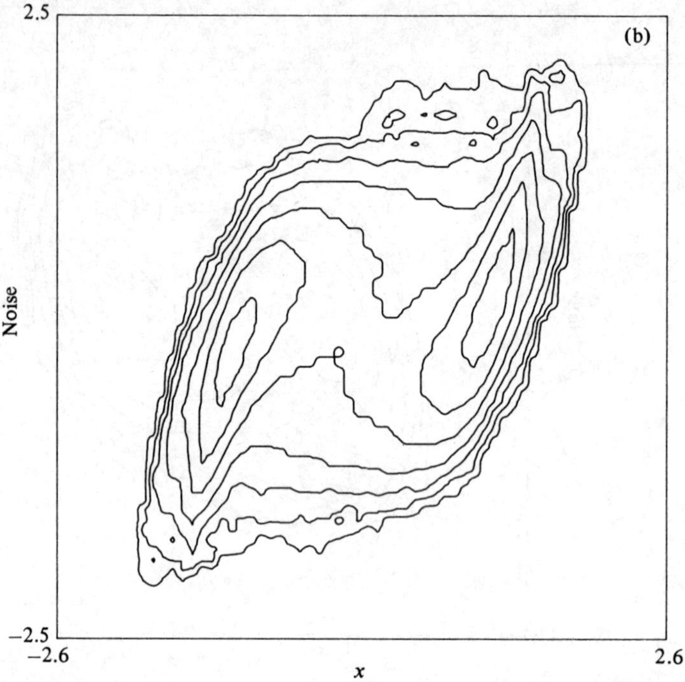
2.5
(b)
Noise
−2.5
−2.6
2.6
x

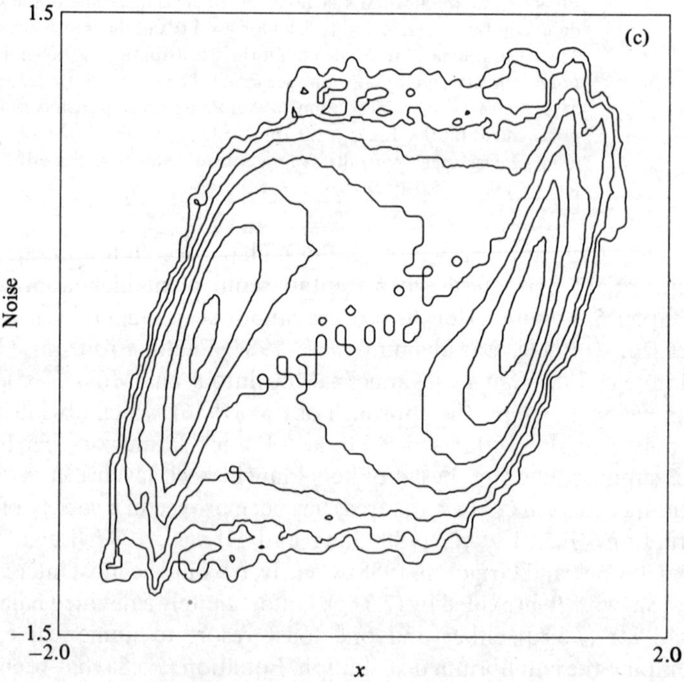
1.5
(c)
Noise
−1.5
−2.0
2.0
x

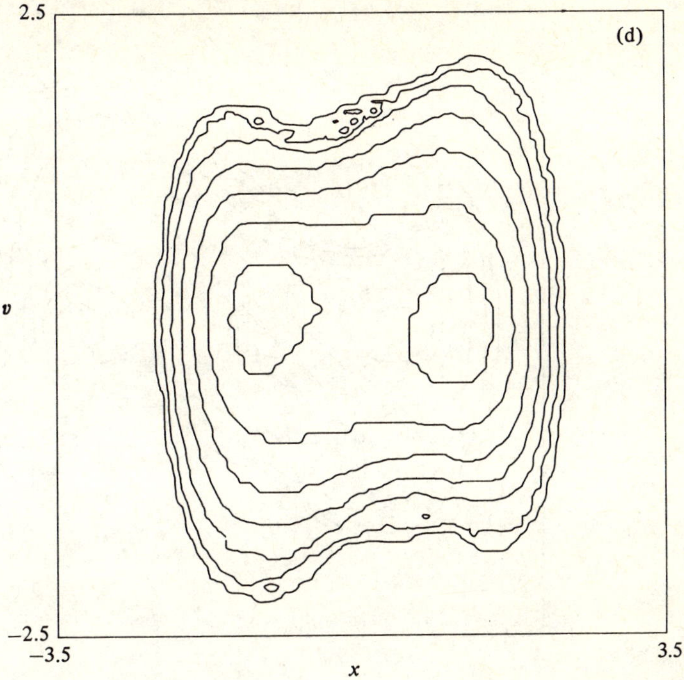

2.5
(d)
v
−2.5
−3.5
3.5
x

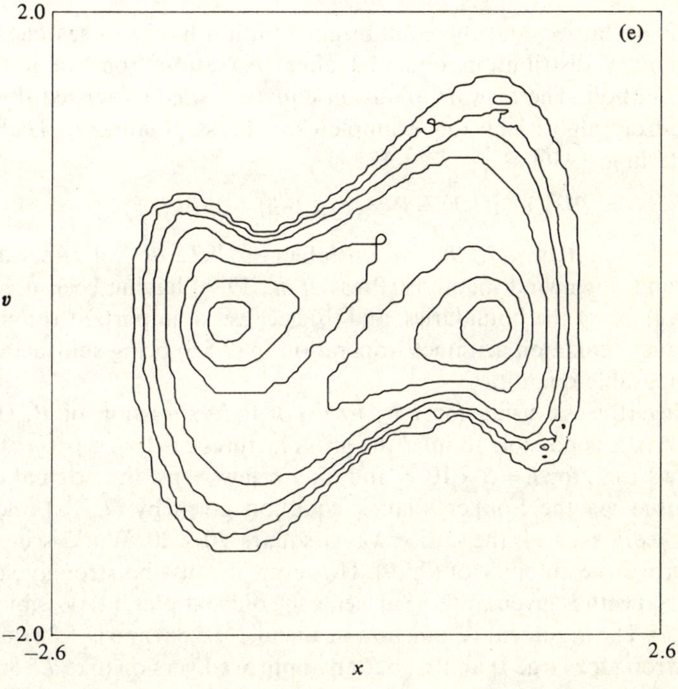

2.0
(e)
v
−2.0
−2.6
2.6
x

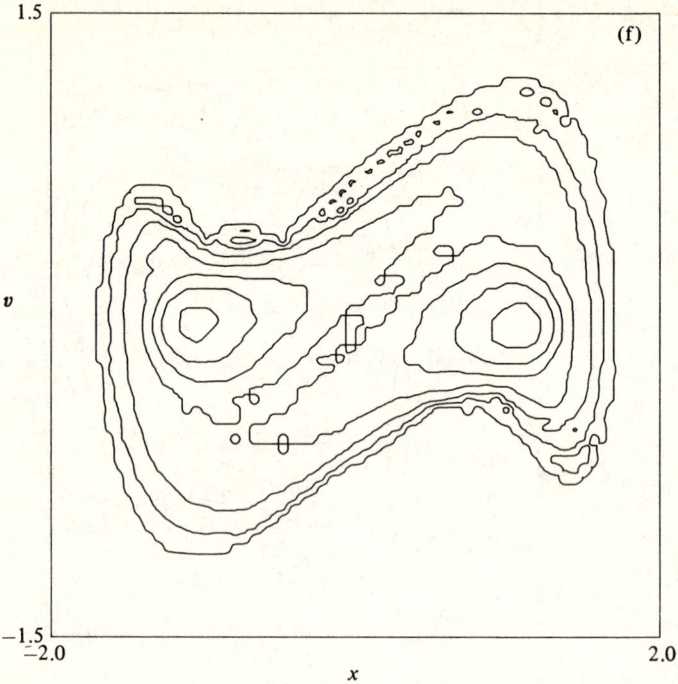

a square lattice, and the equilibrium solution has been searched for as the stationary distribution obtained after relaxation from an initial guessed distribution. The forward diffusion equation was discretized (by use of the leap-frog algorithm: for example, see Press, Flannery, Teukolsky and Vetterling, 1986) as

$$P(x_i, v_J, t+h) = P(x_i, v_J, t-h) + 2h\dot{P}(x_i, v_J, t). \tag{7.5.9}$$

Unfortunately, (7.5.9) is unstable (a) because of free boundaries, upwind/downwind methods (Press *et al.*, 1986) having been used to obtain $P(x, v)$ along the boundaries, and (b) because of important round-off errors, the most common arithmetic operation in (7.5.9) being subtraction of small, comparable quantities.

Nevertheless, using for $P(x_i, v_J, t=0)$ the expression of $P_{eq}(x, v, y)$ from (7.5.6a), it is possible to infer the gross features of $P(x_i, v_J, t \to \infty)$. Figure 7.4 shows that, for $h = 5 \times 10^{-2}$ and after a few steps, the original distribution diffused via the Fokker–Planck equation given by (7.5.8a) and (7.5.8c) is markedly skewed; the lattice was a square 20×20. Work is in progress to improve the stability of (7.5.9). However, it must be strongly stressed that (7.5.8a) with $\tilde{D}$ given by (7.5.8b) yields the only (at present) possible solution of (7.5.7). The argument is somehow academic: (7.5.8a) and (7.5.8b) are known to be wrong for large D and τ_n, but any improved version (like 7.5.8c) cannot be

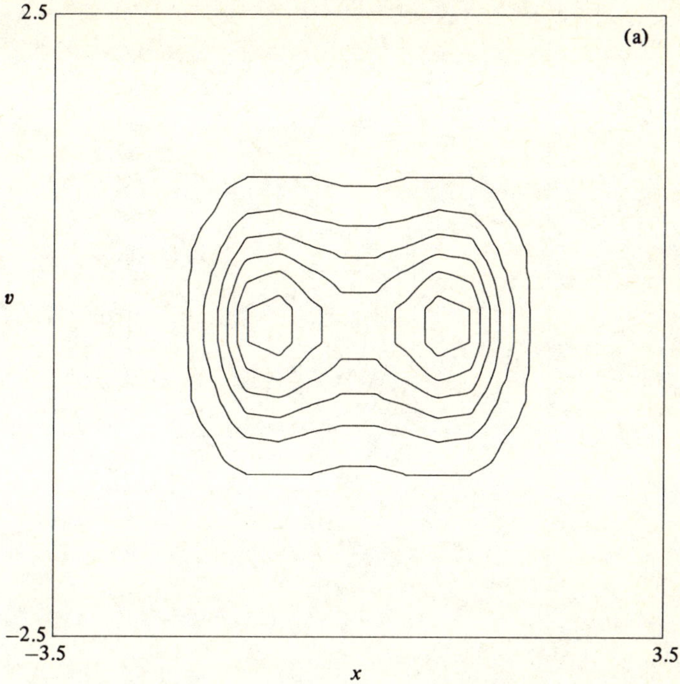

Figure 7.4. Stochastic phase space portraits for the system (7.5.1a, b) (7.5.3c): equilibrium distribution as function of velocity and x. The contours are taken at 1/12, 3/12,…, 11/12 of the maximum. $D = 1.0, \gamma = 1.0, \tau_n = 1.0$.
(a) Equilibrium distribution as from (7.5.6a). (b) Distribution obtained by relaxing via the Fokker–Planck operator of (7.5.8a), starting from the distribution given by (7.5.6a). A marked skewing has been developed.
(c) Equilibrium distribution from the numerical experiment.

closely solved. Indeed, unless (7.5.8a) with $\tilde{D}$ as defined via (7.5.8c) can be solved *analytically*, the author believes that any approximate computer-obtained solution is bound to be useless: as already stressed in Section 7.4, if a computer must be heavily used to extract information, it is probably more straightforward, useful and efficient to integrate directly the corresponding Langevin equation.

The last numerical experiment presented herein is the motion of a linear oscillator subject to multiplicative white or colored noise.

$F(x)$ is in the form

$$F(x) = -\omega_0^2 x \tag{7.5.10a}$$

and

$$g(x) = x. \tag{7.5.10b}$$

It has been shown (Hernández-Machado and Sancho, 1984; Lindenberg and

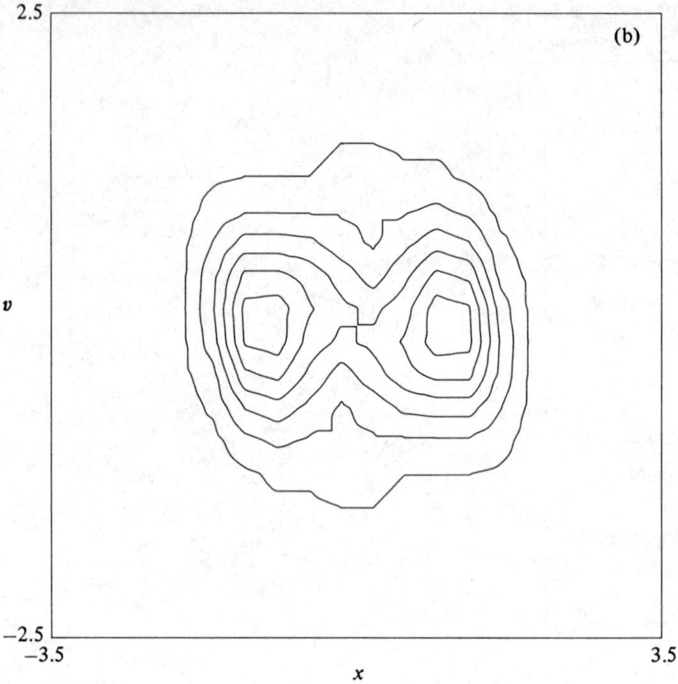

2.5
(b)
v
−2.5
−3.5
3.5
x

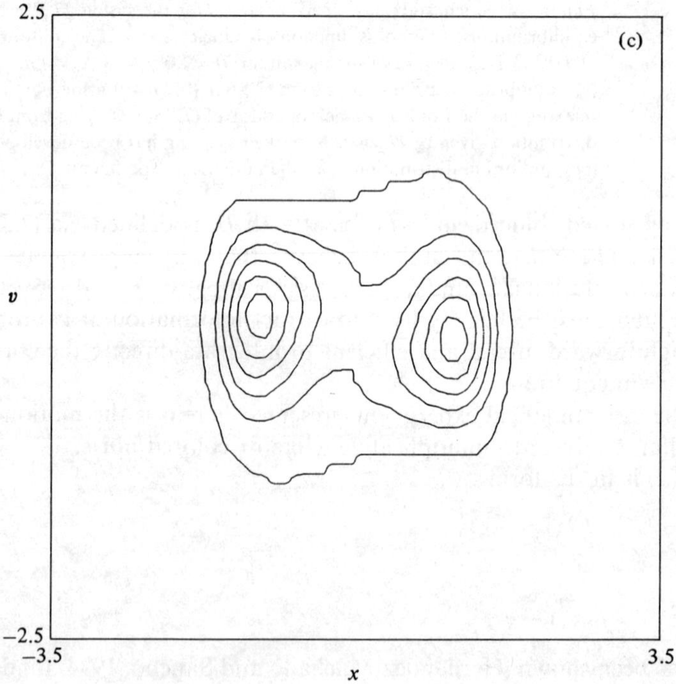

2.5
(c)
v
−2.5
−3.5
3.5
x

West, 1984; San Miguel and Sancho, 1980; West and Lindenberg, 1983) that the system described by (7.5.1a, b) and (7.5.7) becomes unstable at fixed τ_n as D is increased. In Ramshaw and Lindenberg (1986) and West and Lindenberg (1984), the existence of a critical value D_c for D is explained via the lack of a fluctuation–relaxation relation in this system even in presence of simply a white noise forcing: this leads to the interest in observing such an instability even in the simple case of a linear oscillator. In the following Lindenberg and West (1984) will be the basis of the discussion and that paper should be consulted for details and references. It is important first to notice that in the quoted paper an additive noise is added to the right hand side of $\dot{v}$ in (7.5.1a, b). This is because, even though D_c does not depend on the intensity of this additive noise, without the additive forcing the system might get 'stuck' at the origin ($x = 0, v = 0$), where the multiplicative noise would have no effect. Phenomenologically, for D above D_c the system will jig around the origin for a very long time, to suddenly burst away towards very large x and v (and the digital integration must be stopped because of numerical problems). This behavior posed serious computational problems: the system ought to be followed until one such burst occurs (and, if $D < D_c$, a burst will never develop) and, in principle, a burst could happen just a few time steps after the integration is halted: as a matter of fact, data obtained changing either the starting position or the seed used to initialize the random numbers or the value of (x, v) above which the system is assumed to be 'running away' were quite different. The procedure followed ought to be different. The analysis in Lindenberg and West (1984) which leads to the prediction of D_c is based on the study of the eigenvalues of the matrix driving the evolution of the vector w defined as

$$w \equiv (\langle x^2 \rangle, \langle xv \rangle, \langle v^2 \rangle) \tag{7.5.11}$$

(see the quoted paper for more details).

Increasing D, at D_c one of three (initially negative) eigenvalues changes sign, becoming positive for $D > D_c$. The procedure followed to observe the transition stable/unstable has been: to start the system off the equilibrium point; allow a certain time to reach 'thermal equilibrium'; sample (x, v) for a given period; compute the modulus of w defined in (7.5.11) over such a time; sample again a sequence of (x, v); compute again the modulus of w over this second period and compare it to what was found during the first sampling. If the latter value was found to be bigger than the former one, D would be assumed to be above D_c (because of, at least, one positive eigenvalue) and vice versa. This 'recipe' was found to give highly reproducible results against change of starting values of (x, v) and of the seed used to initialize the random numbers. The algorithm was in the form given by (7.5.2a, b), with no additive noise, the starting point being always $x = 1, v = 0$. Runs in the presence of small additive noise gave results consistent with the case without additive noise. A typical result is shown in Figure 7.5: the lower solid curve is the value of D_c at which one of the eigenvalues changes sign; the crosses plus solid line refers to

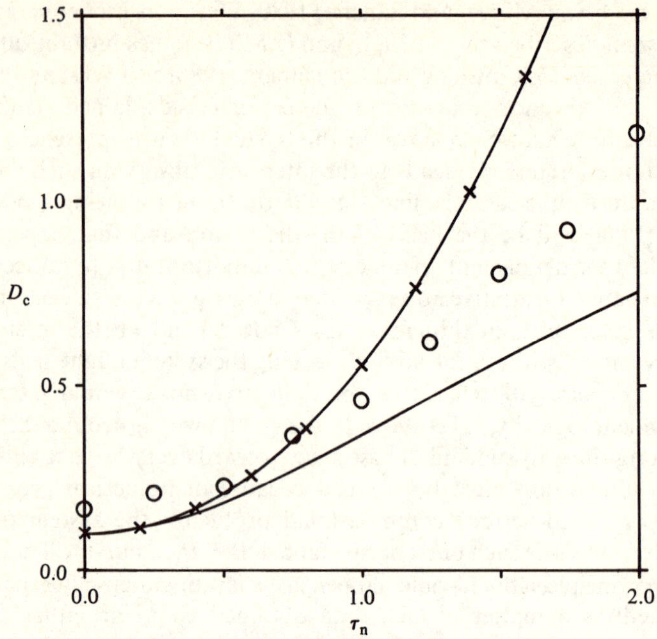

Figure 7.5. D_c as a function of τ_n. The full line is the theoretical value of Lindenberg and West (1984). The full line with crosses is a theoretical prediction based on the instability of the energy envelope (see quoted reference). The circles are the experimental data (integration time step set equal to 10^{-2}).

the D_c at which the energy envelope (for small γ, the energy is an almost conserved quantity and it is possible to write a Fokker–Planck equation for energy-angle variables) becomes unstable. The computer simulation (circles) is clearly in good qualitative agreement with the theoretical predictions.

7.6 Conclusions

An algorithm to numerically integrate sets of stochastic differential equations has been presented and developed to cover the case of multi-dimensional stochastic forcings. Recipes to deal with almost white and nonlinear noisy forcings are also given (see the appendix, Section 7.7).

The examples discussed, covering a wide variety of different systems and situations, demonstrate the utility of digital simulations. Indeed, provided that care has been taken to avoid trivial mistakes (and also to choose the appropriate Ito/Stratonovich calculus in the presence of white, parametric noise), digital simulations are very reliable, closely simulate the theoretical set of stochastic differential equations, and allow us to be quite confident about the 'experimental' result: digital simulations, moreover, are the only possible approach when reproducibility and high precision are important.

There is, however, one major drawback: the constant increase in complexity of the 'interesting' models makes digital simulations more and more greedy in terms of CPU time. Disregarding statistics problems, the time taken for a run increases in proportion to the number of dimensions of the system under study (whereas in analogue simulations it is roughly constant).

The future lies probably with the use of array processors (the present algorithms are in terms of scalar codes) to follow in parallel the evolution of a large number of stochastic trajectories.

7.7 Appendix: The limit of white noise

The algorithms introduced in the previous sections cover most of the situations where a stochastic numerical simulation is to be applied. There are, however, cases when such algorithms are insufficient. It is of interest (for instance, in biological science) to be able to take the white process limit of colored noise. Unfortunately, from an algorithm viewpoint, this limit cannot be taken: two different algorithms must be used for the two different cases. As pointed out, if the noise is multiplicative this forces us to opt either for an Ito or a Stratonovich prescription (but see also Fox, 1988; Fox, Roy and Yu, 1988). In complex situations (for example inertial systems: see Faetti *et al.*, 1984), when it is not clear which of the two should be preferred, it would be most interesting to be able to take the limit of white noise directly in the colored noise algorithm (making the correlation time of the colored noise as small as the time step) to observe towards which of the two different prescriptions the system would 'naturally' evolve. On the other hand, for the given algorithms to be stable, if T is the shortest time scale in the set of stochastic differential equations and h is the integration time step (for the purposes of this section, $T = \tau_n$), the requirement $h/T \ll 1$ must always be met.

Considering the system

$$\dot{x} = f(x) + xy$$

$$\dot{y} = -\frac{1}{\tau_n} y + \frac{(2D)^{1/2}}{\tau_n} \xi(t) \qquad (7.7.1)$$

when $\tau_n \to 0$, if h/τ_n had to be kept small, it would mean that more and more time steps would be needed to integrate $\dot{x}$ (whose time scale will be longer and longer when measured in h units). Furthermore, if h is too small, there will be round-off error problems during the integration of $\dot{x}$. The problem of very different time scales in ordinary differential equations is referred to as a 'stiff problem' (see, for example, Lapidus and Seinfeld, 1971). A 'recipe' to deal with stiff stochastic differential equations is the goal of this appendix, where the stiffness comes from the differential equation describing y in (7.7.1).

The basic idea is to choose a time step h which is suitable for (taking (7.7.1) as an example) integrating $\dot{x}$, letting just $h \leqslant \tau_n$. The equation for y will be integrated with the help of more stable algorithms. We suggest an implicit method (see quoted reference for a description of implicit methods in the theory of ordinary differential equations). For a differential equation of the form

$$\dot{y} = -\frac{1}{\tau_n} y + \frac{(2D)^{1/2}}{\tau_n} \xi(t) \tag{7.7.2}$$

($\xi(t)$ Gaussian with zero average and standard deviation one), it is possible to write with no approximation

$$y(t+h) = \mathrm{e}^{-h/\tau_n} y(t) + \frac{(2D)^{1/2}}{\tau_n} \int_0^h \mathrm{e}^{(s-h)/\tau_n} \xi(s)\, \mathrm{d}s. \tag{7.7.3}$$

Defining

$$Z_\infty(h) = \int_0^h \mathrm{e}^{(s-h)/\tau_n} \xi(s)\, \mathrm{d}s, \tag{7.74}$$

we note that $Z_\infty(h)$ is Gaussian because it is a linear combination of Gaussian variables. As usual, the procedure is now to find a suitable equivalent stochastic variable.

From (7.7.4), (7.2.9) and (7.2.16) it follows that ($q \equiv -h/\tau_n$)

$$\begin{aligned} &\langle Z_\infty(h) \rangle = 0 \langle Z_\infty^2(h) \rangle = \frac{\tau_n}{2}[1 - \mathrm{e}^{2q}] \\ &\langle Z_1(h) Z_\infty(h) \rangle = \tau_n [1 - \mathrm{e}^q] \\ &\langle Z_2(h) Z_\infty(h) \rangle = \tau_n^2 [1 - \mathrm{e}^q (1 - q)]. \end{aligned} \tag{7.7.5}$$

With the help of (7.7.5), a suitable expression for $Z_\infty(h)$ is

$$Z_\infty(h) = \alpha Z_1(h) + \beta Z_2(h) + \gamma\eta, \tag{7.7.6}$$

where η is a Gaussian variable with zero average, standard deviation one, statistically independent of $Z_1(h)$ and $Z_2(h)$, and α, β, γ are computed as

$$\begin{aligned} \alpha &= \frac{12}{h^4}\left[\frac{h^3}{3}\langle Z_1(h) Z_\infty(h)\rangle - \frac{h^2}{2}\langle Z_2(h) Z_\infty(h)\rangle\right] \\ \beta &= \frac{12}{h^4}\left[h \langle Z_2(h) Z_\infty(h)\rangle - \frac{h^2}{2}\langle Z_1(h) Z_\infty(h)\rangle\right] \\ \gamma^2 &= \langle Z_\infty^2 \rangle - \alpha^2 h - \beta^2 \frac{h^3}{3} - \alpha\beta h^2. \end{aligned} \tag{7.7.7}$$

(It is actually simpler to compute α, β, γ directly than to give analytic expressions.)

Not surprisingly, in the limit $\tau_n/h \to \infty$, $Z_\infty(h) \to Z_1(h)$. Equation (7.7.3), via (7.7.4)–(7.7.7), gives the looked-for algorithm. Even if (7.7.3) would be right in the case $h/\tau_n > 1$, in practice a closer look at the expression used to integrate $\dot{x}$ (as in equation (7.7.1) for example) will reveal that the condition $h/\tau_n \leqslant 1$ must be nevertheless satisfied. Equation (7.7.3) can be useful also when dealing with white non-linear forcings. The algorithms devised in Sections 7.2 and 7.3 assumed that the stochastic differential equations were linear in the noisy forcing. When this is not the case, it is possible to substitute for the noisy variable an auxiliary variable whose evolution will be given by something like (7.7.3), adding one more differential equation to the original set. The limit τ_n almost equal to h can now be taken safely, which yields a white noise algorithm for the original set of stochastic differential equations.

Acknowledgements

The author wishes to thank Prof. Frank Moss and Dr Peter McClintock for reading the manuscript and for continuous help and encouragement, Prof. Ian Drummond for making available preprints of his work and discussing problems related to stochastic Runge–Kutta integration schemes, and Dr Paolo Grigolini for a very fruitful correspondence on some theoretical ideas contained in this paper.

This work was supported by the British Science and Engineering Research Council.

References

Arnold, L. 1974. *Stochastic Differential Equations, Theory and Applications*, NY: Wiley-Interscience.

Aronowitz, F. 1977. *IEEE J. Quant. Electron.* **QE-13**, 13.

Borkovec, M., Straub, J. E. and Berne, B. J. 1986. *J. Chem. Phys.* **85**, 146.

Broggi, G., Colombo, A., Lugiato, L. A. and Mandel, P. 1986. *Phys. Rev. A* **33**, 3635.

Büttiker, M., Harris, E. P. and Landauer, R. 1983. *Phys. Rev. B* **28**, 1268.

Casademunt, J., Mannella, R., McClintock, P. V. E., Moss, F. E. and Sancho, J. M. 1987. *Phys. Rev. A* **35**, 5183.

Drummond, I. T., Duane, S. and Horgan, R. R. 1984. *J. Fluid Mech.* **138**, 75.

Drummond, I. T., Duane, S. and Horgan, R. R. 1987. *Nucl. Phys. B. Field Theory and Stat. Phys. B* **280**, 45.

Drummond, I. T., Hoch, A. and Horgan, R. R. 1986. *J. Phys. A* **19**, 3871.

Drummond, I. T. and Horgan, R. R. 1986. *J. Fluid Mech.* **163**, 425.

Faetti, S., Festa, C., Fronzoni, L., Grigolini, P. and Martano, P. 1984. *Phys. Rev. A* **30**, 3252.

Faetti, S., Fronzoni, L., Grigolini, P. and Mannella, R. 1988. Submitted to *Phys. Rev.*

Fox, R. F. 1986. *Phys. Rev. A* **33**, 467.

Fox, R. F. 1988. Submitted to *J. Stat. Phys.*

Fox, R. F., Roy, R. and Yu, A. W. 1988. Submitted to *J. Stat. Phys.*

Fronzoni, L., Grigolini, P., Hänggi, P., Moss, F., Mannella, R. and McClintock, P. V. E. 1986. *Phys. Rev. A* **33**, 3320.

Glorieaux, P., and Dangoisse, D. 1985. *IEEE J. Quant. Electron.* **QE-21**, 1486.
Greensite, H. S. and Helfand, E. 1981. *Bell Syst. Tech. J.* **60**, 1927.
Grüner, G. and Zettl, A. 1985. *Phys. Rep.* **119**, 117.
Hänggi, P., Mroczkowski, T. J., Moss, F. and McClintock, P. V. E. 1985. *Phys. Rev. A* **32**, 695.
Hernández-Machado, A. and Sancho, J. M. 1984. *J. Math Phys.* **25**, 1066.
Klauder, J. R. and Petersen, W. P. 1985. *SIAM J. Numer. Anal.* **22**, 1153.
Knuth, D. E. 1981. *Semi-numerical Algorithms. Vol. 2 of the Art of Computer Programming*, 2nd edn. Reading, MA: Addison-Wesley.
Jung, P. and Hänggi, P. 1987. *Phys. Rev. A* **35**, 4464.
Lapidus, L. and Seinfeld, J. H. 1971. *Numerical Solution of Ordinary Differential Equations.* NY: Academic Press.
Lindenberg, K. and West, B. J. 1984. *Physica* **128A**, 25.
Lugiato, L., Mannella, R., McClintock, P. V. E. and Grigolini, P. 1988. Submitted to *Phys. Rev. A.*
Mandel, P. and Erneux, T. 1985. *IEEE J. Quant. Electron.* **QE-21**, 1352.
Mannella, R., Faetti, S., Grigolini, P., McClintock, P. V. E. and Moss, F. 1986. *J. Phys. A* **19**, L699.
Mannella, R., McClintock, P. V. E. and Moss, F. 1987a. *Phys. Lett.* **120A**, 11.
Mannella, R., McClintock, P. V. E. and Moss, F. 1987b. *Phys. Rev. A* **37**, 721.
Marsaglia, G. and Tsang, W. W. 1984. *SIAM J. Sci. and Stat. Comp.* **5**, 349.
Moss, F., Hänggi, P., Mannella, R. and McClintock, P. V. E. 1986. *Phys. Rev. A* **33**, 4459.
Munakata, T. 1986. *Prog. Theor. Phys.* **75**, 747.
Press, W. H., Flannery, B. P., Teukolsky, S. A. and Vetterling, W. T. 1986. *Numerical Recipes: The Art of Scientific Computing.* Cambridge University Press.
Ramshaw, J. D. and Lindenberg, K. 1986. *J. Stat. Phys.* **45**, 295.
Rao, N. J., Borwankar, J. D. and Ramkrishna, D. 1974. *SIAM J. Control* **12**, 124.
Risken, H. 1984. *The Fokker–Planck Equation, Methods of Solutions and Applications.* Springer Series in Synergetics. NY: Springer.
Sancho, J. M., Mannella, R., McClintock, P. V. E. and Moss, F. 1985. *Phys. Rev. A* **32**, 3639.
Sancho, J. M., San Miguel, M., Katz, S. L. and Gunton, J. D. 1982. *Phys. Rev. A* **26**, 1589.
San Miguel, M. and Sancho, J. M. 1980. *Phys. Lett.* **76A**, 97.
Schneider, T. and Stoll, E. 1978. *Phys. Rev. B* **17**, 1302.
Schöbinger, M., Koch, S. W. and Abraham, F. F. 1986. *J. Stat. Phys.* **42**, 1071.
Smythe, J., Moss, F., McClintock, P. V. E. and Clarkson, D. 1983. *Phys. Lett.* **97A**, 95.
Stratonovich, R. L. 1967. *Topics in the Theory of Random Noise*, vol. II. NY: Gordon and Breach.
Straub, J. E. and Berne, B. J. 1985. *J. Chem. Phys.* **83**, 1138.
Straub, J. E., Borkovec, M. and Berne, B. J. 1985. *J. Chem. Phys.* **83**, 3172.
Straub, J. E., Borkovec, M. and Berne, B. J. 1986. *J. Chem. Phys.* **84**, 1788.
Straub, J. E., Hsu, D. A. and Berne, B. J. 1985. *J. Phys. Chem.* **89**, 5188.
Verlet, L. 1967. *Phys. Rev.* **159**, 98.

Vogel, K., Risken, H., Schleich, W., James, M., Moss, F. and McClintock, P. V. E. 1987a. *Phys. Rev. A* **35**, 463.
Vogel, K., Risken, H., Schleich, W., James, M., Moss, F., Mannella, R. and McClintock, P. V. E. 1987b. *J. Appl. Phys.* **62**, 721.
West, B. J. and Lindenberg, K. 1983. *Phys. Lett.* **95A**, 44.
West, B. J. and Lindenberg, K. 1984. In *Fluctuations and Sensitivity in Non-equilibrium Systems* (W. Horsthemke and D. K. Kondepudi, eds.) *Springer Proceeding in Physics* **1**, 233. Berlin: Springer.

8 Analogue simulations of stochastic processes by means of minimum component electronic devices

LEONE FRONZONI

8.1 Introduction

The study of non-linear stochastic processes is usually characterized by considerable difficulties, thereby rendering it necessary to have recourse to computers for the theoretical predictions to be tested. This in turn involves interminable times of calculation with a great waste of money.

Analogue simulation by means of electronic devices allows us to bypass these difficulties. Although the precision of the analogue technique is modest compared to that of computer calculations, it must be remarked that in many cases it is more significant to have available, immediate results rather than very accurate numerical precision.

It is important to underline that the analogue experiment does not completely replace computer simulation; rather, a comparison of the two techniques, wherever possible, has proven to be a powerful method for studying non-linear stochastic phenomena.

In order to give a short description of electronic simulation, a few basic elements of the techniques will be given, then we shall discuss some significant applications. Finally, we shall describe how it is possible to obtain important conclusions via a joint use of computer and analogue simulation. This has been the object of a recent investigation on the role of coloured noise in some dynamical systems.

8.2 The basis of the analogue simulation via electronic devices

8.2.1 The integration operation

This operation is usually obtained using an RC-circuit; nevertheless, a Miller integrator, assembled in the configuration shown in Figure 8.1(a) is a better choice because of the following advantages:

(a) The high input impedance allows us to add many input signals.
(b) The low output impedance allows us to connect this device to several others without changing the characteristic time of the single integrator.

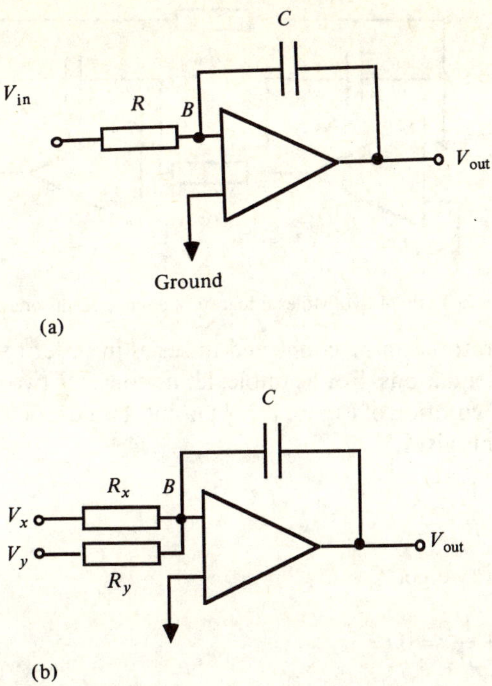

Figure 8.1. (a) Schematic of Miller integrator. (b) Schematic of Miller integrator and adder.

Because of the high input impedance of the Miller integrator, the sum of the currents at point B (see Figure 8.1a) gives a simple differential equation:

$$C\frac{d}{dt}V_{out} + \frac{V_{in}}{R} = 0 \tag{8.2.1}$$

8.2.2 *The addition operation*

It may readily be appreciated that (8.2.1) provides a basis for electronic simulation. In fact, let us suppose that an equation of the form

$$\frac{dz}{dt} = ax + y \tag{8.2.2a}$$

has to be simulated. To add the new term y in (8.2.1) it is sufficient to connect a voltage V_y at the input of the amplifier by means of a suitable resistance, as shown in Figure 8.1(b). The sum of the currents at point B gives

$$\frac{V_x}{R_x} + \frac{V_y}{R_y} + C\frac{d}{dt}V_u = 0 \tag{8.2.2b}$$

which has the same form as (8.2.2a).

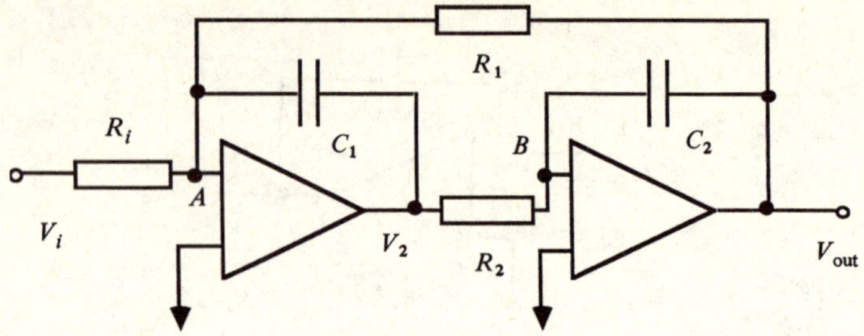

Figure 8.2. Schematic of two Miller integrators connected in series.

Many Miller integrators can be connected in series in order to simulate a system of differential equations. For example, let us consider two integrators assembled in the configuration of Figure 8.2. At point A on the first integrator, the sum of the currents gives

$$C_1 \frac{\mathrm{d}}{\mathrm{d}t} V_2 + \frac{V_i}{R_i} = 0, \tag{8.2.3}$$

and, again, at point B we get

$$C_2 \frac{\mathrm{d}V_{out}}{\mathrm{d}t} + \frac{V_2}{R_2} = 0 \cdot \tag{8.2.4}$$

These are two differential equations with two variables V_{out} and V_2. The two integration times $\tau_1 = R_i C_1$ and $\tau_2 = R_2 C_2$ are the same as those of the single isolated integrator.

Because of the great development of the electronic devices today, it turns out to be easy and inexpensive to perform product and division operations. A typical multiplier is the four-quadrant AD534 made by Analog Devices. Because of the differential nature of the input, we are in a position to assemble 'minimum component devices' so as to simulate the non-linear differential equations. The basic scheme of the AD534 is drawn in Figure 8.3. The transfer function is

$$V_{out} = \frac{(x_1 - x_2)(y_1 - y_2)}{V_0} + (z_1 - z_2). \tag{8.2.5}$$

The parameter V_0 depends on the electric configuration used, normally $V_0 = 10$. By connecting the y-input to the output, a divider is obtained instead of a multiplier. Thus the transfer function reads

$$V_{out} = \frac{10(z_2 - z_1)}{(x_1 - x_2)} + y_1. \tag{8.2.6}$$

In this case, when a divider is based on the use of a multiplier in a feedback-loop, it is important to note that the bandwidth is proportional to the denominator's absolute value.

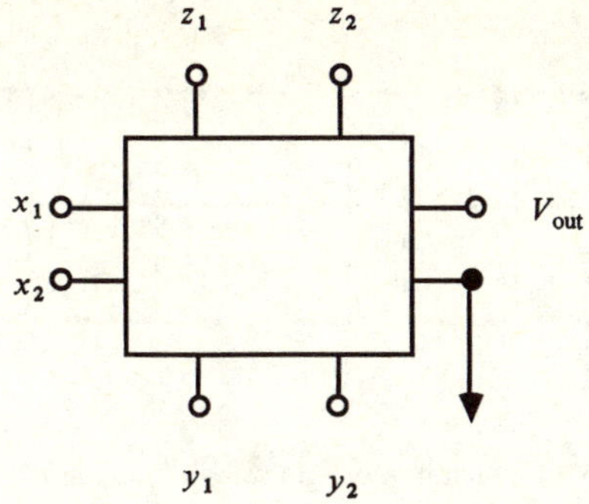

Figure 8.3. Basic schematic of the multiplier Analog Devices AD534.

8.3 A dichotomous noise generator

Many stochastic systems in the presence of dichotomous noise have an exact analytical solution. It is therefore useful to have available a dichotomous noise generator. This allows us to test the accuracy of an analogue simulation. A simple dichotomous noise generator can be made by means of shift-register devices (LFSR) which are inexpensive and easy to obtain (Golomb, 1967). The working principle of this noise generator is based on exclusive-or feedback on the shift register. As a consequence of that, the LFSR generates an *m*-sequence of two-state pulses. The amplitude V of a dichotomous noise as a function of time is shown in Figure 8.4. T_d and T_u denote the times when the out voltage is down and up, respectively. The two times T_d and T_u are affected by an asymmetry resulting from the use of the or-feedback (Tomlinson and Galvin, 1975). The fault can be overcome by alternating the out-exclusive-or with an exclusive-nor feedback (Faetti *et al.*, 1984). The nor-feedback generates a sequence of pulses that are complementary to the or-feedback, so that the asymmetry turns out to be exactly compensated.

In Figure 8.5 a block diagram of a complete dichotomous noise generator is shown: FF is a four-stage flip-flop which drives the two shift-registers SR1 and SR2. The inverse of the frequency of the clock ν_c gives the minimum value of the duration of the sequences of pulses. For this reason the noise generated by means of the shift-register can be considered dichotomous for times that are long compared to the clock time.

A Gaussian noise can be obtained by sending the dichotomous noise to the input of a Miller integrator. The frequency of the clock should be very high compared to the cut-off frequency ν_I of the integrator. For example, with a clock frequency $\nu_c = 100\,\nu_I$ the steady-state equilibrium distribution of $V(t)$, $P(V)$, satisfies the requirement of being a Gaussian function to very large accuracy within a range as wide as five times the noise variance.

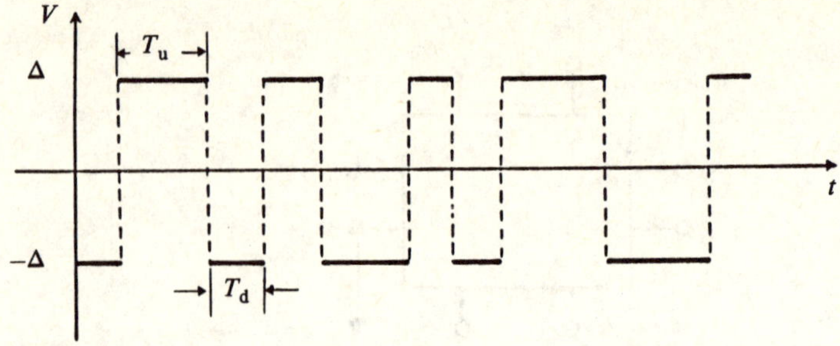

Figure 8.4. Amplitude V of a dichotomous noise as a function of time. Δ denotes the difference $V - \bar{V}$, where $\bar{V}$ is the mean value. T_u and T_d denote the duration times when $V > 0$ and $V < 0$, respectively.

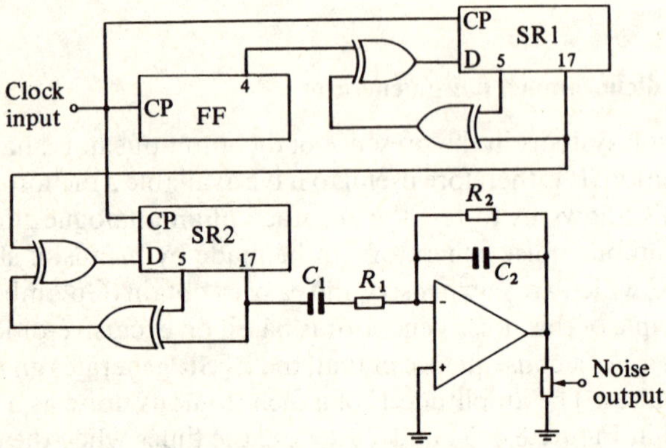

Figure 8.5. Block diagram of the noise generator. The output of the dichotomous noise is available at pin 17 of the shift-register SR2. A Gaussian noise is obtained at the output of the Miller integrator.

8.4 The minimum components technique

The principles on which this technique is based can be summarized in three points:

(1) We have to write the equations in such a way as to reduce to a minimum the number of mathematical operations.

(2) We have to renormalize the equations with respect to time so as to produce characteristic times well within the operating ranges of the electronic devices.

(3) We have to reduce to a minimum the number of electronic components and, whenever it is possible, we must assemble the circuits in only one block. This allows one to reduce the parasitic elements and to reduce the influence of unwanted (and unavoidable) external noise. Note, furthermore,

that using few elements has also the desirable effect of reducing the internal noise.

8.5 The Duffing oscillator in the limit of extremely low friction

A lot of interest has developed in the recent past in the underdamped non-linear oscillator in the presence of a stochastic force. This system plays an important role for testing the well known linear response theory (Kenkre, 1971; Van Kampen, 1971; Visscher, 1974; Weare and Oppenheim, 1974). Our specific problem is that of assessing whether or not the dependence of the maximum of the power spectrum on the viscosity and noise intensity, as predicted by the linear response theory, is correct, even in the extreme underdamped limit as illustrated in Fronzoni, Grigolini, Mannella and Zambon, 1985 (see also Chapter 5, Volume 1).

The equation under investigation is of the form

$$\ddot{x} + \gamma\dot{x} + ax + bx^3 + f(t') = 0, \tag{8.5.1}$$

where $f(t')$ is a Gaussian stochastic force defined by the autocorrelation functions

$$\langle f(t')\rangle = 0 \tag{8.5.2a}$$

$$\langle f(t')f(s)\rangle = (D/\tau_c')\,\mathrm{e}^{-|t'-s|/\tau_c'}, \tag{8.5.2b}$$

where D is the intensity of the noise and $1/\tau_c'$ is its cut-off frequency. In accord with point (1) mentioned above, one should write this equation so as to reduce to a minimum the required mathematical operations. It is thus useful to rewrite (8.5.1) in the form of the following system of two equations:

$$\frac{\mathrm{d}x}{\mathrm{d}t'} = v \tag{8.5.3}$$

$$\frac{\mathrm{d}v}{\mathrm{d}t'} = -\gamma v - (a + b\cdot x\cdot x)\cdot x + f(t'). \tag{8.5.4}$$

It is easy to see that two multiplier elements and two integrators are sufficient to simulate this system.

A possible configuration is shown in Figure 8.6. This circuit is very similar to the circuit of Figure 8.2. The main difference is that the feedback is obtained by using two multipliers in order to get the non-linear term bx^3, while the voltage V_{eq} at the input of the second multiplier gives the constant term 'a'. In fact at point C we have

$$V_{\mathrm{u}} = x(a + bx^2). \tag{8.5.5}$$

By summing up the currents at points A and B, we get

$$\frac{V_1}{R_1} + \frac{V_{\mathrm{c}}}{R} + C_1\frac{\mathrm{d}V_1}{\mathrm{d}t} + \frac{V_{\mathrm{n}}}{R_{\mathrm{n}}} = 0 \tag{8.5.6}$$

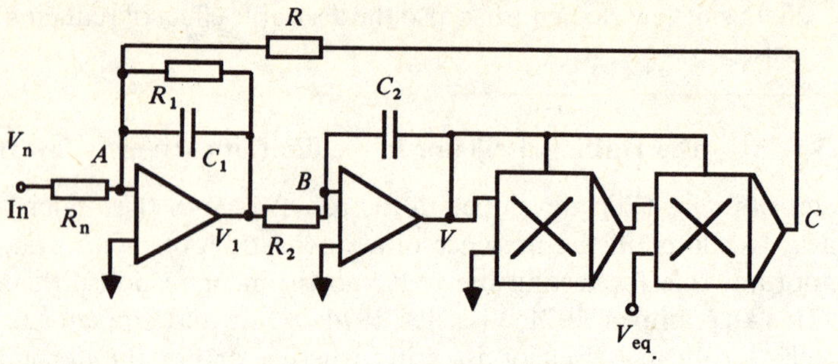

Figure 8.6. The electronic Duffing oscillator. The noise is applied at the input of the first integrator. The voltage V_{eq} gives the linear term in (8.5.1). The variable x corresponds to the voltage V at the output of the second integrator.

$$\frac{V_1}{R_2}+C_2\frac{\mathrm{d}v}{\mathrm{d}t}=0, \tag{8.5.7a}$$

where

$$V_c=\left(\frac{V^2}{10}+V_{eq}\right)\cdot\frac{V}{10} \tag{8.5.7b}$$

$$R_2C_2\frac{\mathrm{d}V}{\mathrm{d}t}=-V_1 \tag{8.5.8}$$

$$RC_1\frac{\mathrm{d}V_1}{\mathrm{d}t}=-V_1\frac{R}{R_1}-\frac{V_{eq}}{10}.V-\frac{V^3}{100}+\frac{R}{R_n}V_n. \tag{8.5.9}$$

By putting $t=R_2C_2t'=RC_1t'$, $v=V_1$ and $x=V$, we then find that (8.5.8) and (8.5.9) are equivalent to the system of (8.5.3) and (8.5.4). Thus, (8.5.8) and (8.5.9) describe the Duffing oscillator driven by a stochastic force when a noise generator is applied at the input of the device.

Note that $\tau=R_2C_2=RC_1$ is the time scaling of the circuit, and $\omega_0=(V_{eq}/10)^{1/2}$ is a renormalized characteristic frequency. In order to understand how this circuit simulates (8.5.3) and (8.5.4) let us make the following remarks:

(a) A large parasitic capacitance would invalidate (8.5.3) and (8.5.4), so particular care is necessary to reduce the number of electric connections. The capacitance elements must be chosen to be much larger than the parasitic elements.

(b) By applying a sine-generator instead of the noise generator we can measure the resonance frequency $\omega_0'=(V_{eq}/10\tau^2)^{1/2}$ and the friction term $\gamma'=R/\tau R_1$. For these measurements to be carried out, it is necessary to apply a very low voltage at the input, the purpose of which is to make the non-linear term bx^2 negligible compared with the linear term a. The direct measurement of the equilibrium voltage x should give the theoretical value $(a/b)^{1/2}$.

(c) In the presence of noise, both the voltage V_n and the noise correlation time τ_c must be scaled so that:

$$f(t) = \frac{V_m(t) R \tau^{1/2}}{R_n} \qquad \tau_c' = \frac{\tau_c}{\tau}. \tag{8.5.10}$$

In order to simplify the theoretical approach to the stochastic Duffing oscillator, we limit ourselves to the case of white Gaussian noise, namely to the case $1/\tau_c' \gg a^{1/2}$, which leads us to replace (8.5.2b) with

$$\langle f(t) f(s) \rangle = 2D\delta(t-s). \tag{8.5.11}$$

The 'Schrödinger like' picture associated with (8.5.1) supplemented by (8.5.11) (and the Gaussian assumption on $f(t)$) leads us to the following Fokker–Planck form:

$$\frac{\partial}{\partial t}\rho(x, v, t) = \mathscr{L}\rho(x, v, t), \tag{8.5.12}$$

where $\mathscr{L}$ is the non-Hermitian Liouvillian operator

$$\mathscr{L} \equiv \left[-v\frac{\partial}{\partial x} + (\omega_0^2 x + bx^3)\frac{\partial}{\partial v} + \gamma\left(\frac{\partial}{\partial v}v + \langle v^2 \rangle \frac{\partial^2}{\partial v^2}\right)\right]. \tag{8.5.13}$$

The general solution of this equation cannot be obtained in terms of an analytical expression. Nevertheless, it is possible to rewrite (8.5.12) in a new form exhibiting the anharmonic strength α, whose explicit expression will be pointed out later in this chapter. Further theoretical details can be derived from Fronzoni *et al.* (1985). Here we limit ourselves to demonstrate the shift of the maximum of the spectrum as a function of α with simple and intuitive arguments.

Let us consider the equilibrium correlation function

$$\phi(t) = \langle x(0)x(t) \rangle / \langle x^2 \rangle. \tag{8.5.14}$$

Then

$$\frac{\mathrm{d}}{\mathrm{d}t}\phi(t) = \left\langle x(0)\frac{\mathrm{d}}{\mathrm{d}t}x(t) \right\rangle \Big/ \langle x^2 \rangle. \tag{8.5.15}$$

On the other hand, from (8.5.4) we have

$$\frac{\mathrm{d}}{\mathrm{d}t}\frac{\langle x(0)v(t) \rangle}{\langle x^2 \rangle} = -\omega_0^2\phi(t) - \gamma\frac{\langle x(0)v(t) \rangle}{\langle x^2 \rangle} - b\frac{\langle x(0)x^3(t) \rangle}{\langle x^2 \rangle}. \tag{8.5.16}$$

The mean field approximation we adopt consists of assuming

$$\langle x(0)x^3(t) \rangle = 3\langle x^2 \rangle \langle x(0)x(t) \rangle, \tag{8.5.17}$$

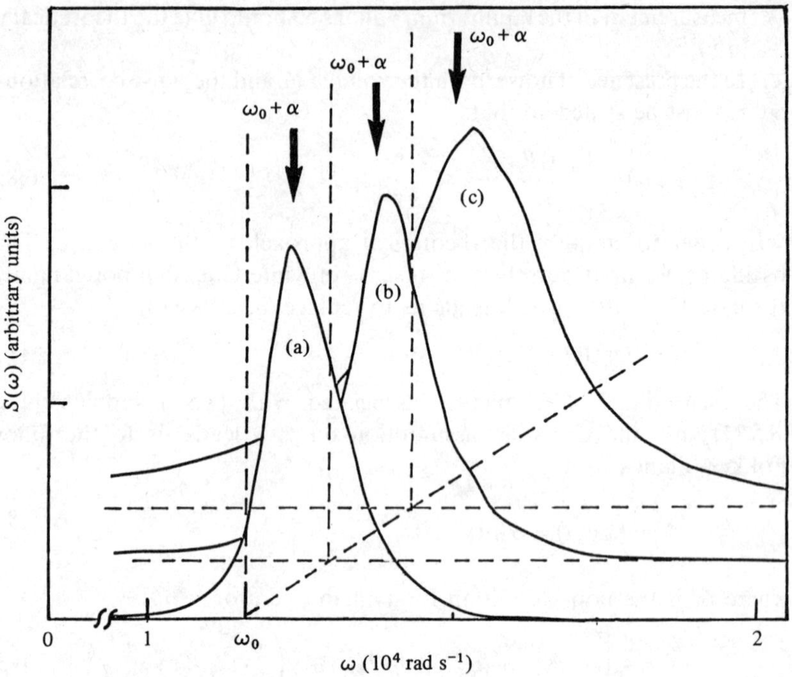

Figure 8.7. The free relaxation spectrum of the variable x for different values of the friction γ. Curves (a)–(c) are for $\gamma = 0.011\ \omega_0$, $\gamma = 0.04\ \omega_0$ and $\gamma = 0.182\ \omega_0$, respectively. The arrows denote the positions of the frequency $\omega_0 + \alpha$, $\omega_0 = 11600\ \text{rad s}^{-1}$ and $\alpha = 0.067\ \omega_0$. From Fronzoni *et al.* (1985).

which is equivalent to the result of statistical linearization (Fronzoni *et al.*, 1985). This results in the renormalized frequency

$$\omega_{\text{eff}}^2 = (\omega_0^2 + 3b\langle v^2 \rangle/\omega_0^2) \tag{8.5.18}$$

or

$$\omega_{\text{eff}} \simeq \omega_0 + 2\alpha, \tag{8.5.19}$$

where

$$\alpha \equiv \frac{3}{4}\frac{b}{\omega_0^3}\langle v^2 \rangle. \tag{8.5.20}$$

However, an improvement on that theory, based on a novel mean field approximation valid in the extremely low friction regime (Fronzoni *et al.*, 1985) predicts that

$$\omega_{\text{eff}} = \omega_0 + \alpha \tag{8.5.21}$$

instead of

$$\omega_{\text{eff}} = \omega_0 + 2\alpha.$$

In conclusion, the two approximations represent high and low friction limits

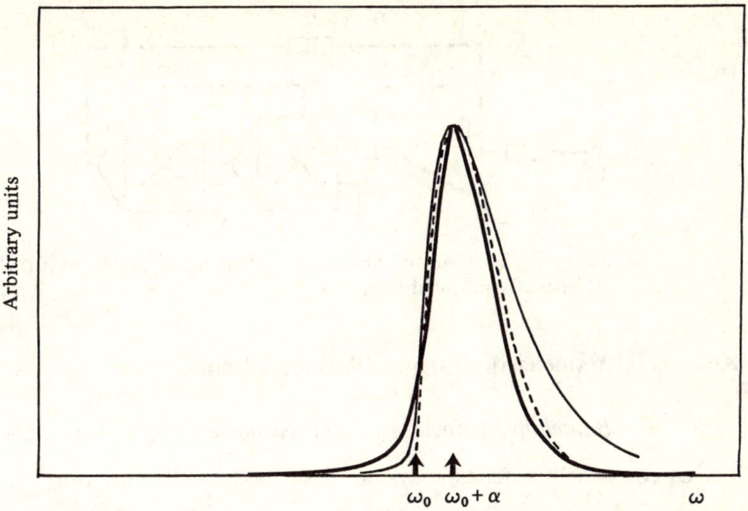

Figure 8.8. The spectrum of the free relaxation of the variable x in the extreme low-friction regime. The solid and dashed lines, obtained from Fronzoni *et al.* (1985), are, respectively, the theoretical prediction (see the quoted paper) and the deterministic spectrum. The result of the analogue simulation is given by the full line. In this experiment $\alpha = 0.067\ \omega_0$, $\gamma = 0.010\ \omega_0$ and $\omega_0 = 11600$ rad s^{-1}.

which indicate a transition from $\Delta\omega_{max} = \alpha$ to $\Delta\omega_{max} = 2\alpha$ when the friction is increased from $\gamma \ll \alpha$ to $\gamma \gg \alpha$.

The analogue results are in a good agreement with the theoretical predictions. In Figure 8.7 we show the free relaxation spectrum of x at different values of the friction γ. The arrows denote the positions corresponding to $\omega = \omega_0 + \alpha$ and they point out the drift of the maximum when γ is increased, corresponding to the predictions of the theory valid in the extreme low friction limit.

In Figure 8.8 a spectrum of the free relaxation of the variable x, obtained by analogue simulation, is given. The friction parameter, γ, in this experiment is very low compared to the parameter 'α' of (8.5.1). Obviously this measurement presents some difficulties.

(a) Because of the low friction, the noise applied at the input was very low; for this reason the electric circuit is particularly sensitive to both the internal noise and any external induced noise.
(b) Low friction means that it takes a long time for equilibrium to be reached, and of course this obliges us to make averages over extremely long intervals of time. Thus the system becomes sensitive to the thermal variations of the electronic components. Because of these limitations, it is important to choose the circuit parameters carefully.

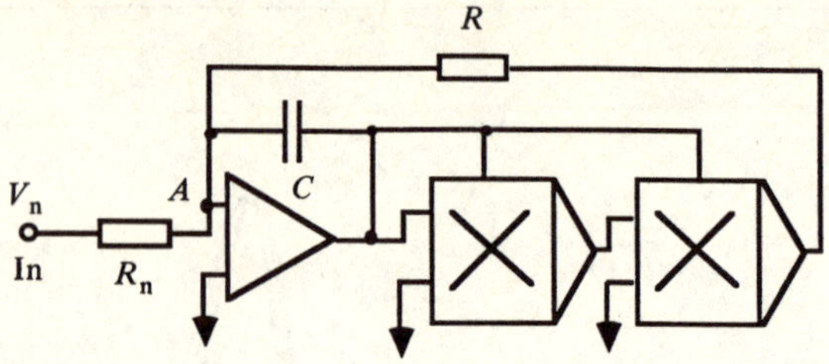

Figure 8.9. Schematic of the analogue simulator of a diffusive particle in presence of a potential $U(x) = bx^4$.

8.6 A stochastic system with coloured noise

8.6.1 Bimodality induced by coloured noise

Let us consider a stochastic system described by the equation

$$\frac{dx}{dt} = -\frac{\partial U}{\partial x} + f(t), \tag{8.6.1}$$

where

$$\langle f(t) f(s) \rangle = (D/\tau_c)\, e^{-|t-s|/\tau_c}.$$

This describes a diffusive particle in the presence of a potential '$U(x)$' and a coloured-stochastic force '$f(t)$'. The value $v_c = 1/\tau_c$ is the cut-off frequency of the noise. For low values of v_c, theoretical predictions are very difficult (Dekker, 1982; Fox, 1983; Fronzoni *et al.*, 1986; Grigolini and Marchesoni, 1985; Hänggi, P., 1978; Hänggi, Marchesoni and Grigolini, 1984; Lindenberg and West, 1983; Lugiato and Horowicz, 1985; Sancho, San Miguel, Katz and Gunton, 1982; Van Kampen, 1976) and experimental results are necessary to test the theory.

A very simple example to illustrate the effect of coloured noise is provided by a potential with the form $U(x) = bx^4$. This means that for the simulation to be carried out, only one Miller integrator is required, and the corresponding design scheme is shown in Figure 8.9. The equation associated with this circuit is

$$\tau \frac{dV}{dt} = -V^3 + \frac{10R}{R_n} V_n, \tag{8.6.2}$$

where $\tau = 10RC$ is the time scaling. The factor of ten appearing in the expression for τ depends on the fact that the first and second multipliers are assembled in that configuration which corresponds to the transfer function, (8.2.5), characterized by $V_0 = 1$ and $V_0 = 10$, respectively.

The statistical distribution, as provided by this analogue simulation, appears to be characterized by two maxima rather than by a single maximum, as would be suggested by a canonical distribution. This is strong evidence for a

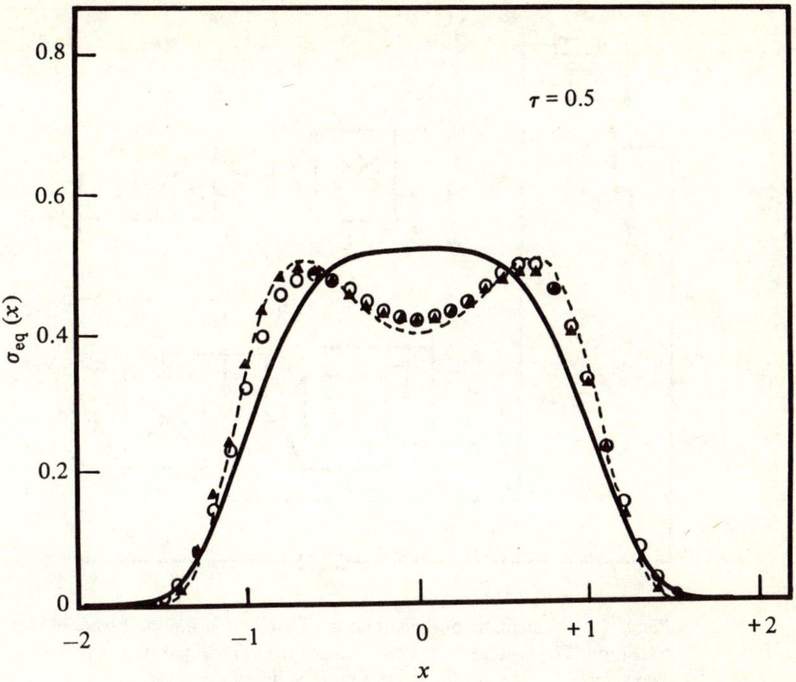

Figure 8.10. Statistical distribution $\sigma_{eq}(x)$ obtained via an analogue simulator whose scheme is shown in Figure 8.9. The full line denotes the result of the mean field approximation. The dotted line denotes the theoretical result of the Fox theory (Fox, 1986). The circles are the results of a digital simulation; the triangles are the results of the analogue simulation. In this case $D = 0.5$ and $b = 1$. From Faetti *et al.*, 1988a.

transition from a one-mode to a two-mode distribution with the increase of τ. In Figure 8.10 the analogue results are compared with two different theoretical predictions (Fox, 1986; Grigolini, 1985; Hänggi, Mroczkowski, Moss and McClintock, 1985) and the results of a digital simulation.

8.6.2 *The effect of external noise on the Freedericksz transition*

Another interesting application of the analogue technique is the simulation of the Freedericksz model (Chandrasekhar, 1977) at the instability threshold. This transition appears in a layer of liquid crystal in the presence of a magnetic field H. Let us consider a layer of liquid crystal with thickness d and uniform orientation. When a magnetic field is applied to the sample, a torque tends to orient the molecules along the field direction. Using the theory of elasticity, we get the approximate equation

$$\lambda\dot{\theta} = -\frac{k\pi}{d^2}\theta + \chi_a H^2(\theta - \tfrac{1}{2}\theta^3), \tag{8.6.3}$$

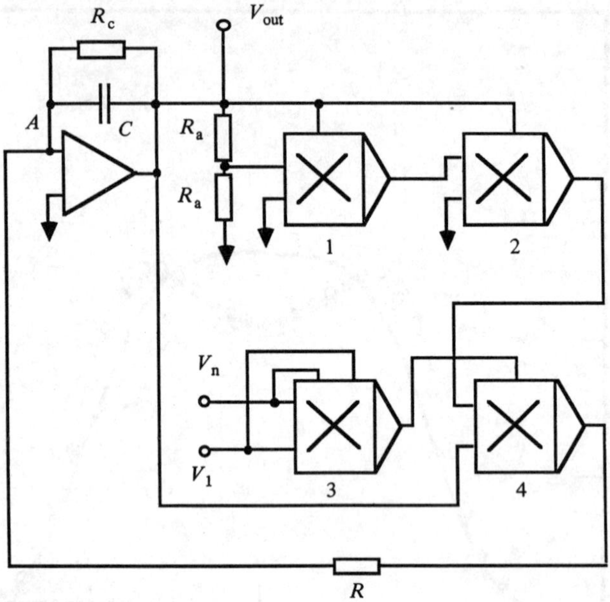

Figure 8.11. Schematic of the circuit used for simulating the Freedericksz transition. The voltage V_1 corresponds to the control parameter H in (8.6.3). The symbol V_n denotes the mean square value of the noise.

where k and λ are, respectively, the bend elastic constant and the friction parameter. χ_a is the magnetic anisotropy, and θ is the mean angle that the molecules make with the direction perpendicular to the magnetic field. From (8.6.3) it appears that when $H_c^2 \simeq k\pi/\chi_a d^2$ a transition occurs which corresponds to the molecular orientation with $\theta \neq 0$.

Fluctuations of the intensity of the magnetic field can change the value of the critical threshold H_c. In this case the theoretical approach presents some difficulties, due to the non-linear stochastic term present in (8.6.3). The theoretical question to answer is which Fokker–Planck equation should be associated with (8.6.3) in the case where the noise is coloured. The analogue experiment has proven to be an important tool to check the theoretical predictions.

From (8.6.3) it turns out that four multipliers are required for the simulation to be carried out. Figure 8.11 shows the schematic of the analogue circuit. The sum of the currents at point A gives

$$C\frac{\mathrm{d}V}{\mathrm{d}t}+\frac{V_4}{R}+\frac{V}{R_c}=0 \tag{8.6.4}$$

and

$$V_4=(V_1-V_n)^2\left(V-\frac{1}{2}\frac{V^3}{100}\right). \tag{8.6.5}$$

Inserting V_4 in (8.6.4), we arrive at the final form

$$RC\frac{dV}{dt} = -\frac{R}{R_c}V + (V_1 - V_n)^2\left(V - \frac{1}{2}\frac{V^3}{100}\right). \tag{8.6.6}$$

which, when the time is scaled, is equivalent to (8.6.3).

The voltage V_1 applied to the input of the 3-multiplier plays the role of the intensity of the magnetic field H. To evaluate the threshold V_c the output of the device was sent to a computer so as to determine a statistical distribution. According to the definition of the threshold for the Freedericksz transition this should take place at the value of V_1 at which the maximum of the statistical distribution $P(x)$ changes from $x = 0$ to $x \neq 0$. Figure 8.12 shows the experimental statistical distribution $P(x)$ at different values of V_1. Figure 8.12(a) corresponds to $V_1 < V_c$, and Figure 8.12(b) corresponds to $V_1 > V_c$.

Sancho, Sagues and San Miguel (1986) provide a prediction of the threshold as a function of the correlation time τ_c of the noise fluctuations. The analogue results (Figure 8.13) seem to be in good agreement with the theory for low values of τ_c (Figure 8.13a) but they move away from theoretical predictions for high values of τ_c (dotted line in Figure 8.13b). This disagreement is not surprising because this theory is an approximate one, but it is difficult to understand whether the disagreement depends on the approximations behind this theory or on the breakdown of the Fokker–Planck equation itself. By using a dichotomous noise rather than a Gaussian one, it has been possible to determine that the main reason for the disagreement is precisely the breakdown produced by the quadratic nature of noise. In fact, when, in (8.6.6), a dichotomous noise is added, the problem has an exact analytical solution (Horsthemke, Doering, Lefever and Chi, 1985). This allows us to test the analogue simulation for very high values of the correlation time τ_c. Comparing these measurements to the predictions of a new theory (Faetti, Fronzoni, Grigolini and Mannella, 1988a; Faetti *et al.*, 1988b), we obtain good agreement, as shown in Figure 8.13(b) (full line).

8.6.3 *Theoretical developments*

The previously described experiments on systems with coloured noise have allowed us to arrive at a new theoretical consideration: we rewrite (8.6.1) and (8.6.3) in a general form

$$\frac{dx}{dt} = \varphi(x) + \psi(x)\xi(t), \tag{8.6.7}$$

where $\xi(t)$ is a genetic coloured noise. The 'Schrödinger like' picture (Grigolini, 1985) corresponding to (8.6.7) reads as

$$\frac{\partial}{\partial t}\rho(x, \xi, t) = (\mathscr{L}_a + \mathscr{L}_b + \mathscr{L}_1)\rho(x, \xi, t), \tag{8.6.8}$$

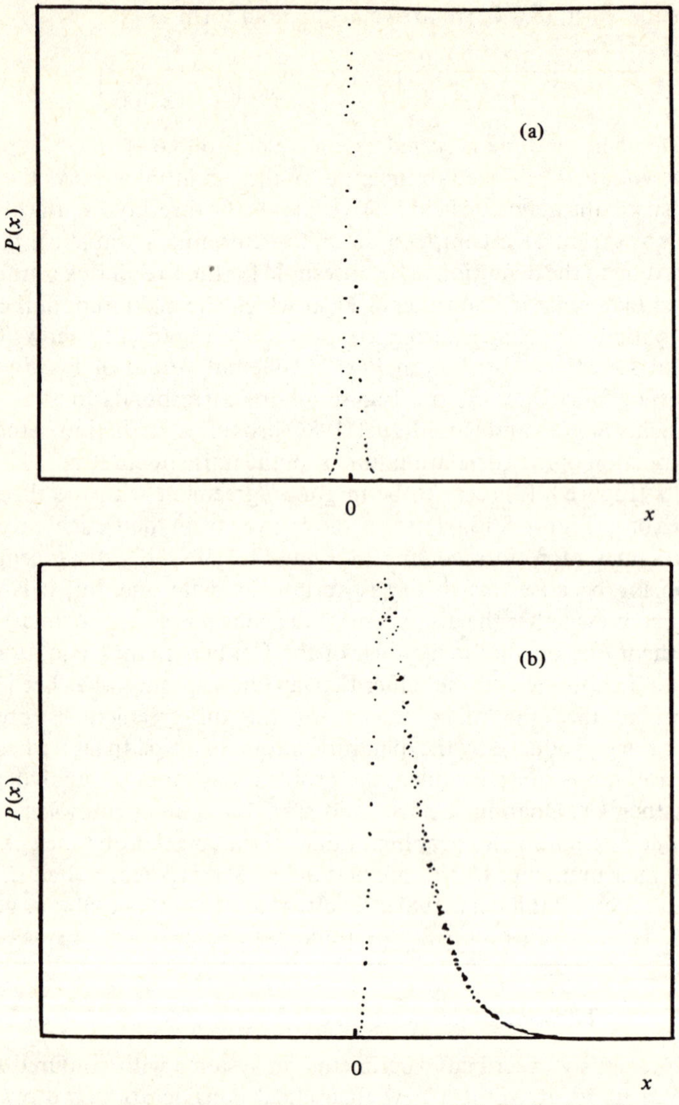

Figure 8.12. Experimental statistical distribution $P(x)$ obtained by the simulation for (a) $V_1 < V_c$ and (b) $V_1 > V_c$, where V_c is the threshold of the Freedericksz transition.

where

$$\mathscr{L}_a \equiv -\frac{\partial}{\partial x}\varphi(x), \quad \mathscr{L}_b \equiv -\psi(x)\xi(t). \tag{8.6.9}$$

$\mathscr{L}_b$ is the operator driving the statistical distribution of the variable ξ. By applying the projection operator method of Grigolini (1986) to the derivation

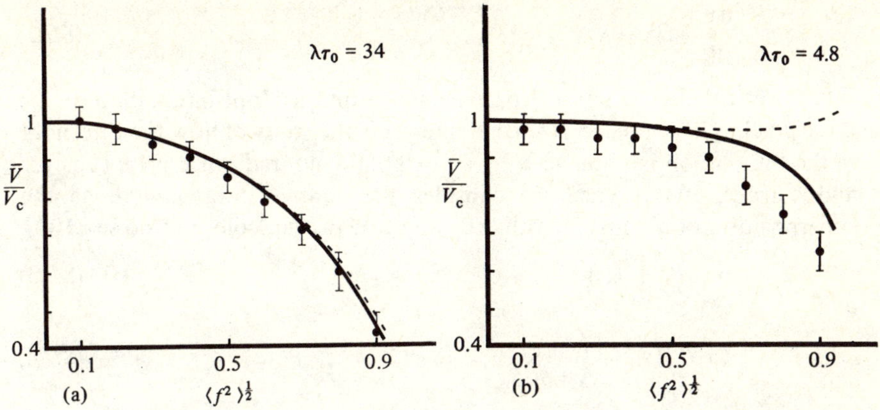

Figure 8.13. The ratio $(\bar{V}/V_c)$ versus the mean square value of the voltage noise. $\bar{V}$ and V_c denote the Freedericksz transition with and without noise, respectively. $\tau_0 = R_c C$ and $\lambda = 1/\tau_c$. (a) Low values of the time correlation (white noise). (b) High values of the time correlation (coloured noise). The dotted line is the theoretical prediction of Sancho, Sagues and San Miguel (1986). The full line is the prediction of Faetti *et al.* (1988b). Circles denote the analogue results.

of an equation of motion for the reduced probability distribution $\sigma(x,t) \equiv \int d\xi\, \rho(x,\xi,t)$ to second-order in $\mathscr{L}_1$ we obtain

$$\frac{\partial}{\partial t}\sigma(x,t) = \mathscr{L}_a\sigma(x,t) + \int_0^t w(\tau)\sigma(x,t)d\tau \tag{8.6.10}$$

where

$$w(t) = -c(t)\frac{\partial}{\partial x}\psi(x)e^{\mathscr{L}_a t}\frac{\partial}{\partial x}\psi(x)e^{-\mathscr{L}_a t} \tag{8.6.11}$$

and

$$c(t) = \langle \xi(t_1)\xi(t_2)\rangle.$$

This Fokker–Planck form is independent of the noise statistics, the dependence on which actually starts being exhibited at the third-order in $\mathscr{L}_1$. Equation (8.6.10) coincides with the so-called 'best Fokker–Planck approximation' (BFPA) at any order in τ.

8.7 A two-dimensional system

8.7.1 *The Brusselator*

Another interesting example is the study of the following system of equations (Nicolis and Prigogine, 1977)

$$\frac{dx}{dt} = A - (1 - B)x + x^2 y \tag{8.7.1}$$

$$\frac{dy}{dt} = Bx - x^2 y. \tag{8.7.2}$$

If B is regarded as a control parameter we find a Hopf bifurcation at the critical value $B_c = 1 + A^2$. A special interest is the study of how the threshold of the bifurcation is influenced by an external coloured noise on B (Lefever and Turner, 1986). Then we consider the control parameter B as the superposition of a constant value B_0 and a Gaussian-coloured noise $f(t)$:

$$B = B_0 + f(t) \tag{8.7.3}$$

and

$$\langle f(t) f(s) \rangle = \left(\frac{D}{\tau}\right) e^{-|t-s|/\tau} \tag{8.7.4}$$

where D characterizes the noise intensity and τ is the correlation time. Using the 'minimum component design philosophy' as discussed in the applications described above it is helpful to write (8.7.1) and (8.7.2) in order to point out the operations required to make the simulation

$$\frac{dx}{dt} = a_1 - bx + z \tag{8.7.5}$$

$$\frac{dy}{dt} = b_1 x - z \tag{8.7.6}$$

$$z = x \cdot x \cdot y \tag{8.7.7}$$

$$b_1 = b_0 + f(t). \tag{8.7.8}$$

These equations suggest the use of only two integrators and two multipliers. In Figure 8.14 the basic scheme of the analogue circuit is shown. The electrical equivalent equations read

$$RC\frac{dx}{dt'} = V_0 - [1 + (B_0 + V_n)]x + x^2 y \tag{8.7.9}$$

$$RC\frac{dy}{dt'} = (B_0 + V_n)x - x^2 y. \tag{8.7.10}$$

Substituting the dimensionless time $t = t'/RC = t'\omega_B$ into these equations we get the same form as that of (8.7.1) and (8.7.2).

8.7.2 *The postponement of the Hopf bifurcations*

The question considered here is what happens when B becomes noisy. The limit cycle consequently becomes a statistical quantity and the bifurcation becomes non-distinct as discussed by Lucke (Chapter 6, Volume 2) and Wiesenfeld (Chapter 7, Volume 2). However, a statistical definition of the new bifurcation point B^*, which occurs in the presence of noise, can be made in

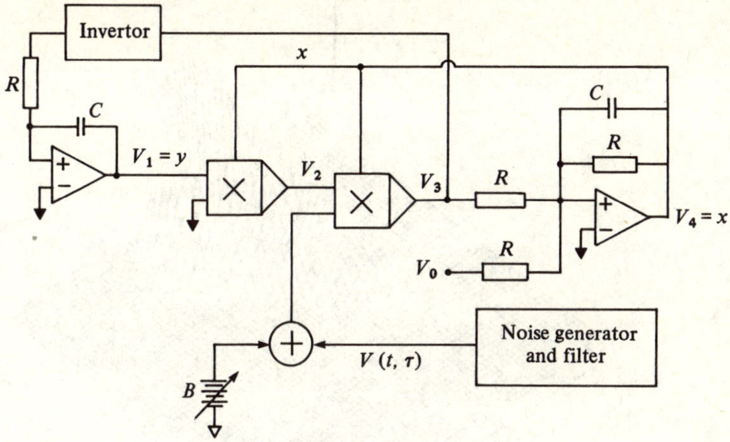

Figure 8.14. Scheme of the analogue circuit of the Brusselator. τ is the correlation time of the noise. The voltage B corresponds to the control parameter B_0 in (8.5.20). From Fronzoni *et al.* (1987).

terms of the topology of the two-dimensional stationary statistical density $P(x, y)$. The behaviour of B^* with D and τ is of interest. Specifically, the Lefever and Turner theory predicts that for $D > 0$, $B^* > B$ (a postponement) for coloured noise, i.e. for $\tau > \omega_B^{-1}$, whereas $B^* < B$ (an advancement) for white noise. The definition of B^* can best be made with actual measured examples of $P(x, y)$ in view. For $B > B_c$, the limit cycle as measured on this device is a closed contour $y(x)$ which is traced out at frequency ω_B. When $D > 0$, this contour becomes stochastic, and the appropriate statistical observable is the stationary two-dimensional density $P(x, y)$. It is shaped like a mountain with an irregularly shaped crater at the top. An example for B well above B_c is shown in Figure 8.15, where (a) shows an isometric view and (b) shows a set of contours of constant probability, which are the cuts made by planes at various heights parallel to the xy plane. As the noise intensity is increased the crater 'floor' rises until it becomes equal in height to the lowest point on the crater rim. This defines the critical stationary density which occurs for a pair of values B^* and D^*. For $B < B^*$ (but still $> B_c$) or for $D > D^*$ the crater 'fails to hold water'. In topological language, the critical density separates the class of all densities which have only simply connected horizontal cross sections from those which have one or more doubly connected cross sections. The pair D^* and B^* then define a phase plane, as shown by the measured results in Figure 8.16. The Lefever–Turner theory predicts both advancements and postponements depending on the value of τ. We observe only postponements, to within the accuracy of our simulation, for all values of $0.1 < \tau < 10$. For a complete discussion and more details about these phenomena readers are referred to Fronzoni, Mannella, McClintock and Moss, 1987.

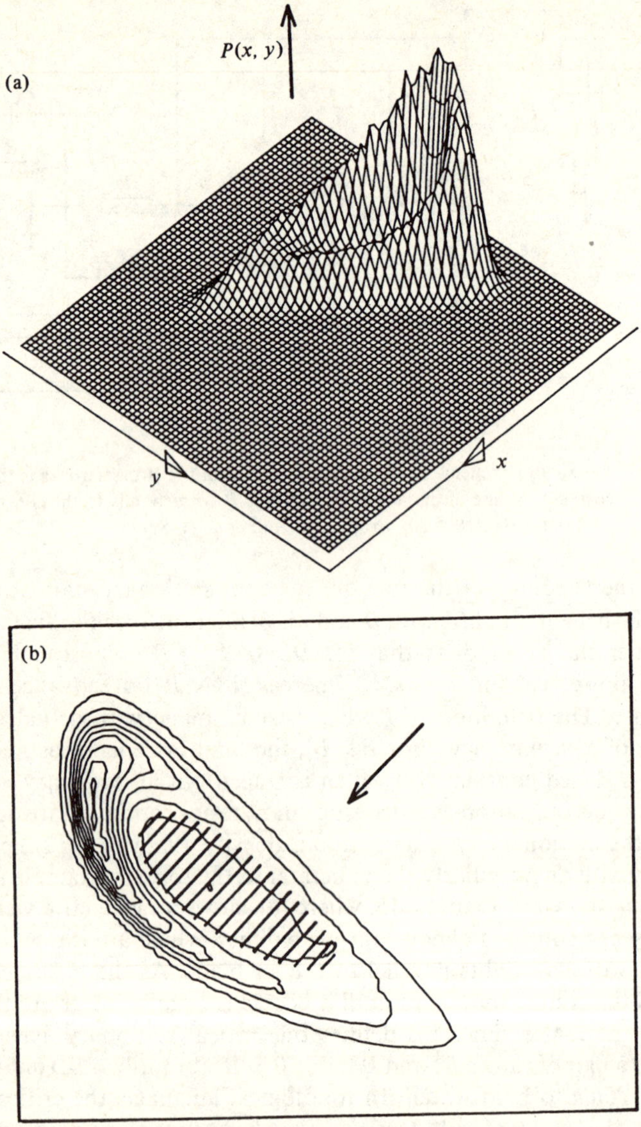

Figure 8.15. Statistical density $P(x, y)$ measured for $B = 2.2$, $D = 0.008$ and $\tau = 0.1$. (a) A three-dimensional view as observed from the direction indicated by the arrow in (b). (b) A plot of contours of constant probability at equally spaced intervals. The cross-hatched area indicates the crater which is concave downward. From Fronzoni *et al.* (1987).

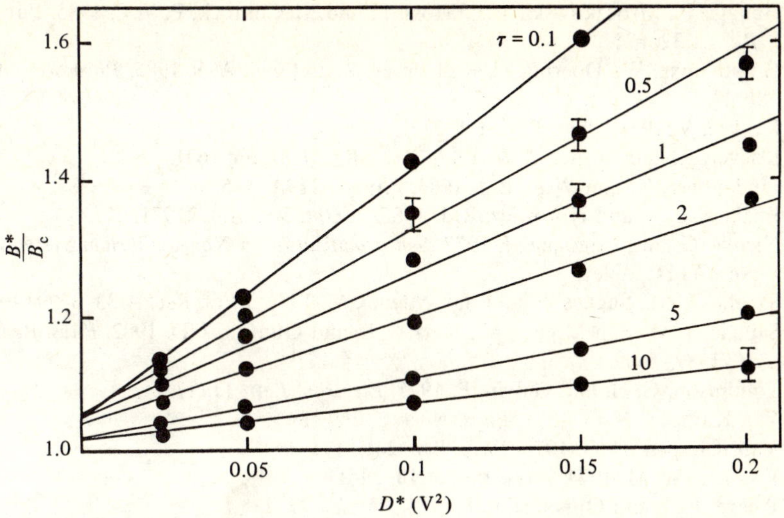

Figure 8.16. The ratio (B^*/B_c) as a function of the noise intensity D for different values of the time correlation τ. B^* and B_c denote the Hopf bifurcation threshold with and without noise, respectively. From Fronzoni *et al.* (1987).

References

Chandrasekhar, S. 1977. *Liquid Crystals.* Cambridge University Press.

Dekker, H. 1982. *Phys. Lett.* **90A**, 26.

Faetti, S., Festa, C., Fronzoni, L., Grigolini, P. and Martano, P. 1984. *Phys. Rev. A* **30**, 3252–63.

Faetti, S., Fronzoni, L., Grigolini, P. and Mannella, R. 1988a. Submitted to *J. Stat. Phys.*

Faetti, S., Fronzoni, L., Grigolini, P., Palleschi, V. and Tropiano, G. 1988b. Submitted to *J. Stat. Phys.*

Fox, R. F. 1983. *Phys. Lett.* **94A**, 281.

Fox, R. F. 1986. *Phys. Rev. A* **33**, 467.

Fronzoni, L., Grigolini, P., Hänggi, P., Moss, F., Mannella, R. and McClintock, P. V. E. 1986. *Phys. Rev. A* **33**, 3320.

Fronzoni, L., Grigolini, P., Mannella, R. and Zambon, B. 1985. *J. Stat. Phys.* **41**, (3/4), 533–79.

Fronzoni, L., Mannella, R., McClintock, P. V. E. and Moss, F. 1987. *Phys. Rev. A* **36**, 834.

Golomb, S. 1967. *Shift Register Sequences.* San Francisco: Holden-Day.

Grigolini, P. 1985. *Adv. Chem. Phys.* **62**, 1–29.

Grigolini, P. 1986. *Phys. Lett.* **119**, 157.

Grigolini, P. and Marchesoni, F. 1985. *Adv. Chem. Phys.* **62**, 29–80.

Hänggi, P. 1978. *Z. Phys.* **B31**, 407.

Hänggi, P., Marchesoni, F. and Grigolini, P. 1984. *Z. Phys.* **B56**, 333.

Hänggi, P., Mroczkowski, T. J., Moss, F. and McClintock, P. V. E. 1985. *Phys. Rev. A* **32**, 625.

Horsthemke, W., Doering, C. R., Lefever, R. and Chi, A. S. 1985. *Phys. Rev. A* **31**, 1123.

Kenkre, V. 1971. *Phys. Rev. A* **4**, 2327.

Lefever, R. and Turner, J. W. 1986. *Phys. Rev. Lett.* **56**, 1631.

Lindenberg, K. and West, B. J. 1983. *Physica* **119A**, 485.

Lugiato, L. A. and Horowicz, R. J. 1985. *J. Opt. Soc. Am.* **2**, 971.

Nicolis, G. and Prigogine, I. 1977. *Self-organization in Nonequilibrium Systems.* New York: Wiley.

Sancho, J. M., Sagues, F. and San Miguel, M. 1986. *Phys. Rev. A* **33**, 3399–403.

Sancho, J. M., San Miguel, M., Katz, S. L. and Gunton, J. D. 1982. *Phys. Rev. A* **26**, 1589–1609.

Tomlinson, G. H. and Galvin, P. 1975. *Electron. Lett.* **11**, 77.

Van Kampen, N. G. 1971. *Phys. Norv.* **5**, 279– 84.

Van Kampen, N. G. 1976. *Phys. Rep.* **24C**, 171.

Visscher, W. M. 1974. *Phys. Rev. A* **10**, 2461.

Weare, J. H. and Oppenheim, I. 1974. *Physica* **72**, 1.

9 Analogue techniques for the study of problems in stochastic nonlinear dynamics

P. V. E. McCLINTOCK and FRANK MOSS

9.1 Introduction

There is nothing new about the use of analogue techniques to find approximate solutions of awkward nonlinear equations. Mechanical and geometrical analogue methods have been used since time immemorial (see, for example, Thom, 1971); electronic analogue computers (Johnson, 1963) have now been available in one form or another for almost half a century. Two important developments of recent years, however, have combined to make it a relatively simple matter to simulate even quite complicated systems usefully on the laboratory bench so that, for many applications, an analogue computer *per se* is no longer required. First, reliable, high quality, inexpensive integrated circuits are now readily available to perform all of the common mathematical operations such as addition and subtraction, multiplication and division, trigonometric functions, integration and differentiation, and so on. Just a few such components, suitably arranged, are normally sufficient to model the particular nonlinear system under study; and the circuits can usually be designed, built and tested for their static response in less than a day. Secondly, the advent in the laboratory of small, relatively inexpensive, digital computers/data-processors has made it possible to carry out quite sophisticated statistical analyses of the circuit response in real time.

Experimental analogue systems of this kind are especially suitable for the study of Langevin equations of the type

$$\dot{x} = h(x) + V_N(t)g(x), \qquad (9.1.1)$$

where h and g are functions of x, and $V_N(t)$ represents external fluctuations imposed on the system (Horsthemke and Lefever, 1984). The use of electrical circuits to study problems of this kind, originally introduced by Landauer (1962) and Stratonovich (1963, 1967), was subsequently developed by a number of workers including, particularly, Morton and Corrsin (1969). The latter authors' experiments on Fokker–Planck systems constitute an early precursor of the analogue studies to be described below. The first application of the technique in its present form was that reported by Sancho, San Miguel, Yamazaki and Kawakubo (1982a).

The motivation underlying the development of both analogue and digital simulation techniques relates to the considerable difficulty (see Volumes 1 and 2) of finding theoretical solutions of stochastic differential equations: only in a minority of cases (e.g. single-variable systems subject to white Gaussian, or to exponentially correlated dichotomous, noise) is it possible to obtain exact analytic solutions. In all other cases approximation is necessary, and this naturally casts a measure of doubt on the accuracy and range of validity of the numerical or semi-analytic approximate solutions that are obtained. It is now well appreciated that approximations which are valid and lead to accurate results under a given set of conditions can nonetheless become extremely poor approximations in a different but possibly even quite close neighbourhood of the parameter space. Consequently, it is highly desirable also to obtain solutions by independent means, and this can conveniently be accomplished either by digital simulation (see Chapter 7) or by the analogue experimental methods to be discussed below (see also Chapters 6 and 8) or, preferably, by both approaches.

In fact, digital and analogue simulation are to a large extent mutually complementary, each method having its own particular advantages and disadvantages. Digital simulation is inherently more accurate, provided, of course, that the appropriate algorithm has been correctly programmed. On the other hand, huge amounts of central processor time (in some cases, many hours) are often required on large computers in order to achieve results of acceptable statistical quality, and this is particularly true of the higher-dimensional systems that are becoming a strong focus of current interest. Analogue experiments, although less accurate, can be performed in a relatively straightforward manner and usually yield results of excellent statistical quality in a matter of minutes. An additional advantage of analogue experiments is that they mimic nature more closely than can ever be the case for a digital simulation. The experimental clarification (Smythe, Moss and McClintock, 1983; Smythe, Moss, McClintock and Clarkson, 1983) of the widespread 'Ito versus Stratonovich' discussion of a few years ago (Van Kampen, 1981b, and references therein) provides an excellent example of what amounts to a very basic distinction between the analogue and digital approaches. On the basis of the analogue technique, it was possible to confirm experimentally the expectation that it is the Stratonovich prescription for the integration of the stochastic differential equation that describes what happens in reality for a real physical system subject to quasi-white noise (see Section 9.3). In performing a digital simulation of the same system, however, it was necessary first to choose to programme either the Ito or the Stratonovich stochastic calculus into the algorithm to be computed; so that the digital simulation approach was, of its nature, innately unable to distinguish between them.

In the light of these considerations it seems in many ways more helpful to regard applications of the analogue technique as amounting to *experiments on real physical systems*, rather than as simulations as such. Although each electronic circuit is contrived, rather than naturally occurring, being specifically designed to model as closely as possible some particular stochastic differential equation or set of equations, it is nonetheless a real physical system in its own right and consequently is obliged to behave as nature prescribes. Furthermore, the circuits necessarily incorporate many of the 'nonidealities,' such as internal noise and nonzero additive constants, that are also found in natural systems. As will become apparent below, this feature can sometimes be helpful in drawing attention to subtle but crucially important differences between particular stochastic differential equations and the natural systems that they are intended to model.

In the next section we present succinct experimental details of the analogue techniques that have been developed. To demonstrate their utility and flexibility we describe in Sections 9.3–9.9 some applications to the study of selected stochastic nonlinear systems; for fuller details of any particular application readers are referred to the original publications. Finally, in Section 9.10, we discuss possible future directions in which the work may develop.

9.2 Experimental technique

Most of the circuits used to date have been based on a relatively small number of standard integrated circuits including, in particular, the LF356 operational amplifier and the Analog Devices AD534 multiplier/divider. The Analog Devices AD639B trigonometric function generator has also been used occasionally for special applications (see Section 9.8 below). The use of these devices limits the upper frequency response of the circuits in practice to a few tens of kilohertz; the lower limit is usually zero, since the circuits are d.c.-coupled. Given the relatively modest maximum frequency requirements, special layouts are not needed and the circuits can usually be constructed on standard plug-in prototype boards. The various mathematical elements are formed in the usual way (see, for example, Horowitz and Hill, 1980) by use of the operational amplifiers in conjunction with suitably chosen resistors and capacitors, but with multiplication/division being effected by means of the AD534. The latter also provides a number of other facilities which can often be exploited to excellent effect, as discussed in Chapter 8.

The circuits are invariably designed to operate in terms of scaled time, such that one second of real-time circuit operation corresponds typically to 100–1000 time units of the model equation. This time-scaling refers to the use of integrator circuit elements (of standard design) whose time constants are made less than unity, and is necessary for two reasons. First, if it were desired to apply pseudo-white external noise to a circuit whose integrator time constant was unity, it would be essential to ensure that the power spectrum of the noise generator remained flat deep into the sub-hertz range. This is not in fact the case for commonly available commercial noise generators. Secondly, there would be the disadvantage that data acquisition periods would be orders of magnitude longer than for the time-scaled circuit: quite apart from inconvenience, this would exacerbate problems of stability and drift of parameter values in the circuit.

The noise voltage V_N used to drive the circuit is exponentially correlated such that

$$\langle V_N(t) V_N(s) \rangle = \frac{D}{\tau} e^{-|t-s|/\tau}, \tag{9.2.1}$$

where τ is the correlation time and D is the noise intensity, so that

$$\langle V_N^2 \rangle = \frac{D}{\tau}. \tag{9.2.2}$$

Because the circuit is operating in terms of scaled time,

$$\tau = \tau_N / \tau_I, \tag{9.2.3}$$

where τ_N is the correlation time (in seconds) of the driving noise voltage and τ_I is the integrator time constant (in seconds) of the circuit. Thus

$$\langle V_N^2 \rangle = \frac{D\tau_I}{\tau_N} \tag{9.2.4}$$

is the relationship that links the mean square noise voltage actually applied to a circuit (of unity scale factor) to the parameter D that most often appears in the stochastic differential equations. In order to perform experiments in the white noise limit where $\tau \rightarrow 0$, and (9.2.1) becomes

$$\langle V_N(t) V_N(s) \rangle = 2D\delta(t-s), \tag{9.2.5}$$

it is necessary to ensure that $\tau_N \ll \tau_I$. In practice, it is usually sufficient that $\tau_N \simeq 0.1\,\tau_I$, in order that the results from analogue experiments on one-dimensional systems agree with the exact, white noise solutions of the Stratonovich Fokker–Planck equation to within a few per cent (which is comparable to the accuracy of the analogue experiment). In certain white noise analogue experiments and theories, particularly the earlier ones, the noise intensity was defined in terms of a parameter σ such that

$$\langle V_N(t) V_N(s) \rangle = \sigma^2 \delta(t-s). \tag{9.2.6}$$

In the interests of consistency with the (more detailed) original publications, to which the reader may wish to refer, the same convention will be used in relevant cases below; the connection between D and σ is simply that $\sigma^2 = 2D$, so that

$$\langle V_N^2 \rangle = \frac{\sigma^2 \tau_I}{2\tau_N}. \tag{9.2.7}$$

A variety of noise generators has been used, but most of the experiments to be described have been based on the Quan-Tech model 420 or Wandel and Goltermann model RG1 instruments. The output of the noise generator is passed through a single-pole active filter with a time constant of τ_N, before being applied to the circuit. The noise intensity, as specified by D or σ, is determined by a direct measurement of $\langle V_N^2 \rangle$ at the output of the filter with the aid of an Analog Devices AD536A true RMS to d.c. converter.

The analogue circuits are usually designed to operate at a scale of unity in volts, so that a change of one volt in the circuit corresponds to a unit change of the relevant variable in the corresponding model equation. In particular cases, however, smaller scales can be used such that unit change in the variable corresponds to less than one volt in the circuit: the D or σ appearing in (9.2.4) or (9.2.7) must then be scaled by an appropriate factor. Operation at a reduced scale factor facilitates studies of system response at high noise intensity while ensuring that voltage swings still lie within the range (*c.* ± 10 V) that can be accommodated by the circuit. The concomitant disadvantage, of course, is that nonidealities such as parameter drift and internal noise in the circuit become correspondingly more important.

The fluçtuating circuit variables, for example x or $\dot{x}$, are digitized typically with 12-bit precision in blocks of 1K–4K samples, which are then analysed to extract and average the particular items of statistical information that are needed. Most of the work to be described below has been based on the use of Nicolet 1080, 1180 and 1280 data-processors, which incorporate two or more internal digitizers for the analogue input signals. They are easily programmed to measure, for example, one-dimensional and two-dimensional statistical densities, moments of any order, time-evolving densities, mean first passage times and distributions of passage times, power spectra and correlation times. The fast internal 24-bit FFT co-processor of the Nicolet 1280 makes it particularly suitable for measurement of the latter two quantities. Data acquisition/averaging is usually continued until the statistical quality of the quantity being measured, as viewed on the CRT display, is judged to be satisfactory: this may involve the averaging of from a few hundred up to several thousand input sweeps, depending on circumstances. The final result may be filed on disk.

The measured, averaged, data can be characterized locally, using electronic cursors to indicate relevant regions on the display. Thus the operator can, for example, locate extrema by least-squares fittings of parabolae or Gaussian curves, or can measure exponents, or integrate correlation functions to determine relaxation times. The results can be plotted in the usual way on an *XY* recorder; contour plots can also be drawn of two-dimensional densities, all directly from the data-processor. Alternatively, the results can quickly be transferred to a mainframe computer (such as a VAX) for more sophisticated analyses or plotting facilities.

9.3 The genetic model equation

The particular motivation for the experimental study (Smythe, Moss and McClintock, 1983; Smythe *et al.* 1983) of the genetic model equation lay in the prediction by Lefever and Horsthemke (1979, 1981) that a noise-induced transition would occur at a critical value of noise intensity. The equation of the genetic model is

$$\dot{x} = \tfrac{1}{2} - x + \lambda_t x(1-x), \tag{9.3.1}$$

where

$$\lambda_t = \lambda + \sigma\xi_t. \tag{9.3.2}$$

λ is a control parameter, and $\sigma\xi_t$ is a Gaussian white noise of variance σ^2 and zero mean as described by (9.2.6). The solution of the corresponding Fokker–Planck equation for the stationary density P_s is

$$P_s = N[x(1-x)]^{-\nu}\exp\left\{-[x(1-x)\sigma^2]^{-1} - \left(\frac{2\lambda}{\sigma^2}\right)\ln\left(\frac{1-x}{x}\right)\right\}, \tag{9.3.3}$$

where $\nu = 1$ or 2 according as to whether the Stratonovich or Ito stochastic calculus, respectively, was used in the derivation. The function (9.3.3) has a single maximum for $\sigma < \sigma_c$, but a double maximum for $\sigma > \sigma_c$, where σ_c defines a critical noise intensity that is dependent on λ, and also on ν: for $\lambda = 0, \sigma_c = 2$ when $\nu = 1$ (Stratonovich) and $\sigma_c = \sqrt{2}$ when $\nu = 2$ (Ito). It is the transition from a single- to a double-peaked density that constitutes the noise-induced transition, for which the associated order parameter can be defined (Lefever and Horsthemke, 1981) as the separation between the peaks.

The system was simulated by Smythe, Moss and McClintock (1983) and Smythe *et al.* (1983) using the electronic circuit shown in Figure 9.1. It functions in the manner outlined in the previous section. All sum and difference operations, and the two integrations, are accomplished with standard (LF356) operational amplifiers; the multiplications are performed with the AD534 integrated circuit. This particular circuit differs from the others to be described in having a slow integrator to stabilize the d.c. working point, which corresponds to the steady state solution of the Langevin equation without noise, as well as the fast integrator of time constant τ_I. The former integrator is not in fact essential, and for certain applications (measurement of nonstationary phenomena) it is inadmissable; it was therefore omitted from all analogue circuits used subsequently. The steady state solution of (9.3.1), with $\sigma = 0$, is given by

$$x_s = (1/2\lambda)[\lambda - 1 + (1+\lambda^2)^{1/2}], \tag{9.3.4}$$

and the circuit could, in practice, be adjusted to follow this relation to an accuracy of $\pm 4\%$ over the full working range of $-7 \leqslant \lambda \leqslant +7$.

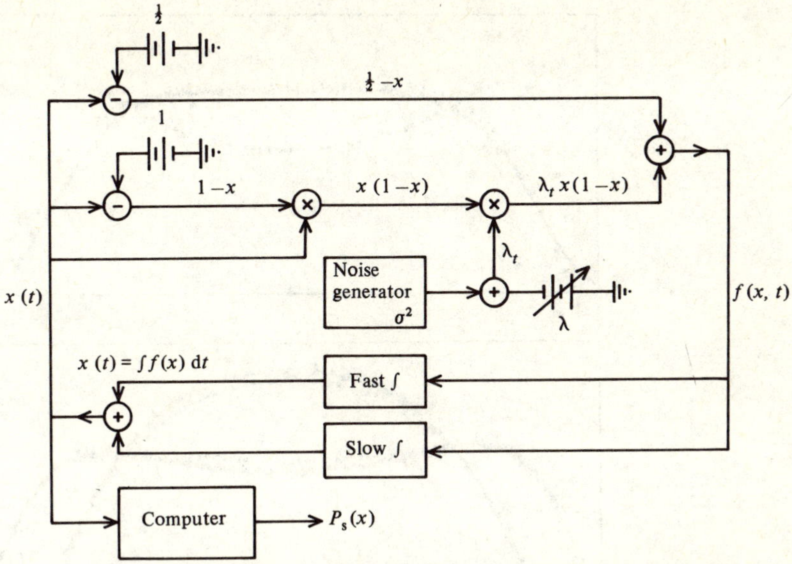

Figure 9.1. Block diagram of the electronic analogue circuit model of the genetic model equation (9.3.1). (Smythe, Moss and McClintock, 1983.)

Some typical density functions measured for the circuit are shown by the continuous curves of Figure 9.2. In Figure 9.2(a), the experimental result refers to $\lambda = 0, \sigma = 2$ which, according to the Stratonovich version of the Fokker–Planck equation (flat-topped dotted line, for $v = 1$) should be the critical noise intensity. This is indeed seen to be the case; and the Ito version of the theory (twin-peaked dotted line, for $v = 2$) is in clear disagreement with the experimental measurements. Figure 9.2(b) shows densities measured for two other values of σ, both above and below σ_c, compared in each case with the Stratonovich theoretical prediction, with which they are seen to be in good agreement.

The phase diagram for the system was mapped out through an experimental determination of the values taken by σ_c for a range of $|\lambda|$; the criterion for $\sigma = \sigma_c$ was taken as the existence of a finite region of nearly zero gradient in the density. The result of this procedure is shown by the open-circle data points of Figure 9.3, where it is compared with the Ito and Stratonovich theoretical predictions (full curves). The measurements are in clear disagreement with the former, but agree with the latter within the systematic experimental error shown by the single bar. On the assumption (see above) that the peak separation m could be identified with an order parameter, measurements were also made of two out of the three classical critical exponents, and were found to be in agreement with the theoretical prediction (Smythe, Moss and McClintock, 1983).

A digital simulation of the system was carried out using an algorithm based

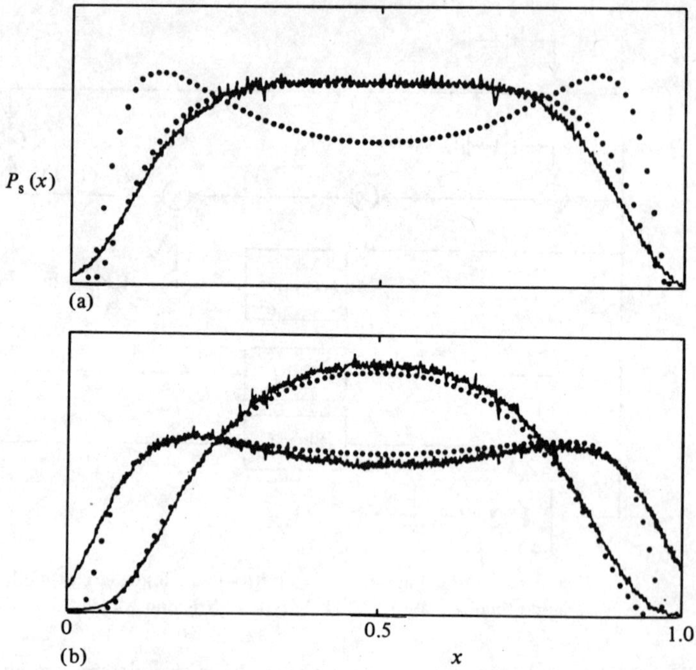

Figure 9.2. Measured density functions $P_s(x)$ for the genetic model (jagged continuous curves) compared with theoretical predictions (filled circles) from (9.3.3). (a) For $\sigma = 2$: Stratonovich calculus, $\nu = 1(\simeq$ flat-topped) and Ito, $\nu = 2$ (twin peaked). (b) For $\sigma = 1.5$ (single maximum) and $\sigma = 2.5$ (double maximum), using the Stratonovich calculus ($\nu = 1$) in both theoretical curves. (From Smythe, Moss and McClintock, 1983.)

on the Ito stochastic calculus. It was found (Smythe *et al.*, 1983) to yield the results shown by the filled circles of Figure 9.3. These are clearly in good agreement with the Ito ($\nu = 2$) version of (9.3.3) but, of course, in disagreement with the experimental results from the analogue circuit. This outcome was very much in agreement with the earlier theoretical discussion given by Van Kampen (1981a, b). As is well known, the 'Ito/Stratonovich' degeneracy (in the predictions of white noise theories) was lifted by the first successful consideration of coloured noise (Blankenship and Papanicolaou, 1978; Wong and Zakai, 1965), which recovered the Stratonovich result in the white noise limit. In the analogue circuit, 'white' noise is really coloured (i.e. real) noise but with $\tau_N \ll \tau_I$, so that the results of such experiments would be expected to agree with the Stratonovich calculus. In fact, in a digital simulation of the genetic model using the Ito algorithm (J. Smythe and F. Moss, 1983, unpublished observations), it was observed that including even an unsophisticated simulation of noise with a very small but nonzero correlation time significantly pushed the results toward the Stratonovich solution.

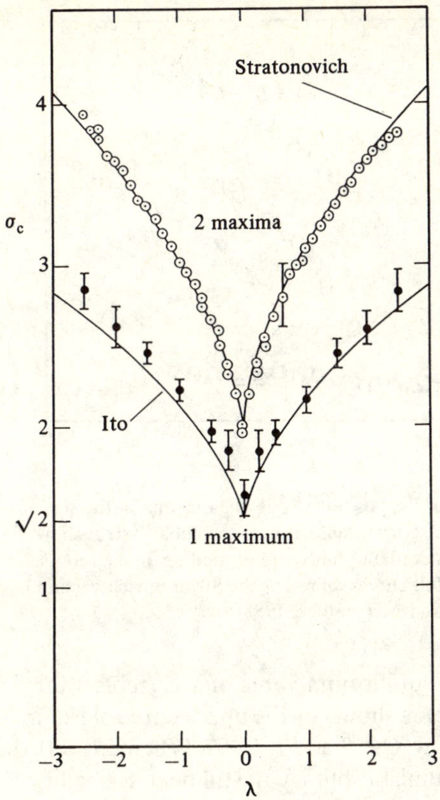

Figure 9.3. The phase diagram for the genetic model. The measured experimental data (open circles) are in agreement with theory (upper full curve) based on the Stratonovich calculus ($v = 1$ in 9.3.3). The filled circles, representing the results of a digital simulation based on the Ito algorithm, are in good agreement with the Ito version (lower curve, calculated with $v = 2$ in 9.3.3) of the theory. (From Smythe *et al.*, 1983.)

9.4 The cubic bistable system

The cubic bistable is the simplest example of a system that exhibits a continuous instability; it can be considered as a generic model representative of first order phase transitions which deterministically show hysteresis and bistability in the presence of noise. It may be written

$$\dot{x} = -x^3 + \lambda_t x^2 - Qx + R = f(x), \tag{9.4.1}$$

where

$$\lambda_t = \lambda + \xi(t) \tag{9.4.2}$$

is the noisy control parameter, $\xi(t)$ represents fluctuations of zero mean, and Q and R are constants. With $\xi(t) = 0$ and appropriate choices of Q and R, there is

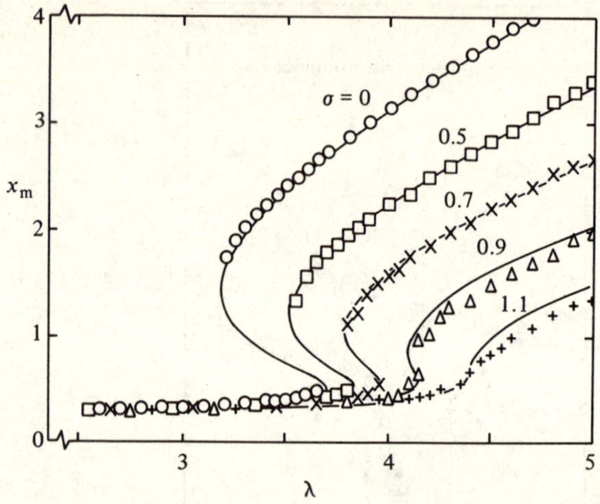

Figure 9.4. Plots of the positions x_m of the maxima of the stationary densities measured for the electronic simulator of the cubic system shown in Figure 9.5. The measured values (data points) are plotted against λ for various noise intensities σ. The full curves represent the Stratonovich version of (9.4.5) (Robinson, Moss and McClintock, 1985).

a range of λ for which the equilibrium value of x is treble-valued (two stable roots and one unstable one) as shown by the upper curve of Figure 9.4. For the experiments discussed below, $Q = 3$ and $R = 0.7$. When $\xi(t) \neq 0$, the system is, of course, no longer deterministic, but it can still be described by its statistical density $P(x)$.

The main interest of such systems in the present context lies in their somewhat counter-intuitive response to the multiplicative fluctuations of the control parameter: the transition from the lower to the upper root can often be shifted by the modulation towards larger values of the control parameter, i.e. *stochastic postponement* of the transition. Effects of this kind have already been observed, or are to be anticipated, in such diverse areas as: quantized turbulence in superfluid helium (Chapter 1 of this volume and references therein; Moss and Welland, 1982); electronics (Kabashima and Kawakubo, 1979); optics (Bowden, Ciftan and Roble, 1981); liquid crystals (Brand and Schenzle, 1980; Kai, Kai, Takata and Hirakawa, 1979; Kawakubo, Yanagita and Kabashima, 1981; San Miguel and Sancho, 1981); chemical equilibria (Roux, de Kepper and Boissonade, 1983); and tumour immunology (Lefever and Horsthemke, 1979).

Welland and Moss (1982) considered the particular case where $\xi(t)$ in (9.4.2) represents Gaussian white noise of variance σ as defined by (9.2.6) so that

$$\langle \xi(t)\xi(s) \rangle = \sigma^2 \delta(t-s). \tag{9.4.3}$$

Following the approach developed by Horsthemke and Lefever (1984; and references therein), they were able to show that the stationary density should be

$$P_s(x) = \frac{N}{x^2(v + 1/\sigma^2)} \exp\left[\frac{2}{\sigma^2}\left(\frac{-\lambda}{x} + \frac{Q}{2x^2} - \frac{R}{3x^3}\right)\right], \tag{9.4.4}$$

where N is a normalizing constant and $v = 1$ or 2 depending on whether one chooses to use the Stratonovich or Ito version of the white noise Fokker–Planck equation. An interesting feature of (9.4.4) is that it predicts a shift with σ of the positions x_m of the extrema of $P_s(x)$, which are given by

$$(1 + v\sigma^2)x_m^3 - \lambda x_m^2 + Qx_m - R = 0. \tag{9.4.5}$$

Plots of $x_m(\lambda)$ for different values of σ are shown by the full curves of Figure 9.4 for the Stratonovich version of the equation. It is clear that when considered in terms of the maxima of $P_s(x)$, i.e. in terms of the most probable values of x, an increase of σ tends to stabilize the system in its lower state, corresponding to a stochastic postponement of the transition: behaviour that is, of course, highly reminiscent of that observed in some of the experiments mentioned above. Furthermore, as σ is varied with fixed λ, the system can evidently undergo a variety of noise-induced transitions, where the density changes from unimodal to bimodal form, or vice versa.

In order to establish whether or not behaviour of this kind occurs in reality, in a well-characterized system governed by (9.4.1)–(9.4.3), an analogue experiment was carried out by Robinson, Moss and McClintock (1985). A block diagram of the electronic circuit is shown in Figure 9.5, and some typical densities obtained from it are plotted in Figure 9.6. It is evident that, when σ is gradually increased while keeping all other parameters constant, the density broadens, shifts, becomes bimodal and finally, for very large σ, converges towards a delta function close to the lower root of the deterministic ($\sigma = 0$) version of the cubic system (9.4.1), (9.4.2). Detailed comparisons (McClintock and Moss, 1985) between the shapes of the densities in Figure 9.6 and those predicted by (9.4.4) yield excellent agreement provided that the Stratonovich ($v = 1$) version of the equation is used. The loci of the maxima in the experimental densities, shown by the points in Figure 9.4, are also in good agreement with the corresponding Stratonovich theoretical curves.

The same electronic circuit has also been used for measurements of the correlation time (Sancho, Mannella, McClintock and Moss, 1985) of the system, a quantity that is complementary to the stationary density. The correlation function of the system within the parameter range of interest can be markedly nonexponential but, following Hernández-Machado, San Miguel and Sancho (1984), a relaxation time T can be defined as

$$T = \int_0^\infty \mathrm{d}s\, C_2(s)/C_2(0) \tag{9.4.6}$$

where

$$C_2(s) = \langle \delta x(t + s)\delta x(t) \rangle \tag{9.4.7}$$

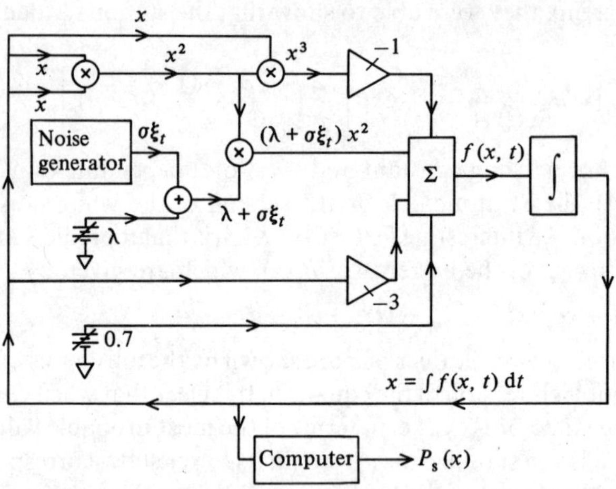

Figure 9.5. Block diagram of the electronic analogue circuit model of the cubic bistable system (9.4.1) (Robinson, Moss and McClintock, 1985).

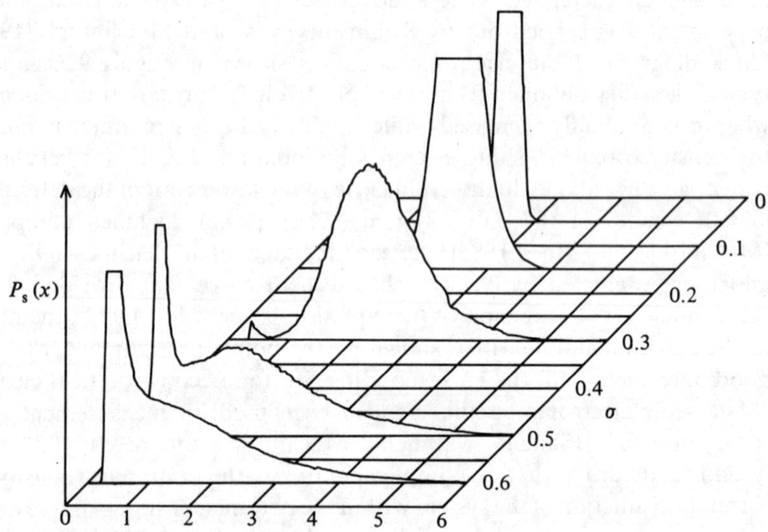

Figure 9.6. Examples of stationary density functions $P_s(x)$ measured with $\lambda = 3.6$ for the electronic simulator of the cubic system (9.4.1) shown in Figure 9.5. As the variance σ^2 of the noise is increased the density broadens, shifts, becomes bimodal and finally, for very large σ, converges towards a delta function near to the lower root of the deterministic ($\sigma = 0$) equation. (Robinson, Moss and McClintock, 1985).

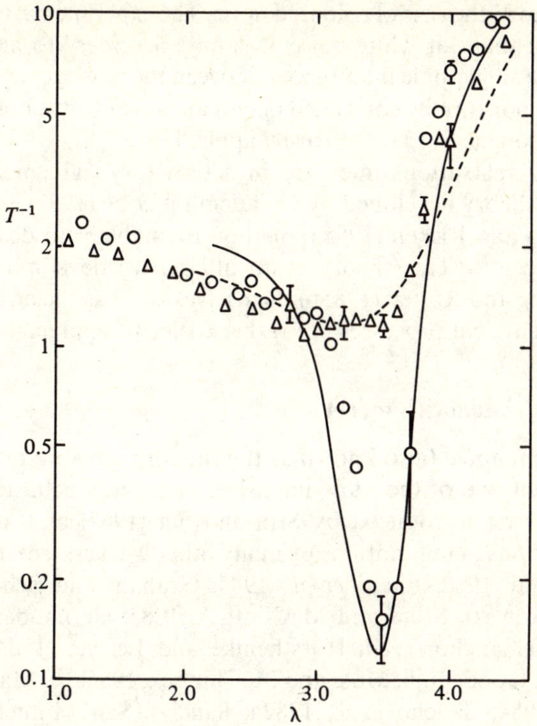

Figure 9.7. Inverse relaxation times T^{-1} measured as a function of λ for the electronic simulator of the cubic system shown in Figure 9.5, with $\sigma = 0.5$ (circles) and $\sigma = 1.0$ (triangles).

and

$$\delta x(t) = x(t) - \langle x(t) \rangle. \tag{9.4.8}$$

When the measured values of the inverse relaxation time T^{-1} are plotted either against λ for a fixed σ, or against σ for fixed λ within the bistable region, a well-defined minimum is found in each case: as an example Figure 9.7 shows $T^{-1}(\lambda)$ for two values of σ. The full and dashed curves (Sancho *et al.*, 1985) represent calculations based on the theory of Jung and Risken (1985a), and can be seen to be in good agreement with the experimental data. The minima may be construed as representing the characteristic critical slowing down that occurs in the vicinity of the noise-induced transition, which was discussed above purely in relation to the changes in the density $P_s(x)$.

The relaxation time of the cubic system is also being studied in the presence of coloured noise. (Casademunt *et al.*, 1987), such that (9.4.3) must be replaced by

$$\langle \xi(t)\xi(t') \rangle = (D/\tau)\exp\{-|t-t'|/\tau\} \tag{9.4.9}$$

to describe Ornstein–Uhlenbeck noise. The theory is, of course, a great deal more complicated in the case of coloured noise. The experiments, however, are no more difficult than for white noise: it is only necessary to alter the time constant τ_n of the single-pole filter placed between the noise generator and the circuit, so that exponentially correlated Gaussian noise of any chosen $\tau = \tau_N/\tau_I$ (where τ_N is no longer $\ll \tau_I$) can then be applied.

The measured relaxation times are in satisfactory agreement with the predictions of a theory developed by Casademunt *et al.* (1987) extending the white noise Jung and Risken (1985a) method to enable it to deal with non-Markovian processes. This theory is an alternative development to that proposed by Jung and Risken (1985b) themselves and has the advantage that the underlying physical processes are rather easier to appreciate.

9.5 The Stratonovich model

The Stratonovich model (also known as the random growing rate model, or RGRM) has been one of the most intensively studied stochastic nonlinear systems of all. First introduced by Stratonovich (1967) and subsequently studied in its various manifestations by many other workers (see, for example: Brenig and Banai, 1982; Faetti *et al.*, 1984; Graham and Schenzle, 1982; Hänggi, Mroczkowski, Moss and McClintock, 1985; Hernández-Machado, San Miguel and Sancho, 1984; Horsthemke and Lefever, 1984; Jung and Risken, 1985a, b; Kondepudi, Moss and McClintock, 1986a, b; Mannella *et al.*, 1986; Risken, 1984; Sancho *et al.*, 1982a; Sancho, San Miguel, Katz and Gunton, 1982b) the stochastic differential equation describing the model may be written

$$\dot{x} = dx - bx^3 + xV_N(t), \tag{9.5.1}$$

where $V_N(t)$ can represent either white (9.2.5) or colored (9.2.1) multiplicative noise, and d and b are constants. In what follows, we will assume $d = b = 1$, except where otherwise indicated. Alternatively, the noise term may appear additively, in which case (9.5.1) becomes

$$\dot{x} = dx - bx^3 + V_N(t). \tag{9.5.2}$$

In relation to real macroscopic physical systems there are two points that must be borne carefully in mind. First, there will always be a small additive constant on the right hand sides of (9.5.1) and (9.5.2) (Kondepudi, Moss and McClintock, 1986a; Moss, 1984); secondly, even in the case of (9.5.1) where the noise term is multiplicative, there will also be some weak additive noise to be taken into account (Brand, 1984). As will become clear below, these two small, easily overlooked deviations from the idealized forms of (9.5.1) and (9.5.2) can sometimes be exceedingly important in determining how the corresponding real physical systems behave in practice. The Stratonovich model is readily simulated by application of the techniques outlined in Section 9.2: many aspects of its behaviour have been investigated by such means. A variety of

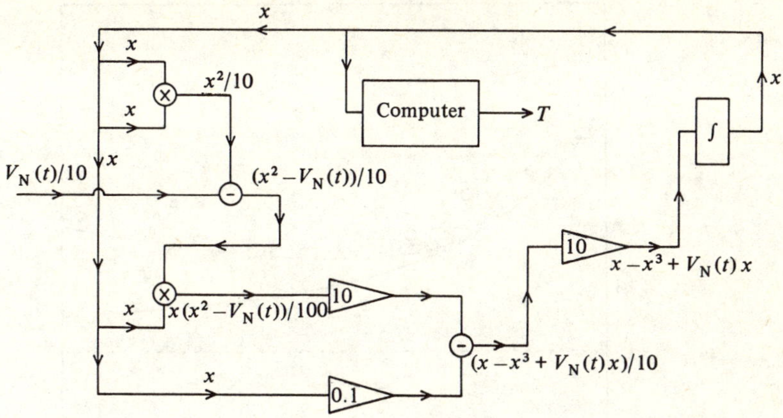

Figure 9.8. Block diagram of an electronic simulator for the Stratonovich model (9.5.1) with multiplicative external noise $V_N(t)$ (Mannella *et al.*, 1988).

circuit arrangements are possible, one example of which (Mannella, Faetti, Grigolini and McClintock, 1988) is shown in Figure 9.8. (This particular circuit was designed to operate at large noise intensities: the overall scale factor is unity, but intermediate scaling down is used in order to minimize the voltage excursions of the noise term.)

In one of the first experiments of this kind, Yamazaki and Kawakubo (reported in Sancho *et al.*, 1982b) measured the one-dimensional density $P(x)$ and were thereby able to confirm the existence of a noise-induced transition: they found that, above a critical noise intensity, the maximum in $P(x)$ at finite x disappears and is replaced by a singular maximum at $x=0$ (Schenzle and Brand, 1979).

Partly stimulated by the existence of the noise-induced transition, there has been a great deal of interest in the behavior of the relaxation time T of (9.5.1) as defined by (9.4.6)–(9.4.8). It is now known (Jung and Risken, 1985a) that, in fact, T^{-1} falls monotonically with increasing D from its deterministic ($D \to 0$) value of two towards an asymptotic ($D \to \infty$) limit of $2/\pi$. Measurements of $T^{-1}(D)$ for the circuit of Figure 9.8 (Mannella, McClintock and Moss, 1987; Mannella *et al.*, 1988) have shown, however, that T^{-1} at first falls but then passes through a shallow minimum and rises again at large D, as shown by the data points of Figure 9.9.

These results confirmed those of a preliminary experiment (Mannella *et al.*, 1986) and were not inconsistent with an earlier analogue simulation by Faetti *et al.* (1984). It can be shown (Mannella *et al.*, 1986, 1988) that the apparent discrepancy between theory and experiment can be resolved if explicit account is taken of the small additive constant and additive noise in (9.5.1) mentioned above which, in reality, should be written as

$$\dot{x} = x - x^3 + xV_N(t) + V'_N(t) + g, \tag{9.5.3}$$

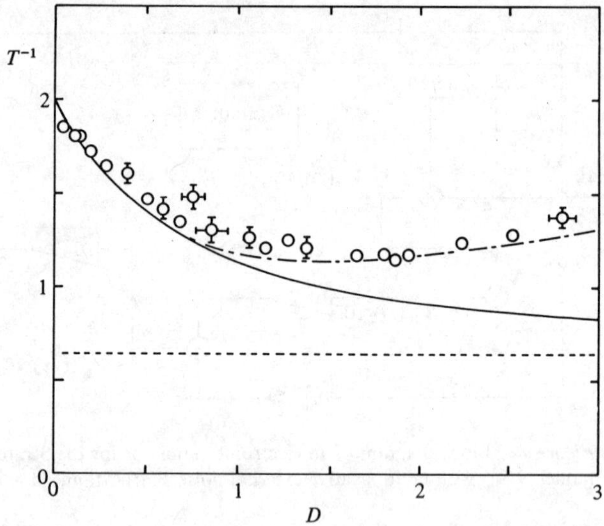

Figure 9.9. Reciprocal relaxation times T^{-1} measured for the Stratonovich model simulator shown in Figure 9.8, plotted as a function of noise intensity D (points), compared with the calculation (full curve) by Jung and Risken (1985a). The dashed curve indicates the limiting ($D \to \infty$) value of T^{-1} for (9.5.1) with $d = b = 1$; the dash–dot curve shows the calculated $T^{-1}(D)$ for (9.5.3) with $Q = 0$, $g = 5 \times 10^{-3}$ (Mannella *et al.*, 1988).

where g is the small additive constant

$$\langle V'_N(t) V'_N(s) \rangle = 2Q\delta(t - s) \tag{9.5.4}$$

and

$$\langle V_N(t) V'_N(s) \rangle = 0 \tag{9.5.5}$$

so that the two white noises given by (9.2.5) and (9.5.4) are entirely uncorrelated. The main consequence of finite $V'_N(t)$ is to cause the system to jump back and forth between the positive and negative wells: something that cannot occur at all in the case of (9.5.1) where the noise is purely multiplicative. To facilitate a meaningful comparison with theory, the data of Figure 9.9 refer only to realizations of $x(t)$ where such jumps did *not* take place. The theoretical behaviour of $T^{-1}(D)$ for (9.5.3) with a realistic value of g (dash–dot curve in Figure 9.9) is entirely different from that of (9.5.1) where $g = 0$ (full curve), but is remarkably close to the measured data. It may be concluded that, although the monotonic decrease of $T^{-1}(D)$ with D predicted for (9.5.1) is formally correct, it is not a phenomenon that occurs in real physical systems for which $g \neq 0$ and $Q \neq 0$, so that they must in reality be described by (9.5.3) yielding the broad minimum in $T^{-1}(D)$ exemplified by the data and dash–dot curve of Figure 9.9.

Two-dimensional densities for the Stratonovich model with additive (9.5.2) coloured (9.2.1) noise have been measured (Moss and McClintock, 1985) and found to exhibit a number of interesting features. The circuit used was similar

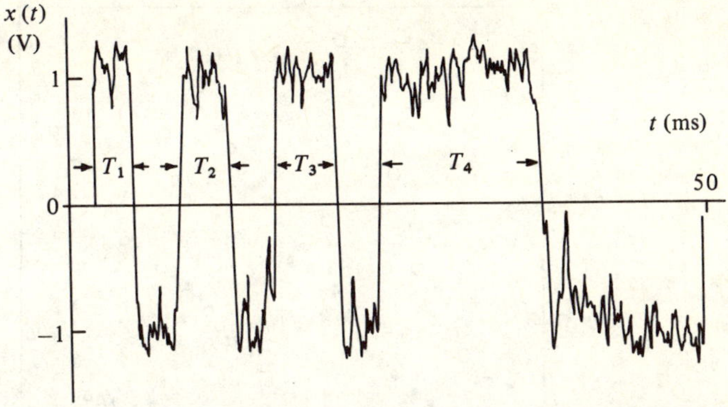

Figure 9.10. An example time function measured for an electronic simulator of the Stratonovich model (9.5.2) with additive coloured noise. The times $T_1, T_2, T_3 \ldots$ represent sojourn times in the potential well at $x = +1$ (Hänggi *et al.*, 1985).

to that of Figure 9.8, except that the noise was added just prior to the integrator, rather than to the term in x. Both the stochastic phase portraits $P(x, \dot{x})$ and noise portraits $P(V_N, x)$ were studied as a function of the dimensionless noise correlation time τ defined by (9.2.3). In both cases, a striking 'skewing' effect was observed in the shapes of the density peaks, behaviour that was in excellent accord with theoretical predictions by Jung and Risken (1985b). The Stratonovich model can, of course, be regarded as the overdamped limit of the double-well Duffing oscillator. It was not particularly surprising, therefore, that a very similar skewing effect was also found in the latter system (Moss, Hänggi, Mannella and McClintock, 1986), which will be discussed in more detail below, in the next section.

As already mentioned, an important effect of the additive noise in (9.5.2) is to cause transitions to take place between the two potential wells. There will, of course, also be the small additive constant g as in (9.5.3) but, provided that this can be made very small compared to the separation $(d/b)^{1/2}$ of the minima, its effect on the transition rate will be negligible. A typical time function $x(t)$, demonstrating the hopping between the wells, is shown in Figure 9.10: the periods T_1, T_2, $T_3 \ldots$ represent *sojourn times* (closely related to the *exiting times* of Pontryagin, Andronov and Vitt, 1933) in the well at $x = +1$. It is straightforward, for any given values of D and τ, to measure the distribution of sojourn times $P(T)$ and thence to determine the mean sojourn time $\langle T \rangle$. Hänggi *et al.* (1985; and see Chapter 9, Volume 1) developed an approximate Fokker–Planck-type equation showing that $\langle T \rangle$ would be expected to increase exponentially with τ at constant D; experimental measurements of $\langle T \rangle$ for the analogue circuit (Figure 9.11) were found to be in excellent agreement with this prediction.

The Stratonovich model also constitutes a simple and convenient model system for studies of the influence of noise on branch selectivity at a bifurcation

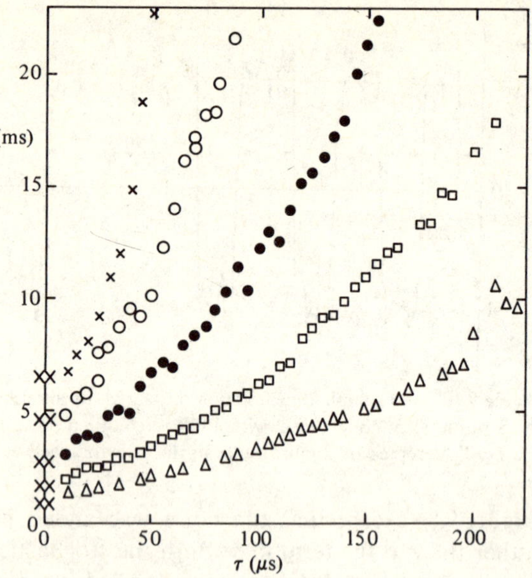

Figure 9.11. Values of the mean sojourn time $\langle T \rangle$, measured for $x(t)$ time functions similar to that of Figure 9.10, as functions of the (unscaled) noise correlation time τ, for several values of the dimensionless noise intensity D. Symbols: ×, 0.073; ○, 0.083; ●, 0.114; □, 0.153; △, 0.212. The theoretical white noise ($\tau \to 0$) limits, with no adjustable parameters, are indicated by XX (Hänggi *et al.*, 1985).

point. Equation (9.5.2) may be re-written in the form

$$\dot{x} = -x^3 + \lambda x + g + V_N(t), \tag{9.5.6}$$

where the small symmetry breaking constant g has been included explicitly, $V_N(t)$ can be either white or coloured noise, $b = 0$, and the parameter λ can be swept in time from a negative to a positive value. When $V_N = 0$, $g = 0$ and λ is negative, there is only one stable solution at $x = 0$; as λ passes through zero and becomes positive, this solution becomes unstable and the system bifurcates into stable solutions $x = \pm\sqrt{\lambda}$ with equal probability. A finite value of g will have the effect of selecting one of these two stable states as λ passes through the origin. Even in the presence of large amplitude noise such that $g \ll [\langle V_N^2 \rangle]^{1/2}$, it has been predicted that g will still be an efficient state selector, provided that the rate at which λ is swept through the origin is slow enough (Kondepudi and Nelson, 1984, 1985). This remarkable assertion has been tested, and verified, through experimental measurements of the time evolving density of x, $P(x, t)$ for the analogue circuit (Kondepudi, Moss and McClintock, 1986a, b; Moss, Kondepudi and McClintock, 1985). Details of these investigations are given in Chapter 10 of Volume 2.

9.6 The double-well Duffing oscillator

Brownian motion in a double-well potential, first treated by Kramers (1940), has since been studied by numerous other workers on account of its inherent interest and wide applicability. Recent theoretical developments have been reported by Jung and Risken (1985b) and Voigtlaender and Risken (1985), and the topic has been reviewed by Hänggi (1986). The Duffing oscillator subject to external noise, which can be used as an archetypal model of double-well systems, may be written

$$\dot{x} = v \tag{9.6.1}$$

$$\dot{v} = x - x^3 - \gamma v + V_N(t), \tag{9.6.2}$$

where γ is a damping constant and the additive external noise $V_N(t)$ can be either colored (9.2.1) or white (9.2.5). In the limit of large damping, (9.6.1) and (9.6.2) reduce to the Stratonovich model (9.5.2) of the preceding section.

For coloured noise in the underdamped regime, (9.6.1) and (9.6.2) with (9.2.1) constitute a three-dimensional (3-d) system whose properties are in practice rather difficult to calculate. Matrix continued fraction (MCF) expansion (Risken, 1984) may be used to solve the relevant Fokker–Planck equation, but yields a matrix of infinite dimension that must be truncated and inverted numerically, a necessarily approximate procedure. In fact, the matrix in question can only be evaluated with sufficient accuracy on ordinary computers for $d = 2$ systems: for (9.6.1) and (9.6.2) this would imply *either* white noise *or* taking the overdamped limit. Even for $d = 2$ systems, huge amounts of central processor time ($\sim$ one hour) are needed on standard mainframe computers. In these circumstances, the analogue experimental approach offers clear advantages in terms of speed and economy.

By addition of a second integrator to a circuit similar to that of Figure 9.8 (but with additive rather than multiplicative noise), it is straightforward to incorporate the inertial term necessary to model (9.6.1) and (9.6.2). Two such circuits are described by, for example, Fronzoni *et al.* (1986) who studied the influence of the noise colour τ and intensity D, and the damping constant γ, on the single-dimensional density $P(x)$: see also Chapter 8, this volume. The results were compared with predictions based on an approximate, nonlinear Fokker–Planck-type equation. Good agreement between experiment and theory was obtained over a wide range of D for small τ; excellent agreement was obtained, even for large values of τ, for relatively small D.

The analogue experiments were extended by Moss *et al.* (1986) to measure the stochastic phase portraits of the system, of which two examples are shown in Figure 9.12. The upper pair of drawings, marked (a), provide, respectively, a perspective plot (left) and a contour plot (right) of the same joint statistical density $P(\dot{x}, x)$ for noise with a relatively short dimensionless correlation time $\tau = 0.25$. The lower pair of drawings refer to identical physical conditions, except for a substantial increase in the noise correlation time, to $\tau = 4$. Several

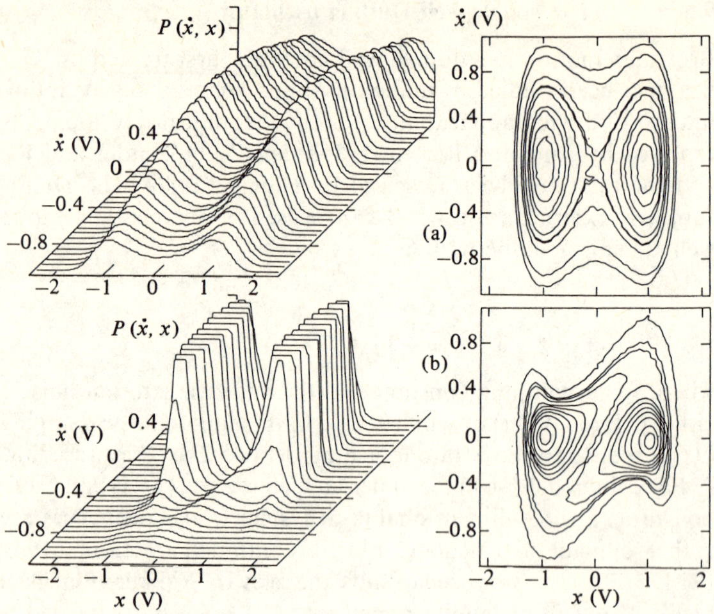

Figure 9.12. Stochastic phase portraits measured an electronic simulator of the double-well Duffing oscillator (9.6.1), (9.6.2) with $\gamma = 1$, $D = 0.3$ and: (a) $\tau = 0.25$; (b) $\tau = 4.0$. The vertical scale is arbitrary. The amplitudes of the contours of constant probability (right hand plots) vary: (a) from 256 to 3.5×10^3 with two contours near the separatrix at $\simeq 1.2 \times 10^3$; (b) 512 to 32×10^3 with separatrix at $\simeq 4 \times 10^3$ (Moss *et al.*, 1986).

features of these results are of interest, and particularly so given that some of them had not been predicted in advance of the measurements, and even now still lie beyond the practical capacity of current theoretical approaches. The perspective plots clearly show a narrowing of the peaks and an increasing depth of the valley separating them as τ is increased, an effect which is predicted in current approximate theory (Hänggi *et al.*, 1985). The contour plots exhibit an increasing asymmetry which tends to skew the valley towards an axis running from lower left to upper right, an effect which only very recently has been analytically predicted (see Chapter 8 of Volume 1). It is also interesting to note the emergence of a third maximum in the $P(x)$ cross-sections near $x = -1.2, \dot{x} = -0.5$ (and, of course, near $x = 1.2, \dot{x} = 0.5$). This latter feature was not observed either in the MCF calculations (Jung and Risken, 1985b) or in the analogue simulations (Moss and McClintock, 1985) of the 2-d equivalent system (i.e. the Stratonovich model discussed above), although the skewing effect was clearly evident. Digital simulations by Mannella (see Chapter 7, this volume) have confirmed the existence of the third maximum, but yielded $P(\dot{x}, x)$ portraits of relatively poor statistical quality even after the expenditure of hours of central processor time on a VAX11/780.

It is apparent that, at present, analogue experiments offer the only effective means by which the stochastic phase portraits of higher-dimensional systems can be established.

The same analogue circuit has also been used for measurements (Dykman *et al.*, 1988) of the power spectral density of the system when driven by a random force, in this case with V_N Gaussian and pseudo-white (9.2.5). For the work in question, the damping constant γ was made as small as possible consistent with the stability of the system but, otherwise, the circuit remained as described by Fronzoni *et al.* (1986). Analytic calculations of the power spectra to be expected for the system have been presented by Onodera (1970) and, more recently and in greater detail, by Dykman, Soskin and Krivoglaz (1985).

A typical set of power spectra $Q(\omega)$ measured for the analogue circuit are shown in Figure 9.13(a), (b) for increasing values of noise intensity, defined in terms of a parameter $\beta = D/2$. The two peaks in $Q(\omega)$ at finite frequency can be readily understood in physical terms: the high frequency peak, present even for very small values of β, corresponds to motion within either one of the potential wells; the lower frequency peak, which only appears for relatively large β, corresponds to large amplitude oscillations that take in both wells. For very large β, the energy in the system becomes so great that the intra-well oscillation peak disappears entirely.

The feature that is not, perhaps, so intuitively obvious is the third peak in $Q(\omega)$, at $\omega = 0$. This was missed by Onodera (1970) but predicted by Dykman, Soskin and Krivoglaz (1985) on the basis of an analytic theory (for the nondissipative system with $\gamma = 0$) and by Voigtlaender and Risken (1985) on the basis of numerical calculations that allowed for finite values of γ. The peak in question arises from the slowing down of the system in the vicinity of the central potential maximum. It can be regarded (G. Nicolis, 1986, private communication) as the residual effect of the homoclinic phase space orbit that can occur in the dissipation-free ($\gamma = 0$) system in the absence of stochasticity ($V_N(t) = 0$), which would give rise to a delta function at $\omega = 0$. In the present case, where $\gamma \neq 0$ and $V_N(t) \neq 0$, the combined effects of dissipation and stochasticity broaden the delta function to take the form shown.

The analytic theory of Dykman *et al.* (Dykman, Soskin and Krivoglaz, 1985; Dykman *et al.*, 1988) provides a satisfactory quantitative fit to the experimental measurements of $Q(\omega)$ over a wide range of ω. The dotted and dashed lines in Figure 9.13 represent calculated values referring, respectively, to the dissipative and nondissipative versions of the theory. The fact that all of the experimental spectra are shifted slightly towards larger ω is a consequence of the imperfect modelling of (9.6.1) and (9.6.2) by the circuit. The measured loci of the maxima at finite ω, and of the minimum between them, and of the minimum between the inter-well maximum and the $\omega = 0$ peak, are shown by the data points in Figure 9.14: the circled data and crosses indicate minima and maxima, respectively. The full and dashed curves represent, respectively, the dissipative and nondissipative forms of the theory. Again the level of

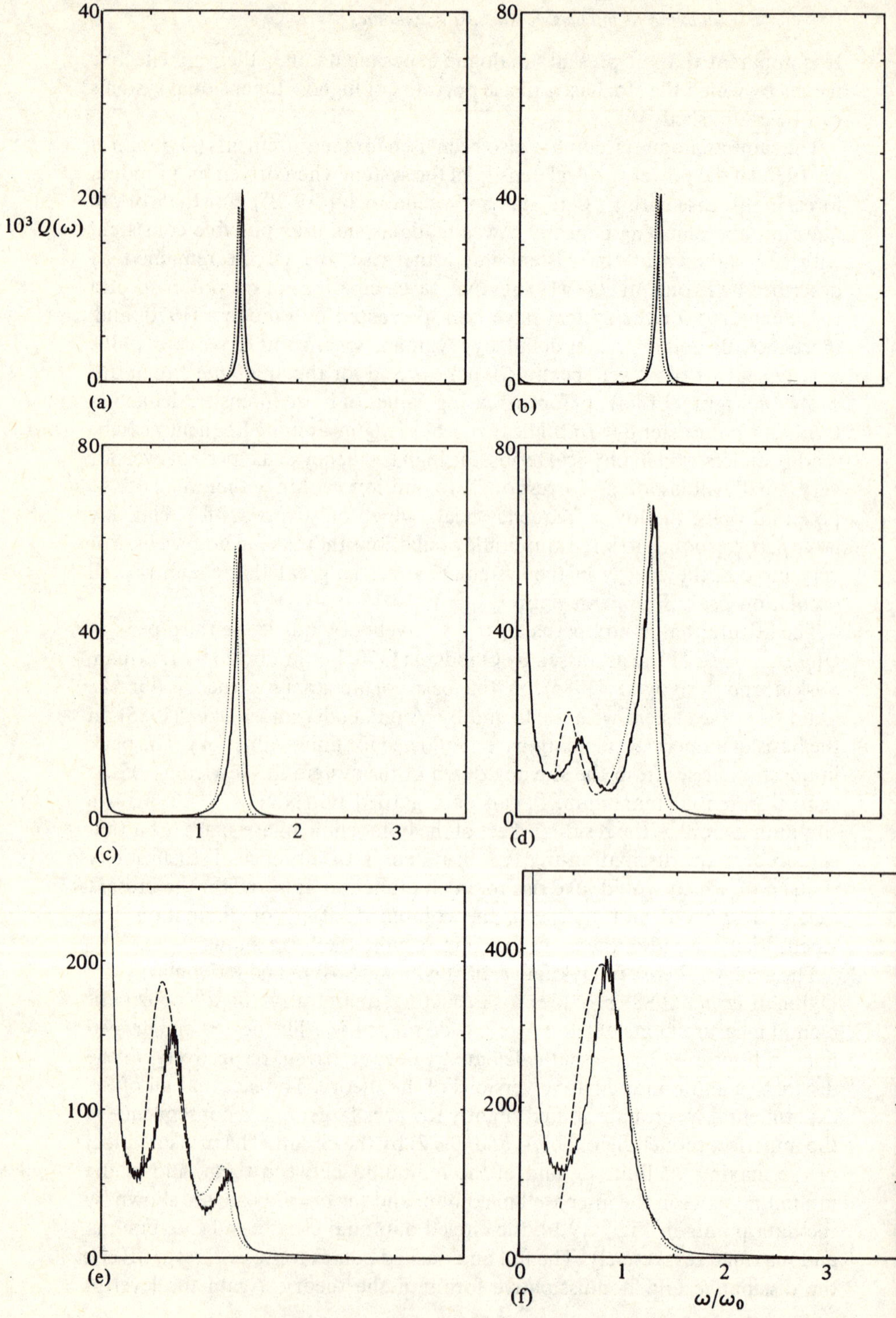

Figure 9.13. Power spectral densities $Q(\omega)$ measured on an electronic simulator of the double-well Duffing oscillator (9.6.1), (9.6.2) subject to quasi-white noise of intensity: (a) $\beta = 0.0055$; (b) 0.0142; (c) 0.0291; (d) 0.06; (e) 0.149; (f) 0.369.

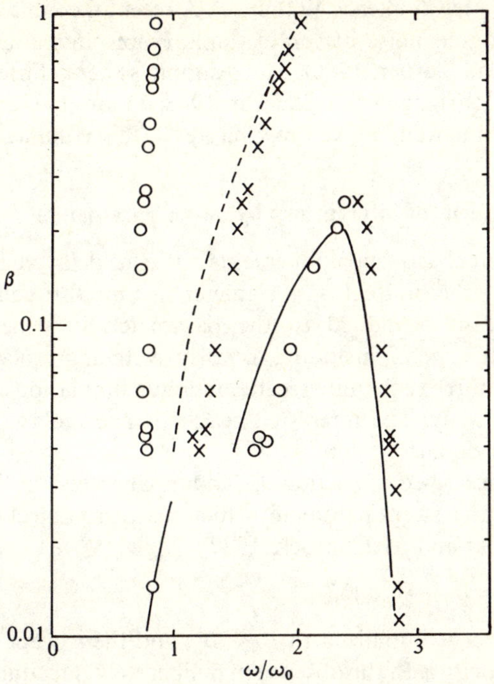

Figure 9.14. Loci of the extrema x_m of the power spectra measured for an electronic simulator of the double-well Duffing oscillator (9.6.1), (9.6.2) as a function of the intensity D of the quasi-white external noise: the circled data represent minima of $Q(\omega)$ and the crosses represent maxima. The curves represent calculated values: the full curve is for a calculation taking account of dissipation; the dashed curve represents a calculation in the nondissipative approximation (Dykman *et al.*, 1988).

agreement between the theory and the analogue measurements is strikingly good.

The fact that the existence of the three maxima in $Q(\omega)$ predicted by Dykman, Soskin and Krivoglaz (1985) has been confirmed both by numerical calculations (Voigtlaender and Risken, 1985) and also by experiments on a real physical system (the analogue circuit) means that analogous behaviour may confidently be anticipated in a very wide range of noisy bistable systems occurring in nature.

9.7 The postponement of Hopf bifurcations by multiplicative noise

Stochastic postponements resulting from parametric (multiplicative) noise at pitchfork bifurcations and at first order instabilities have been discussed in Section 9.4. Such an effect seems to be ubiquitous in nonlinear dynamics, occuring at bifurcations even in discrete maps (Fronzoni, Mannella, McClintock and Moss, 1987; and see Chapter 6, Volume 2). A recent theoretical treatment of a parametric coloured noise driven example Hopf bifurcation has been presented (Lefever and Turner, 1986). The example system chosen for study was the 'Brusselator' (Prigogine and Lefever, 1968; Tyson, 1973). A complete discussion of this simulation is given in Chapter 8, this volume.

9.8 Postponement of bifurcations by swept parameters

Closely related to stochastic postponements are the delayed bifurcations induced by sweeping the bifurcation parameter in time through its critical value. Noise may also be added to the parameter, in which case the applications are quite general, including to noisy switching processes (Moss, 1984). In this section three such quite recent analogue simulation experiments are described very briefly. The interested reader is referred to the original literature for greater detail.

The Brusselator, described in Section 9.7 and in Chapter 8 of this volume, has been operated with a swept parameter which was also subject to coloured noise (Fronzoni, Moss and McClintock, 1987):

$$B \to B(t) = B_0 + vt + V_N(t). \tag{9.8.1}$$

The control parameter is initialized to $B_0 \ll B_c$, and then swept linearly for times $t > 0$ at the velocity v. In the absence of deliberately introduced external noise, i.e. for $V_N = 0$, large postponements $\Delta B = B^* - B_c$ are observed. Here 'large' means that typical values of ΔB are considerably larger than the deterministic change in B during one oscillation period of the Brusselator, i.e.

$$\Delta B_{\text{typical}} > v\omega_B^{-1}. \tag{9.8.2}$$

It turns out that even very small noise destroys this postponement in the sense that, if B^* is determined by a convenient definition based on the growth of limit cycle amplitude, the location becomes impossible to determine in the presence of noise. In this case 'small' means that typically $(\langle V_N^2 \rangle)^{1/2} \ll \Delta B^*$. These results are qualitatively similar to some preliminary theory on the postponement of Hopf bifurcations (Erneux and Mandel, 1986; Erneux and Reiss, 1984; Wallet, 1986).

Naturally, a very fruitful area in which to apply swept parameter theory is laser switching (see also Chapter 12, Volume 2), especially so since transients in noisy lasers have recently been studied experimentally (Roy, Yu and Zhu, Chapter 4, this volume; Zhu, Yu and Roy, 1986). The system studied here and in the theory of Broggi, Colombo, Lugiato and Mandel (BCLM) (1986) is an extremely simple model for a ring laser with additive noise:

$$\dot{x} = x[-1 + A/(1 + x^2)] + V_N(t), \tag{9.8.3}$$

which is assumed to operate on a perfectly tuned single mode with homogeneous broadening. Here $\langle x^2 \rangle$ represents the mean intensity of the laser field. The control parameter A was swept in time,

$$A \to A(t) = A_0 + vt, \tag{9.8.4}$$

from an initial value A_0. Even though simple, such models are useful for studies of the macroscopic behaviour. In the BCLM calculations, white noise was assumed and the statistical distribution of (postponed) bifurcation times $W(t)$ was obtained from numerical integrations of the Fokker–Planck equation analogous to (9.8.3). Deterministically, the laser transition was postponed to a value $A^* > A_c$ by the swept parameter, and it was observed that added noise decreased the postponement, i.e. the maxima of the $W(t)$ were shifted toward smaller values of A^* (but still larger than the deterministic bifurcation point A_c). Analogue simulations of this system, wherein the $W(t)$ were measured directly, agreed quite well with the white noise theory (Mannella, McClintock and Moss, 1987) and have recently been extended to include colored noise (Mannella, Moss and McClintock, 1987), a case for which the Fokker–Planck equation is impossible to deal with in an exact analytical theory and exceedingly difficult numerically. A linearized bifurcation theory, modified to account for colored noise, has recently been reported (Van den Broeck and Mandel, 1987).

Finally, swept parameter postponements have been observed and measured with an electronic circuit model of the iterated map

$$x_{n+1} = 1 - \lambda x_n^2 \tag{9.8.5}$$

where $\lambda \to \lambda(t) = \lambda_0 + vt$. The cascade of subharmonic (period doubling) bifurcations leading to chaos is by now very well known and highly celebrated (see, for example, Collet and Eckman, 1980). A recent theory (Kapral and Mandel, 1985) predicted both the magnitude of the postponements and a scaling law for the cascade, but for a somewhat different map. Measurements on the map (9.8.5) were in excellent agreement within the theory and hence indicated the universality of the predicted scaling behaviour (Morris and Moss, 1986).

Of course, maps which are noisy, such as (9.8.5) with either additive or multiplicative noise, are of high current interest (see Chapters 4–6, Volume 2).

9.9 Systems with periodic potentials

The Langevin equations of example systems with periodic forcing and additive noise are

$$\frac{1}{\gamma}\ddot{\phi} + \dot{\phi} = a + b \sin \phi + V_N(t), \tag{9.9.1}$$

which in the limit of a very large damping constant γ becomes

$$\dot{\phi} = a + b \sin \phi + V_N(t). \tag{9.92}$$

The system is exposed to a constant force a plus a period term. Such systems are qualitatively different than those discussed previously in this chapter since, even in the absence of noise, if a $\geqslant b$ 'running' or unbounded solutions exist. It is useful to view the potential, $V(\phi) = -a\phi + b\cos\phi$, as a 'tipped washboard'. The running solutions then lead to trajectories which can wander over an unlimited number of cycles of the potential.

Three actual devices which exhibit unbounded solutions are the Josephson tunnel junction (Magerlein, 1978) and the electronic phase locked loop (Bak and Pedersen, 1973; Cirillo and Pedersen, 1982; D'Humieres, Beasley, Huberman and Libchaber, 1982; Henry and Prober, 1981) which are modelled by (9.9.1) and the ring laser gyroscope (Chow, *et al.*, 1985) which obeys (9.9.2). These applications, chiefly the first, have stimulated interest in analogue simulations of the dynamics dating back to the 1960s (Werthamer and Shapiro, 1967); however, it is a fact that (9.9.1) becomes chaotic if a time periodic forcing, e.g. $c\sin\omega t$, is added in addition to or instead of the noise, which chiefly accounted for the burst of interest in analogue simulations dating from the early 1980s and continuing even now (Alstrom, Jenson and Levinsen, 1984; Alstrom and Levinsen, 1985; Ben-Jacob, Braiman, Shainsky and Imry, 1981; Cirillo and Pedersen, 1982; D'Humieres *et al.*, 1982; Kao, Huang and Gou, 1986; Seifert, 1983; Yeh and Kao, 1982).

In order to simulate systems with periodic potentials in the running state, one must arrange to return the trajectory repeatedly to a necessarily finite number of samples of the potential. It is not unreasonable to sample a very large number of connected cycles when making digital simulations, but analogue simulators, by contrast, are usually able to 'look at' only one to four actual cycles. An escaping trajectory must therefore be returned to or confined within this sample. There are (at least) two methods for accomplishing this: (1) the sign of the constant force a can be reversed each time the trajectory crosses a boundary (Magerlein, 1978); or (2) at either boundary $\pm\phi_b$, a constant $2\phi_b$ can be subtracted from $x(t)$ (Kao, Huang and Gou, 1986). Method (1) is not suitable for stochastic problems, since the trajectory can escape the interval $[-\phi_b, \phi_b]$ with a probability which depends on the noise intensity. Method (2) is a means of imposing periodic boundary conditions assuming that v is periodic in $2\phi_b$, i.e. $v(\phi) = v(\phi \pm 2\phi_b)$. Both methods assume unity normalization, i.e. that $\int_{-\phi_b}^{\phi_b} d\phi\, P(\phi) = 1$, where $P(\phi)$ is the stationary probability density.

An analogue simulator of (9.9.2) is shown in Figure 9.15. It functions according to the same principles as described previously, except for two modifications.

The first modification is the subcircuit marked PBC, which imposes periodic boundary conditions at the extremities of two cycles of the potential. The circuit is not shown in detail, but its operation is straightforward. A pair of voltage comparators is operated at the output of the integrator, and they are triggered in turn, respectively, as $\phi \to \pm\phi_b = \pm 2.5\,\mathrm{V}$. The scale is chosen as a

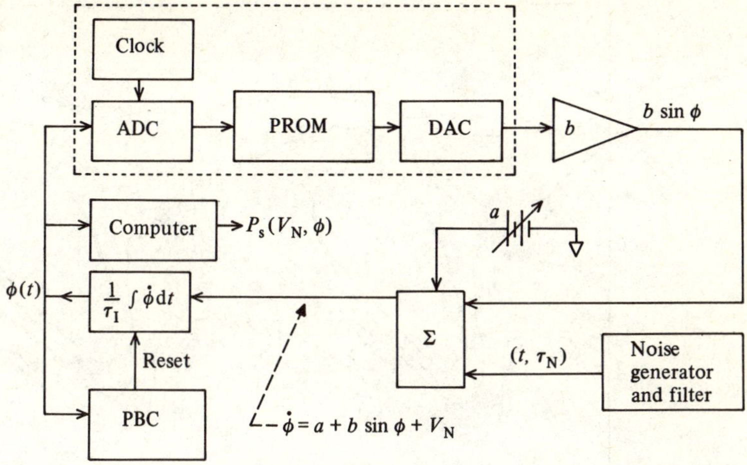

Figure 9.15. Electronic circuit model for simulation of a Langevin equation with a periodic potential. The circuit is a closed loop system which dynamically approximates (9.9.2). Its operation is discussed in the text. (From Vogel *et al.*, 1987a.)

convenience since it corresponds to the full scale range of the analogue-to-digital converter (ADC) inputs of the Nicolet data-processor. The boundaries are at $\pm\phi_b = \pm 2\pi$ so that two full cycles of the potential are observed. When a boundary is crossed, and consequently the appropriate comparator is triggered, an analogue switch is closed which resets the integrator. The integrator is simply a good quality operational amplifier with an input resistor R and feedback capacitor C (so that the integrator time constant $\tau_I = RC$). Since the output voltage of the integrator is $\phi(t) = Q(t)/C$, all that is required to reset it when $Q(t) = \pm\phi_b$ is to reverse the sign of the charge Q. The analogue switch accomplishes this by momentarily connecting C to a voltage source which charges it with $2Q$ of the opposite sign. It is important that the resetting process is achieved in a time that is small compared to any other time-scale of importance in the simulation. Usually this means that it must be small compared to the τ_N which corresponds to the quasi-white noise limit.

The second modification is shown by the subcircuit enclosed by the broken line on Figure 9.15. It is necessary in order to simulate $\sin\phi$ where ϕ is an input voltage. Of course, this operation can be accomplished very simply by using the Analog Devices AD639 Universal Trigonometric Function chip. The subcircuit shown in Figure 9.15 can accomplish this by essentially storing the $\sin\phi$ function in a programmable, read only memory (EPROM) which then functions as a 'look-up table'. The input voltage $\phi(t)$ is converted in the ADC to digital information which is compared with the contents of the look-up table. The matching entry at each interval Δt, the digitizing time, is then passed to the digital-to-analogue converter (DAC) which then outputs $\sin\phi(t)$. Of course, it is necessary that $\Delta t \ll \tau_N$, and that the storage and bit capacity of the

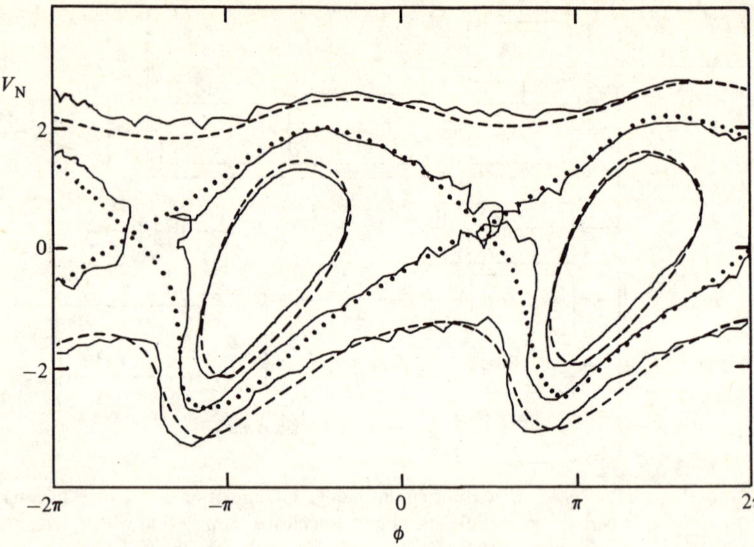

Figure 9.16. Contour plots of the noise portrait $P_s(V_N, \phi)$ measured for the periodic potential simulator of (9.9.2) shown in Figure 9.15 (full curves) compared with theoretical predictions (dashed and dotted lines) at close parameter values. For both experiment and theory $a = b = 1.0\,\mathrm{V\,s^{-1}}$ and the noise correlation time $\tau = 1.0$; the theory is for noise intensity $D = 1$ and the experiment for $D = 0.92 \pm 0.06\,\mathrm{V}^2$. (From Vogel *et al.*, 1987a.)

EPROM be large enough to resolve sin ϕ. In one system already realized an 8 bit, 3 MHz ADC (the AD578LD) outputs to a pair of 4K by 8 bit EPROMs (the MMC27C32-35) followed by a DAC (the AD565AKD). The resulting system, with the ADC operated by a 1 MHz clock, has an analogue input to analogue output voltage frequency response of 10 kHz $\pm$ 0.1 dB.

Of course, the advantage of the more complicated second method is its versatility, *Any* function – not just trigonometric functions – can be programmed into the EPROM, and can be made periodic by the PBC. For example, a periodic, parabolic potential in the noise-driven system (9.9.2) has been studied by comparing the power spectra with the stationary statistical densities (M. James, 1986, unpublished MS Thesis, University of Missouri at St. Louis).

Circuits of the type described have been used to study (9.9.2) with coloured noise, for comparisons with and tests on the accuracy of solutions of the associated two-dimensional Fokker–Planck equations obtained by means of infinite matrix continued fraction expansions (Risken, 1984). The solutions so obtained are the two-dimensional stationary densities $P(V_N, \phi)$, i.e. the noise portraits, and these were also measured using the circuit (Vogel *et al.*, 1987a). An example noise portrait for coloured noise with $\tau = 1$ is shown in Figure 9.16 as a set of three contours of constant probability with the measurements, shown by the solid (noisy) lines compared to the theory shown by the broken and dotted lines. The lowest (fully open) contour was separated from the two highest (closed) contours by one order of magnitude in statistical weight. The only adjustable constant used to fit the data to the theory was the normalization of the joint density $P(V_N, \phi)$, and the fit was made at the separatrix shown by the dotted line. In addition to the now familiar skewing effect, one can easily discern those contours (the open ones) which statistically favour running trajectories destined to intersect the boundaries and those (closed) ones which favour trajectories bound to the pair of potential wells located near $\pm\pi$. In a ring laser gyroscope application (Vogel *et al.*, 1987b) the mean velocity $\langle\dot{\phi}\rangle$ averaged over both the noise and in time was calculated, measured and simulated digitally.

9.10 Conclusions and future directions

The projects described above represent a rich variety of timely and challenging topics in nonlinear dynamics. The two most dominant characteristics of analogue simulations as quasi-experimental approaches to such problems will probably determine their future applications.

First, being real macroscopic systems, they often immediately reveal sometimes hidden defects in idealized theories. Because of this alone, electronic circuit modelling of proposed theoretical systems will continue to be a useful supplement to digital simulations and actual experimental realizations.

The second characteristic is speed of data acquisition, which is revealed immediately in studies on higher-dimensional systems. Three-dimensional, noisy systems, for example those with two dynamical dimensions and one resulting from a nonzero noise correlation time, are already expensive in terms of CPU time to simulate even with modest accuracy on ordinary digital computers; and accurate matrix continued fraction solutions for such systems cannot be obtained with anything less than a super-computer. Related to speed is another advantage of analogue simulation, especially evident in studies on complex systems: very complicated dynamics can be rapidly 'scanned' over large regions of parameter space. This aspect is especially useful when used in conjunction with digital simulations and actual experiments to show the investigator 'where to look'.

Finally, one should mention the versatility of the hybrid analogue–digital–analogue forcing system discussed in Section 9.9, as a factor in indicating future directions for research. Since any potential can be programmed into the EPROM, one can immediately imagine some exotic higher-dimensional systems, for example the three-dimensional system

$$\dot{x} = F(x) + V_N(t), \qquad (9.10.1)$$

where $F(x)$ is a 'spatially colored' noise, and $V_N(t)$ is the usual time correlated noise. Such a system, which can also be easily simulated digitally, might be useful for studying classical models of thermally excited conduction in one-dimensional disordered conductors.

References

Alstrom, P., Jenson, M. H. and Levinsen, M. T. 1984. *Phys. Lett.* **103A**, 171–4.

Alstrom, P. and Levinsen, M. T. 1985. *Phys. Rev. B* **31**, 2753–8.

Bak, C. K. and Pedersen, N. F. 1973. *Appl. Phys. Lett.* **22**, 149–50.

Ben-Jacob, E., Braiman, Y., Shainsky, R. and Imry, Y. 1981. *Appl. Phys. Lett.* **38**, 822–4.

Blankenship, G. and Papanicolaou, G. C. 1978. *SIAM J. Appl. Math.* **34**, 437–76.

Bowden, C. M., Ciftan, M. and Roble, H. R., (eds). 1981. *Optical Bistability*, New York: Plenum.

Brand, H. R. 1984. *Prog. Theor. Phys.* **72**, 1255–7.

Brand, H. R. and Schenzle, A. 1980. *J. Phys. Soc. J.* **48**, 1382–3.

Brenig, L. and Banai, N. 1982. *Physica D* **5**, 208–26.

Broggi, G., Colombo, A., Lugiato, L. A. and Mandel, P. 1986. *Phys. Rev. A* **33**, 3635–7.

Casademunt, J., Mannella, R., McClintock, P. V. E., Moss, F. E. and Sancho, J. M. 1987. *Phys. Rev. A* **35**, 5183–90.

Chow, W. W., Gea-Banacloche, J., Pedrotti, L. M., Sanders, V. E., Schleich, W. and Scully, M. O. 1985. *Rev. Mod. Phys.* **57**, 61–104.

Cirillo, M. and Pedersen, N. F. 1982. *Phys. Lett.* **90A**, 150–2.

Collet, P. and Eckman, J-P. 1980. *Iterated Maps on the Interval as Dynamical Systems.* Basel: Birkhäuser.

D'Humieres, D., Beasley, M. R., Huberman, B. A. and Libchaber, A. 1982. *Phys. Rev. A* **26**, 3483–96.

Dykman, M. I., Mannella, R., McClintock, P. V. E., Moss, F. and Soskin, S. M. 1988. *Phys. Rev. A.* **37**, 1303–13.

Dykman, M. I., Soskin, S. M. and Krivoglaz, M. A. 1985. *Physica* **133A**, 53–73.

Erneux, T. and Mandel, P. 1986. *SIAM J. Appl. Math.* **46**, 1–15.

Erneux, T. and Reiss, E. L. 1984. *SIAM J. Appl. Math.* **44**, 463–78.

Faetti, S., Festa, C., Fronzoni, L., Grigolini, P. and Martano, P. 1984. *Phys. Rev. A* **30**, 3252–63.

Fronzoni, L., Grigolini, P., Hänggi, P., Moss, F., Mannella, R. and McClintock, P. V. E. 1986. *Phys. Rev. A* **33**, 3320–7.

Fronzoni, L., Mannella, R., McClintock, P. V. E. and Moss, F. 1987. *Phys. Rev. A* **36**, 834–41.

Fronzoni, L., Moss, F. and McClintock, P. V. E. 1987. *Phys. Rev. A* **36**, 1492–4.

Graham, R. and Schenzle, A. 1982. *Phys. Rev. A* **26**, 1676–85.

Hänggi, P. 1986. *J. Stat. Phys.* **42**, 105–48.,

Hänggi, P., Mroczkowski, T. J., Moss, F. and McClintock, P. V. E. 1985. *Phys. Rev. A* **32**, 695–8.

Henry, R. W. and Prober, D. E. 1981. *Rev. Sci. Instr.* **52**, 902–14.

Hernández-Machado, A., San Miguel, M. and Sancho, J. M. 1984. *Phys. Rev. A* **29**, 3388–96.

Horowitz, P. and Hill, W. 1980. *The Art of Electronics*. Cambridge University Press.

Horsthemke, W. and Lefever, R. 1984. *Noise Induced Transitions*. Berlin: Springer.

Johnson, C. L. 1963. *Analog Computer Techniques*, 2nd edn. New York: McGraw-Hill.

Jung, P. and Risken, H. 1985a. *Z. Phys. B* **59**, 469–81.

Jung, P. and Risken, H. 1985b. *Z. Phys. B* **61**, 367–79.

Kabashima, S. and Kawakubo, T. 1979. *Phys. Lett.* **70A**, 375–6.

Kai, S., Kai, T., Takata, M. and Hirakawa, K. 1979. *J. Phys. Soc. J.* **47**, 1379–80.

Kao, Y. H., Huang, J. C. and Gou, Y. S. 1986. *J. Low Temp. Phys.* **63**, 287–305.

Kapral, R. and Mandel, P. 1985. *Phys. Rev. A* **32**, 1076–81.

Kawakubo, T., Yanagita, A. and Kabashima, S. 1981. *J. Phys. Soc. J.* **50**, 1451–6.

Kondepudi, D. K., Moss, F. and McClintock, P. V. E. 1986a. *Physica* **21D**, 296–306.

Kondepudi, D. K., Moss, F. and McClintock, P. V. E. 1986b. *Phys. Lett.* **114A**, 68–74.

Kondepudi, D. K. and Nelson, G. W. 1984. *Phys. Lett.* **106A**, 203–6.

Kondepudi, D. K. and Nelson, G. W. 1985. *Nature* **314**, 438–41.

Kramers, H. A. 1940. *Physica* **7**, 284–304.

Landauer, R. 1962. *J. Appl. Phys.* **33**, 2209–16.

Lefever, R. and Horsthemke, W. 1979. *Bull. Math. Biol.* **41**, 469–90.

Lefever, R. and Horsthemke, W. 1981. In *Nonlinear Phenomena in Chemical Dynamics* (C. Vidal and A. Pacault, eds.), p. 120. Berlin: Springer.

Lefever, R. and Turner, J. W. 1986. *Phys. Rev. Lett.* **56**, 1631–4.

McClintock, P. V. E. and Moss, F. 1985. *Phys. Lett.* **107A**, 367–70.

Magerlein, J. H. 1978. *Rev. Sci. Instr.* **49**, 486–90.

Mannella, R., Faetti, S., Grigolini, P. and McClintock, P. V. E. 1988. *J. Phys. A: Math. Gen.* **21**, 1239–52.

Mannella, R., Faetti, S., Grigolini, P., McClintock, P. V. E. and Moss, F. 1986. *J. Phys. A: Math. Gen.* **19**, L699–704.

Mannella, R., McClintock, P. V. E. and Moss, F. 1987. *Phys. Lett.* **120A**, 11–14.

Mannella, R., Moss, F. and McClintock, P. V. E. 1987. *Phys. Rev. A* **35**, 2560–6.

Morris, B. and Moss, F. 1986. *Phys. Lett.* **118A**, 117–20.

Morton, J. B. and Corrsin, S. 1969. *J. Math. Phys.* **10**, 361–8.

Moss, F. 1984. In *Workshop on Chaos in Nonlinear Dynamical Systems* (J. Chandra, ed.), *SIAM Proceedings*, pp. 119–29. Philadelphia.

Moss, F., Hänggi, P., Mannella, R. and McClintock, P. V. E. 1986. *Phys. Rev. A* **33**, 4459–61.

Moss, F., Kondepudi, D. K. and McClintock, P. V. E. 1985. *Phys. Lett.* **112A**, 293–6.

Moss, F. and McClintock, P. V. E. 1985. *Z. Phys. B-Condensed Matter* **61**, 381–6.

Moss, F. and Welland, G. V. 1982. *Phys. Rev. A* **25**, 3389–92.

Onodera, Y. 1970. *Prog. Theor. Phys.* **44**, 1477–99.

Pontryagin, L., Andronov, A. and Vitt, A. 1933. *Zh. Eksp. Theor. Fiz* **3**, 165–80 (of which an English translation appears as an appendix to Volume 1).

Prigogine, I. and Lefever, R. 1968. *J. Chem. Phys.* **48**, 1695–700.

Risken, H. 1984. *The Fokker–Planck Equation*. Berlin: Springer.
Robinson, S. D., Moss, F. and McClintock, P. V. E. 1985. *J. Phys. A: Math. Gen.* **18**, L89–94.
Roux, J. C., de Kepper, P. and Boissonade, J. 1983. *Phys. Lett.* **97A**, 168–70.
Sancho, J. M., Mannella, R., McClintock, P. V. E. and Moss, F. 1985. *Phys. Rev. A* **32**, 3639–46.
Sancho, J. M., San Miguel, M., Katz, S. L. and Gunton, J. D. 1982b. *Phys. Rev. A* **26**, 1589–609.
Sancho, J. M., San Miguel, M., Yamazaki, H. and Kawakubo, T. 1982a. *Physica* **116A**, 560–72.
San Miguel, M. and Sancho, J. M. 1981. *Z. Phys. B* **43**, 361–72.
Schenzle, A. and Brand, H. R. 1979. *Phys. Rev. A* **20**, 1628–47.
Seifert, H. 1983. *J. Low. Temp. Phys.* **50**, 1–19.
Smythe, J., Moss, F. and McClintock, P. V. E. 1983. *Phys. Rev. Lett.* **51**, 1062–5.
Smythe, J., Moss, F., McClintock, P. V. E. and Clarkson, D. 1983. *Phys. Lett.* **97A**, 95–8.
Stratonovich, R. L. 1963. *Topics in the Theory of Random Noise*, vol. I. NY: Gordon and Breach.
Stratonovich, R. L. 1967. *Topics in the Theory of Random Noise*, vol. II. NY: Gordon and Breach.
Thom, A. 1971. *Megalithic Lunar Observatories*. Oxford: Clarendon Press.
Tyson, J. J. 1973. *J. Chem. Phys.* **58**, 3919–30.
Van den Broeck, C. and Mandel, P. 1987. *Phys. Lett.* **122A**, 36–8.
Van Kampen, N. G. 1981a. *J. Stat. Phys.* **24**, 175–87.
Van Kampen, N. G. 1981b. *J. Stat. Phys.* **25**, 431–42.
Vogel, K., Risken, H., Schleich, W., James, M., Moss, F. and McClintock, P. V. E. 1987a. *Phys. Rev. A* **35**, 463–5.
Vogel, K., Risken, H., Schleich, W., James, M., Moss, F., Mannella, R. and McClintock, P. V. E. 1987b. *J. Appl. Phys.* **62**, 721–3.
Voigtlaender, K. and Risken, H. 1985. *J. Stat. Phys.* **40**, 397–429.
Wallet, G. 1986. *Ann. Inst. Fourier (Grenoble)* **36**, 157–84.
Welland, G. V. and Moss, F. 1982. *Phys. Lett.* **89A**, 273–5.
Werthamer, N. R. and Shapiro, S. 1967. *Phys. Rev.* **164**, 523–35.
Wong, E. and Zakai, M. 1965. *Ann. Math. Stat.* **36**, 1560–4.
Yeh, Y. J. and Kao, Y. H. 1982. *Phys. Rev. Lett.* **49**, 1888–91.
Zhu, S., Yu, A. W. and Roy, R. 1986. *Phys. Rev. A* **34**, 4333–47.

Index